T0239336

A VERTICAL EMPIRE

A VERTICAL EMPIRE

History of the British Rocketry Programme

Second Edition

C. N. Hill

Formerly Charterhouse, UK

Imperial College Press

Published by

Imperial College Press
57 Shelton Street
Covent Garden
London WC2H 9HE

Distributed by

World Scientific Publishing Co. Pte. Ltd.
5 Toh Tuck Link, Singapore 596224
USA office: 27 Warren Street, Suite 401-402, Hackensack, NJ 07601
UK office: 57 Shelton Street, Covent Garden, London WC2H 9HE

British Library Cataloguing-in-Publication Data
A catalogue record for this book is available from the British Library.

A VERTICAL EMPIRE
History of the British Rocketry Programme
(2nd Edition)

Copyright © 2012 by Imperial College Press

All rights reserved. This book, or parts thereof, may not be reproduced in any form or by any means, electronic or mechanical, including photocopying, recording or any information storage and retrieval system now known or to be invented, without written permission from the Publisher.

For photocopying of material in this volume, please pay a copying fee through the Copyright Clearance Center, Inc., 222 Rosewood Drive, Danvers, MA 01923, USA. In this case permission to photocopy is not required from the publisher.

ISBN-13 978-1-84816-795-7
ISBN-10 1-84816-795-4
ISBN-13 978-1-84816-796-4 (pbk)
ISBN-10 1-84816-796-2 (pbk)

Printed in Singapore by World Scientific Printers.

Contents

Acknowledgements

This book would not have been possible without the help given by very many people.

Ed Andrews, the Central Services Manager of Westcott Venture Park.

Alan Bond of Reaction Engines.

Roy Dommett CBE, of the RAE and DERA, who was involved in much of the work detailed in this book, and whose sharp and percipient comments have thrown light on many of the ideas outlined.

Wayne Cocroft of English Heritage for his help and assistance with the Spadeadam and High Down sites.

Andy Davis for the photograph of the VC 10 as a Skybolt carrier.

Guy Finch for his encyclopaedic knowledge of aircraft, Blue Streak and the rocket interceptors.

Professor Edward James, who set me on this search following an interview when I applied for his MA course in Science Fiction at Reading University, and after reading his book *Science Fiction in the Twentieth Century*, where he notes that Dan Dare 'gave a whole generation of British boys... a totally false impression that Britain was going to dominate the space race.'

James Macfarlane of Airborne Engineering Limited, Westcott Venture Park.

Doug Millard, Space Curator at the Science Museum, who with great kindness started me on my research by allowing me access to his filing cabinet. He also was the first to put the idea in my mind: why do you want to launch satellites anyway?

Kate Pyne, official historian at the AWRE, Aldermaston, for answering blundering questions with tact.

Dave Wright, who has pursued Blue Streak with dogged perseverance, and without those endless telephone conversations this book would not have been possible. Many of the ideas outlined in this book originated from him. Thanks too to his wife Lesley for her patience!

The staff at the Public Record Office in Kew, the ELDO section of the Historical Archives of the European Union at the European University Institute of Florence, the Coventry History Centre and the Science Museum at Wroughton.

Thanks also to David Cheek of GKN Aerospace, Susan Kinsella, Tom Lukeman, Sean Potter and Barrie Ricketson, for their help and suggestions. Any mistakes are entirely due to me.

Images and copyright:

Thanks to GKN Aerospace for supplying images. Also images from the Defence Evaluation Research Agency: © (British) Crown Copyright, 2000 Defence Evaluation and Research Agency, reproduced with the permission of the Controller Her (Britannic) Majesty's Stationery Office.

The image of the 'underground launcher' on page 122 is by kind permission of English Heritage, and is copyright English Heritage. It was drawn by Allan Adams, and I am grateful to Wayne Cocroft for his help in obtaining the image.

Chapter 1

Introduction

It has been said that Britain acquired an empire in a fit of absent-mindedness. It might also be said that it acquired a rocketry programme in a similar fit of absent-mindedness. The UK space programme, or rocketry programme, has always been so low key that the public perception is that the UK has never even had a space programme. Yet for a time in the late 1950s and throughout the 1960s, the programme was technically as advanced as any in the world. If it did not achieve the high profiles of Sputnik, Vostok or Apollo, it is in the main because the projects were less ambitious, subject to much greater financial restrictions, and had a more modest goal. Most of the work was driven by the needs of the military. This was true too in the USA and USSR, but there the civilian effort also became caught up in the Cold War propaganda battles. Kennedy's cry to arms '... to put a man on the moon before this decade is out ...' had no resonances in the UK, and the motives that drove many of the other projects in the US were also very often military in origin, even if they have been used in civilian guise. GPS began as a way for nuclear submarines to fix their position so they could launch their missiles more accurately.

It must be admitted at the outset that almost all the work described here began life as a military project designed to obliterate cities and their inhabitants. The biggest project of all described in this book is Blue Streak, whose sole purpose was to launch hydrogen bombs at the USSR. It was only later that its application to a satellite launcher was seized upon as a political fig leaf for an embarrassed Government, and even then many of the potential satellites might well have been military. Likewise, Blue Steel was intended to deliver megaton warheads. Black Knight was a research vehicle whose initial function was to act as a test bed for Blue Streak and to research re-entry vehicles for nuclear warheads. Black Arrow and Skylark were the only major projects discussed here whose applications were intended to be solely civilian and scientific.

In the end, though, the British work on rocketry and satellite launchers died, mainly as a consequence of lack of funding, political vacillation and a perceived lack of need either for satellites or other forms of space research, whether military or commercial. Although now there is a developing and thriving international commercial market for the launching of communications satellites in particular, the British rocketry programme is certainly now completely dead and there is no prospect of resurrection. All the engineers with any relevant experience have retired long ago. All the infrastructure has disappeared. It is ironic that the systems that were built and tested in the 1960s, and then abandoned, could have been commercially successful in the 1980s and 1990s. It was, perhaps, a penalty paid for being too early in the field.

To understand the story fully, we have to go back more than half a century, to the early days of the Cold War. During the Cold War era, the USA and USSR were driven by ideological pressures that the UK did not experience. Each feared the other and their systems of government. In addition, when it came to development and production of hardware, they had vastly greater resources than the UK. Indeed, the USSR can be said to have 'lost' the Cold War in the sense that it was driven into final collapse in part by the demands of the military and space programmes on its shaky economy. In some sense that was true for Britain as well: after Blue Streak, there was little further attempt to develop a purely indigenous deterrent system. Since the mid-1960s, the deterrent has been maintained at minimal expense.

Politically, missiles and the nuclear threat meant very different things to the UK compared with the USA and USSR. The UK had no hope of 'winning' a nuclear war, given its limited geography (no-one else did, but there was a perception among some in America that a nuclear war was 'winnable'). America and Russia were driven by a paranoid fear that the one was intent on the other's destruction, and the ideologies of the two were so far apart as to be virtually irreconcilable, despite ideas of 'peaceful co-existence'.

The UK had no such geopolitical or ideological dynamic. It had a considerable interest in the state of Europe and the Continental balance of power, as it always has had, but that interest was to be subsumed into NATO, whose purpose, as its first Secretary General, Lord Ismay, put it, was 'to keep the Americans in, the Russians out, and the Germans down'. The UK had also suffered tremendous economic damage in the Second World War, from which it took a long time to recover. In addition to the expenses of maintaining a far flung Empire, it also had to provide an army of occupation for Germany. One of the problems of wanting to be a Great Power is taking on the burdens and expenses of Great Power status, which Britain was less and less able to do after the war. And then the nuclear factor entered into the equation.

The story of the development of the Bomb is a complicated one, but most of the theoretical and practical work was carried out by European émigrés, backed up by American know-how and resources. The UK sent many of its atomic scientists to America. The US and UK agreed to pool information, an agreement that was to fall foul of a later Act of Congress, the McMahon Act. But the British need for nuclear weapons in the immediate post-war period was not that pressing, since the only country that possessed such weapons at that time was America, Britain's closest ally. Possession of nuclear weapons by the UK would have been useful for the influence they may have carried, but were not at that stage essential to the strategic balance, and would not have had much military significance. They have always been weapons of mass destruction, aimed more at cities than at armies.

As earlier noted, Britain's interests were in her Empire and in Europe. In neither of these areas were nuclear weapons necessary or desirable. But that picture changed in 1949 with the explosion of the first Russian nuclear device. This was to be the first of the many scares that the Soviet Union was reaching parity with or overtaking the West technologically. The need for a British nuclear device now became that much more pressing since the Soviet Union was now perceived to be the most likely candidate for hostilities within the foreseeable future. Then came all the various nuclear scenarios that were so to bedevil military and political planners. In what circumstances would the UK need to use such weapons? In what circumstances might they be used on the UK? NATO doctrine held that an attack on one was an attack on all, but there was always the unspoken fear – would America risk nuclear annihilation for the sake of London? Or Bristol, or Birmingham? No one wanted to find out, and, fortunately, we never did.

Another factor, which should not be discounted, was that, as mentioned earlier, Britain still regarded herself as one of the leading Powers. If the other two had the Bomb, then Britain should have a Bomb too, not from any intrinsic merit of ownership, but so as to keep a seat at the Top Table. The 'nuclear club' was a club she felt she could not afford to be excluded from, yet could only just afford to join.

So work on a British nuclear device began very soon after the war. Soon Britain would have a working device. But there was the problem common to all three powers as to how the Bomb would be delivered. In the early post-war period, there was no alternative to the bomber, and the UK had produced some excellent jet bomber designs in the V bombers – the Valiant, Victor and Vulcan, which were to give sterling service to the UK for many years. Indeed, the Operational Requirement was issued at the end of the war, and nearly 40 years later, Vulcans were used in the Falklands conflict in the bombing role, with Victors in the tanker role.

It was realised in the early 1950s that with the increase in sophistication of missile defences, the V bombers, or bombers in general, would be increasingly vulnerable. Certainly it was expected that the likes of Moscow and other major cities would be surrounded by rings of guided weapons that could take out all but the most major bombing offensive – hence the issue of Operational Requirement OR 1132 in September 1954 for a stand-off missile, which would become Blue Steel. In 1954, the principal problem for such weapons was guidance over a long distance of flight (accuracy decreases with time of flight), and with that in mind, Blue Steel was designed with an operational range of 100 nautical miles. This would keep the bomber clear of Moscow and its attendant defences, although still leaving them with a large amount of hostile territory to cross.

At the same time, the Americans were working on various air-breathing long-range missiles, precursors of the later cruise missile. Ballistic missiles were being worked on by von Braun's team, and by Convair under Brossart, but neither technology had advanced sufficiently to produce an effective weapons system that could deliver a nuclear device over a range of some thousands of miles. In 1954, Duncan Sandys of the UK and Charles Wilson of the US signed an agreement to share information on the development of ballistic missiles. By 1955, technology, particularly in guidance, had advanced far enough for serious design work to begin on a UK ballistic missile, Blue Streak, with a range sufficient to reach Moscow (the criterion for any UK nuclear delivery system) and beyond. At the same time, a parallel programme, called Black Knight, was also started to carry out some of the basic research, particularly on re-entry vehicles. And America began work on a much longer range missile, Atlas.

At that time, thermonuclear warheads were much more massive than they would subsequently become, and so all the early missiles designed by the US, by the USSR, and by the UK turned out to be far larger than was in the end necessary. This was to have important consequences as far as the Soviet Union and Sputnik were concerned. The enormous ballistic missile that had been developed by the Russians turned out to be much more effective as a satellite launcher. Neither the UK nor the US had designed anything quite as big as the Russian R-7, or Semiorka. Western politicians, often technically ignorant themselves and with axes to grind, assumed that these immensely powerful Russian boosters meant the Russians were that much further ahead in technology. In effect the reverse was true. The West had not built such large rockets because they were not necessary once lighter warheads had been developed.

	Height/ft	Mass/lb	Thrust/lb
Semiorka R-7	98	588,000	874,000
Atlas E	92	260,000	385,000
Blue Streak*	~70*	198,000	270,000*
Thor	65	110,000	150000

* In its probable configuration if it had been deployed operationally.

All the early Western missiles such as Blue Streak, Thor, Atlas and Titan I, were designed to use kerosene and liquid oxygen as fuels, as did the first Soviet designs. Solid fuel rockets had not yet sufficient size or range given the weight of the warheads of the 1950s. (Minuteman and Polaris were designed assuming warheads would get lighter.) Such large rockets were also very vulnerable to a first strike attack, so would have to be stored in and fired from underground storage silos, hardened against nuclear attack. This added considerably to the expense of the system, and meant in addition that the missile and silo complex itself became a target.

All these missiles were close contemporaries in conception. Where they differed was that America and Russia pressed ahead with development despite the cost.

Development was carried on with Blue Streak as fast as funds allowed, although the whole project was bedevilled throughout its life by Treasury reluctance to release the necessary money. It could be said of the whole history of Blue Streak from 1955 to 1970 that the technical will was there, the political will was there intermittently, and the financial will was never there. It is astonishing how well the morale of those involved with the project stood up in the face of such political and financial uncertainty.

But in 1957 came the shock of Sputnik. The psychological effect on the Americans was considerable, and Atlas, among others, became a crash programme. In more than just the defence field, the US felt it had been overtaken. This led, among other things, to a massive effort in science and technical education. Its effect on British opinion was very much more muted. Britain did not see itself in any technological race, and was not perturbed by the thought of a satellite orbiting overhead. In the US, it was felt almost as an invasion of the country. Britain had suffered bombing of London as early as 1916, but the US had never experienced hostile aircraft in its skies. Sputnik was perceived in those terms.

Curiously enough, the Rand Corporation (and the RAE) had been undertaking studies into reconnaissance satellites, and had recognised that one legal problem might be that a satellite orbiting over another country may be taken as an invasion of the other country's airspace. This is one of the reasons why the first planned US satellite was intended to be perceived as entirely civilian and part of the 1957 Geophysical Year. Sputnik had resolved this problem at a stroke. The Russians were in no position now to claim invasion of their airspace by American reconnaissance satellites.

Back in the UK, by 1958 the Black Knight rocket programme, intended to provide a lot of the basic research for Blue Streak, was up and running. It would yield a lot of useful information for the UK and the US on the physics of re-entry vehicles, necessary for any ballistic missile system, and also for studies into possible defences against them. The first flight of Blue Streak was planned for 1960, when the decision was taken by the Macmillan Government to cancel the system for military purposes. The reasons for this are complex and will be explored further in the Blue Streak chapters. In the same way that the USSR was eventually driven out of the arms race, so too was the UK, becoming increasingly reliant on the US for delivering its deterrent.

Mainly, I suspect, to minimise the political damage that ensued from the decision, it was announced that although Blue Streak had been cancelled as a weapons system, work would still continue, albeit at a much reduced rate, on developing a satellite launcher based on the missile. At least £60 million (all costs are given as of the period and not corrected for inflation), if not more, including large sums at Woomera by the Australians, had been spent on the project by this stage. A design, which would be known as Black Prince from the Saunders Roe brochure, or more inelegantly in official papers as the BSSLV (Blue Streak Satellite Launch Vehicle), had been under consideration for some time. It would have used Blue Streak as the first stage together with the proven technology of Black Knight as the second stage. Again, though, the major problem was money: one source mentioned that the development costs would amount to half the annual UK university budget, which even given the relatively small university sector in the UK at the time, gave pause for thought. And although the US military had found many uses for satellites, there was not the same perceived need by the UK military, particularly since British Intelligence had access to a good deal of the US information. Although the scientific community would have liked to launch various satellites (a stellar ultraviolet telescope was a favourite project), there were not the funds available in the civilian science research budget. Hence the UK was in danger of building a satellite launcher with no satellites to launch.

The decision was then taken to involve other nations in the project in the hope of sharing the costs. The Old Commonwealth countries were not interested, or lacked the finances and resources. France might be interested, but there was also the opportunity for France to acquire much needed data relevant to its own ballistic missile programme, which led to some difficulties and embarrassment. In the end, the European Launcher Development Organisation, ELDO, was born with little enthusiasm from many of its members. And the ELDO launcher ran into considerable criticism almost from the start, being widely perceived as unnecessary and based on obsolete technology.

The latter criticism was unfounded, although the much slower pace of development in the cash-strapped UK meant that the US tended to be there first. But Blue Streak remained irredeemably tarnished by its cancellation for military purposes. It had, however, the potential to be the equivalent of almost any American launcher until the Saturn vehicles. ELDO was both a political failure and a technical failure. Blue Streak itself performed almost flawlessly, but the same could not be said of the French and German upper stages. One of the reasons for this problem was that the other European countries were a good deal less experienced than Britain; another was that putting together a vehicle designed and built by three different teams of engineers in three different countries, speaking three different languages, was no mean feat. ELDO and its launcher died, never to be resurrected.

And what of the other project, Black Knight? After 22 successful firings, the project was declared at an end. But while the UK was still a member of ELDO, a decision was taken to proceed with an alternative, much smaller satellite launcher, and this would be based on Black Knight. The new design was called Black Arrow.

Two test vehicles were flown, one successful and one not, and then an orbital attempt failed by a small margin. On 29 July 1971, the announcement was made that Black Arrow was cancelled. However, the fourth vehicle was subsequently fired and achieved orbit on 28 October 1971, and that, effectively, was the end of rocketry in the UK. Skylark launches would continue for another 34 years, but there was little further development of the vehicle.

Space science has continued, and the UK has always been successful at building satellites. Indeed, Surrey Satellite Technology Ltd (SSTL) has been one of the major success stories of the past decade. It is also perhaps an exemplar of what the Treasury was maintaining – that if there is money to be made in space, then let private business get on with it.

Project Names

In the 1950s, the fashion in the UK was to give many of the military projects two word code names, the first of which was a colour: thus Orange Herald, Blue Streak, Yellow Sun, Red Duster, Violet Club, Green Flax and so on. A good code name should reveal nothing about the nature of the project. However, Yellow Sun for an H bomb was a bit of a giveaway, since the sun is a gigantic fusion reactor (or perhaps not: Mark 1 was not what is commonly understood by a 'hydrogen bomb', so perhaps there was an element of double bluff).

The rocketry projects covered in this book are Blue Steel, Blue Streak, Black Knight, Black Prince, Black Arrow and the various rocket interceptors. The 'Black' designations were applied, albeit unofficially, to research projects without a direct military application; indeed, Black Arrow was entirely civilian, but was named by extension from Black Knight, as probably was Black Prince.

The Rocket Propulsion Establishment (RPE) at Westcott, Buckinghamshire, produced many solid fuel rocket motors. A Superintendent who had been in charge of the Establishment had been a keen ornithologist, and so all the motors produced there were named after birds: Raven, Rook, Cuckoo, Waxwing etc.

Rocket Fuels

Liquid fuel rockets need a fuel and an oxidant. The UK used kerosene as a fuel almost exclusively, although a lot of work was done on liquid hydrogen, but sadly no use was made of this work. HTP (High Test Peroxide) was the most common oxidant: this was an 85% solution of hydrogen peroxide (H_2O_2) in water. The hydrogen peroxide decomposes to steam and oxygen on a catalyst (silver mesh gauze) and kerosene injected into the resultant hot gases ignited spontaneously. However, Blue Streak used engines (see Figure 1 below) licensed from the US and re-engineered, which burned oxygen and kerosene. The measure of the effectiveness of a rocket motor or fuel combination is called Specific Impulse. This was always written in documents of the period as S.I., but nowadays is usually written as I_{sp}. This can be defined in many different ways: one is thrust times burn time divided by mass of fuel burnt. Another way of looking at it is the thrust obtained from each pound of fuel burned per second. The HTP/kerosene combination had a relatively low S.I., around 210–220 at sea level. Oxygen/kerosene gives an S.I. of around 245 at sea level. The effectiveness of a rocket motor is increased at high altitude or in vacuum. This is because the thrust from a rocket engine derives from the pressure difference between the pressure inside the combustion chamber and the pressure outside. In a vacuum there is no outside pressure. Thus S.I. is sometimes quoted at sea level and sometimes in vacuum. Vacuum S.I. is typically 10–15% higher than sea level. Hydrogen/oxygen is the most effective combination of all, reaching S.I.s of at least 400. This means double the thrust for the same weight of fuel (and burn time). Against that, there are weight penalties in the use of liquid hydrogen, since it has a very low density and needs large tanks. It is also very cold, boiling at −253 °C or 20 K, so the tanks usually need extra insulation, which in turn implies a further weight penalty.

Figure 1. Rolls Royce RZ 2 rocket motors in a test stand at Spadeadam.

Units

The past being another country, they did things differently. I have kept almost entirely to the units of the day. Thus feet and inches. 1 metre is 39 inches. Weight was used where today we would (I hope) use mass. Mass is difficult to define, but can be thought of as the inertia of a body. Weight is the force of gravity on that mass. 1 pound, abbreviated lb (derived from Latin libra), is 0.45 kg.

More annoyingly, force was also expressed in lb, meaning in this context the weight of an object with a mass of 1 lb. Modern usage takes the unit of mass as 1 kg and of force as 1 N. The force of gravity on 1 kg is 9.8 N on Earth at sea level. Thus rocket thrust was defined in pound force, sometimes abbreviated to lbf, or plain lb, which would now be written as 0.45 kg × 9.8 N = 4.4 N. A metric tonne is 1000 kg, an imperial ton is 2,240 lb. By a lucky coincidence they are almost identical.

Pressure was usually given in pounds per square inch, psi, or atmospheres. 'Atmosphere' is still sometimes used today under the name 'bar' (and by extension, the millibar or mb found on weather charts). The modern SI unit of pressure is the newton per square metre (N/m^2) or pascal (Pa). 1 atmosphere or 1 bar is approximately 15 psi or 100,000 Pa (100 kPa).

Orbit heights are usually given in nautical miles above the Earth's surface. A nautical mile is, strictly speaking, one second of arc of a Great Circle on the Earth, or around 2000 yards. A statute mile is 1,760 yards. Thus a nautical mile is a little less than 2 kilometres.

After all this, the merits of the metric system appear obvious!

Acronyms and Abbreviations

AWE – Atomic Weapons Establishment, situated in Aldermaston, Berkshire, and responsible for the development of British nuclear weapons. Now the AWRE or Atomic Weapons Research Establishment.

BND(SG) – British Nuclear Deterrent (Study Group). An *ad hoc* group set up to report on the deterrent, and which recommended the cancellation of Blue Streak as a military weapon.

BSE – Bristol Siddeley Engines. Armstrong Siddeley merged with Bristol to form Bristol Siddeley Engines, which was later taken over by Rolls Royce.

CGWL – Controller of Guided Weapons and Electronics. A senior post within the Ministry of Supply. The Controller was responsible for overseeing guided weapons projects.

DRPC – Defence Research and Policy Committee. A high level committee which recommended particular lines of research and development.

ELDO – European Launcher and Development Organisation.

GW Department – Guided Weapons Department at the Royal Aircraft Establishment (RAE). Responsible for missile development.

HTP – High Test Peroxide, a mixture of 85% hydrogen peroxide and 15% water.

JIC – Joint Intelligence Committee. Provides advice to the Cabinet related to security, defence and foreign affairs

OR – Operational Requirement. This was the specification for a new weapon. The requirement would be issued by the Air Staff, and the Ministry of Supply was responsible for circulating the requirement to the various aircraft manufacturers, choosing the winning design and overseeing the development of the project.

RAE – Royal Aircraft Establishment, situated in Farnborough, Hampshire, and responsible for Government-directed research into aeronautics.

RPE – Rocket Propulsion Establishment, situated in Westcott, Buckinghamshire, and responsible for research into rocket motors.

S.I. – Specific Impulse. One of the measures of rocket motor performance. The higher the S.I., the better the performance.

Places

The Royal Aircraft Establishment (RAE) at Farnborough

The RAE at Farnborough, Hampshire, provided the guiding hand behind most of these projects. Together with the Rocket Propulsion Department (RPD), later the Rocket Propulsion Establishment (RPE), based at Westcott, Buckinghamshire, RAE carried out all the preliminary research for the ballistic missile, and also initiated and oversaw the Black Knight programme. It was also responsible for all the further studies that culminated in Black Arrow.

RAE was a large and wide spread organisation in the 1950s and 1960s; the Guided Weapons (GW) Department was responsible for all the early work on Blue Streak and Black Knight. In January 1962 Space Department was created, taking over much of the work of the GW Department. Space Department was responsible for a good deal of the UK's work with ELDO, for the later re-entry experiments, Skylark, Black Arrow, and the technology satellites Prospero and Miranda.

In 1988 the RAE was renamed the Royal Aerospace Establishment, and in 1991 the RAE was merged into the Defence Research Agency (DRA). In 1995 the DRA and other Ministry of Defence organisations were merged to form the Defence Evaluation and Research Agency (DERA). In 2001 DERA was part-privatised, resulting in two separate organisations, the state-owned Defence Science and Technology Laboratory (DSTL), and the privatised company QinetiQ.

Rocket Propulsion Establishment (RPE) at Westcott

Like other post-war establishments such as Harwell and Aldermaston, Westcott was originally an RAF airfield before being taken over in 1946 as the Guided Projectile Establishment. In 1947 its status changed to the Rocket Propulsion Department of RAE, and then in 1958 it became the Rocket Propulsion Establishment.

A good deal of work was done on solid motors (all named after birds, since the Superintendent was a keen ornithologist), but in addition, there was a good deal of early work done on liquid motors using HTP involving German engineers who had come to Britain at the end of the war. These were the Alpha, Beta and Gamma series of motors. The Delta was a low-key liquid oxygen/kerosene programme, which became rather pointless once Rolls Royce began development of the RZ 1 and RZ 2 chambers. There was also a very considerable liquid hydrogen programme, producing some sophisticated chambers which could almost have been fitted directly into a rocket stage without much further work. In addition, P site was used by Rolls Royce in some early firings of the RZ 1 and RZ 2 before Spadeadam became available.

One of the major projects of the 1950s was the 1/6th scale silo, which must have been quite an impressive sight when the motors were fired inside the tube. Nothing remains of the site today, although there are still hexagonal pieces of concrete which formed the walls of the model silo to be found to this day.

Figure 2. Test stands for rocket motors at Westcott as seen in 2010.

Spadeadam

Spadeadam in Cumbria was chosen as the site to test and develop the Blue Streak engines, and here Blue Streak vehicles were assembled for static firing before shipment to Australia. Rolls Royce ran the site as an agency.

The engine test area at Prior Lancy Rigg consisted of four concrete stands into which the engines could be mounted for test firing. Three remain, copied from a Rocketdyne design used at their California Santa Susana Field Laboratory site; the fourth has been demolished. This last and lost stand seems to have been built to a different design, using an innovative application of pre-stressed concrete to contain liquid oxygen spills.

Two static firing stands themselves stood at Greymare Hills and were large enough to accommodate a full Blue Streak missile. All firings were controlled from command centre bunkers connected to the stands by tunnels or surface cabling ducts.

After the demise of ELDO, Spadeadam was taken over by the RAF. Its primary purpose today is to provide a location for teaching of electronic warfare to RAF and other NATO aircrew.

Woomera

Woomera was, of course, the test range, jointly funded by the UK and Australia. It was run by the WRE (Weapons Research Establishment) based at nearby (relatively!) Salisbury, South Australia. The airfield at RAAF Edinburgh was used for the V bombers during Blue Steel trials and for transport generally. The Range was first established in 1947 under the Anglo-Australia Defence Project and the Woomera Prohibited Area (WPA) was declared for the purposes of 'testing war materials'. It was sometimes referred to as the 'Joint Project'. Most of the early work was devoted to surface-to-air missiles, which would result in the Bloodhound, Thunderbird and Sea Slug missiles. A Blue Streak launch site was built at Lake Hart, with the Black Knight launch sites not far away. Later, a Black Knight launch site was adapted for the Black Arrow satellite launcher.

Work began on a Blue Streak launch site set into the side on a canyon, with the intention of building a silo ('underground launcher') around it, but with the military cancellation this was abandoned.

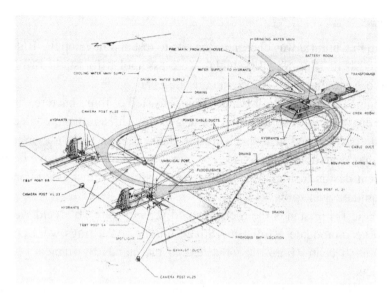

Figure 3. Black Knight launch area at Woomera. There were two gantries, one of which was later adapted for the Black Arrow satellite launcher.

High Down

This was the test site for Saunders Roe, where Black Knight and Black Arrow vehicles were taken for static firing before being taken out to Woomera for

Figure 4. The High Down site on the Isle of Wight.

launching. It was situated by the old Needles Battery, on the top of the chalk cliffs overlooking the Needles lighthouse.

Political and Administrative Matters

Most of the projects covered in this book started life in a military guise, and so the procurement process, as it would be called today, needs to be examined.

Often a project might have its origins in the Defence Research Policy Committee, or DRPC. Thus in 1953, German rocket scientists who had gone to work in Soviet Russia returned to their homeland, and were then debriefed. The DRPC considered the debriefings, and concluded that the UK should begin research on both ballistic missiles and also investigate possible defences against them. This can be seen as the beginnings of what became Blue Streak.

When a need for a particular weapon had been clarified, the Ministry of Defence or the Air Ministry issued an Operational Requirement, or OR. Thus Blue Streak was OR 1139, the warhead Orange Herald was OR 1142, and so on. The OR could be very specific about some of the requirements: thus for aircraft it might specify range, altitude, speed, maximum weight, and so on. Then the OR would be circulated to various firms, who would produce appropriate designs. The Ministry would then evaluate rival designs and award the contract to a particular manufacturer. Development was the responsibility of the Ministry of Supply, who dealt directly with the firms concerned. When the winning design had been selected, it would look after the timetable, finances and so on for the project.

The major problem was that the Ministry of Supply was not the end user, nor did it benefit or suffer directly from the success or failure of the project. A considerable amount of rancour developed between the Ministry of Supply and the Air Staff as a consequence.

As a result, Blue Streak files in the Public Record Office can be found in various different forms: Supply files, Defence files and Air Ministry files. Often material here is duplicated, as the same set of minutes of meetings were circulated to all relevant Ministries. A further complication is that the names given to a project by the Ministry and by the firm can sometimes be different: thus the Saunders Roe SR53 is known in Ministry files as the F138D.

There is a strong perception that this cumbersome bureaucracy did nothing to speed up projects, and that it would have made considerably more sense to give the function of overseeing development and production to the Ministry who would be the end user. This situation was never helped by continual Defence Reviews, changes of policy, and Treasury oversight. Whilst the latter three are obviously necessary, they can also cast doubts on projects which cannot have helped when it came to producing enthusiastic efforts to get the relevant project in service as swiftly and effectively as possible. Relations between firms and Ministries could also be difficult. Given directions as they were from Whitehall, the firms acted almost as government agencies at times, free from normal commercial pressures. They were often at the mercy of the vagaries of changes in defence policy. Thus, as Saunders Roe was gearing up to produce 27 prototypes of the P177, together with work on Black Knight and the SR53, the work force approached 4,000, only to be cut back drastically as a result of the cancellation of the P177. This is always a problem when a company relies too much on government work.

Politics

Blue Streak was effectively in the hands of three Ministries: Defence, Supply and the Air Ministry, whilst the whip hand was held by a fourth, the Treasury. The set up seems Byzantine to modern eyes.

The Air Ministry

The Air Ministry, set up in 1919 to oversee the RAF, was represented in Cabinet by the Secretary of State for Air. Although both Winston Churchill and Harold Macmillan had held the post, during the 1950s and 1960s the position was occupied by ministers who were not destined for greater things, the office holders being respectively Lord De L'Isle and Dudley, Nigel Birch, George Ward, Julian Amery and Hugh Fraser. The post was abolished in 1964 when the Air Ministry was absorbed into the Ministry of Defence.

There was controversy over the procurement process for what would today be called weapons systems, such as Blue Streak. When the Air Staff had decided on the specification for a new project, they would issue an Operational Requirement (OR). The project would often have come through the DRPC – Blue Streak is a good example. In 1953, the DRPC had been studying missile development and had decided that both a ballistic missile and a defence against a ballistic missile should be investigated. This led to the issuing of the OR for Blue Streak in 1955 (and also one for an anti-ballistic missile defence, but this did not get very far).

The Air Staff might have issued the OR, but it was up to the Ministry of Supply to circulate the requirement to industry, take in the proposals, evaluate them and issue the contract to a particular firm. It would then follow the project through to service entry.

The Ministry of Supply

The design and production of aircraft became the concern of the Ministry of Aircraft Production in May 1940. In April 1946 the Ministry of Aircraft Production was dissolved and its powers transferred to the Ministry of Supply, whose primary duty was the furnishing of supplies and the carrying out of research design and development for the services.

Firstly, this lead to problems in that the Ministry of Supply was responsible for developing aircraft, but at the same time, it would not be the end user, and thus lacked the incentive to overcome obstacles, and to speed the process along. Secondly, it did not have to operate the obsolescent material that the prototypes would replace, and so here too lacked that final sense of urgency. A third criticism was its industrial policy: projects were often not allocated to firms on the basis of their ability to carry them out, but often given to firms who were short of work in order to keep them busy. Sometimes the rationale behind some of the decisions was hard to fathom. Blue Steel was given to Avro, who had no experience whatsoever in guided weapons and had to set up a division from scratch – a process which must have cost a year or so of development time.

Reginald Maudling was Minister of Supply from 1955 to 1957. He has this to say about the Ministry in his autobiography:

> When Anthony Eden became Prime Minister in 1955, he promoted me to Minister of Supply, which was my first full Ministerial post … It was a strange Department, and the target of a good deal of criticism, much of it justified. It was supposed to be concerned mainly with the supply of munitions to the three Services, and this was a large part of the routine work of the Department, but in addition it had responsibility for the aircraft production industry generally. The Government exercised a great deal of influence over the industry because, with the scale of modern projects and the vast amount of research expenditure involved, the industry had to rely heavily on the Government for contracts and for support. In addition, the Ministry of Supply was responsible for the Royal Aircraft Establishment at Farnborough, a quite remarkable institution, upon which industry relied heavily for scientific and technical support.
>
> Inevitably we got caught in the middle in all disputes that went on between manufacturer and consumer. This was particularly true in the field of military aircraft, with the Air Force always demanding more from the manufacturers and complaining they were not getting their requirements met, while the manufacturers were saying that they were doing all that was possible and the RAF were asking too much. Relations between the Ministry of Supply and the Air Ministry were not ideal, and indeed I had from time to time considerable battles with Nigel Birch, who was then Secretary of State for Air … I came to the conclusion during the time I was there that the system was a bad one and that the interposition of a third party between customer and supplier, rather than acting as a pacifying agent, merely exacerbated

argument. I did, in fact, recommend the abolition of the Ministry of Supply and when Harold Macmillan asked me to continue in that job when he became Prime Minister I naturally refused, because it seemed absurd to continue as Minister in charge of a Department whose existence I did not think was justified.[1]

Sir Frank Cooper, one of the senior Civil Servants of the time (among many other posts, Permanent Secretary at the Ministry of Defence from 1976 until 1982), had this to say about the Ministry of Supply in the context of the TSR 2, although his strictures could be applied more generally:

> There was no doubt that relations with the Ministry of Aviation /Supply and the Air Ministry went from bad to worse and that these poor relations spread increasingly to the Ministry of Defence as a whole. The breach itself was of long-standing. The basic cause was lack of trust, particularly as regards the information received by the Air Ministry. The trust was lacking because the Procurement Ministry stood between the Air Ministry as the customer, and industry as the supplier. Moreover nothing seemed to arrive at the right time and at the right price, let alone with the desired performance. The lack of trust was exacerbated by the financial arrangements under which the Ministry of Supply/Aviation recovered production costs from the Air Ministry but was left with the research and development costs. Hence, there was no clear objective against which the supply department could assess performance and value.

Ministry of Defence

From 1946 to 1964 five Departments of State did the work of the modern Ministry of Defence: the Admiralty, the War Office, the Air Ministry, the Ministry of Aviation and an earlier form of the present Ministry of Defence. These departments merged in 1964, apart from the defence functions of the Ministry of Aviation which were merged into the Ministry of Defence in 1971.

The main purpose of the Ministry of Defence in the early 1950s was to co-ordinate the three services. At this point in its history it did not have the powers that it would later have. Duncan Sandys was appointed Minister in 1957 by Macmillan, and was given much extended powers. He is remembered for his 1957 Defence White Paper, which was (unfairly) blamed for the demise of a large part of the aircraft industry. It would be more correct to say that Sandys was the first to articulate changes that were inevitable and probably overdue.

The Ministry of Defence did have considerable influence on policy by means of the extremely powerful DRPC. It was this committee that decided general defence policy needs and hence which projects should proceed. The cancellation of the rocket interceptors was a direct consequence of a change in policy initiated by the DRPC, and the requirement for a ballistic missile originated with the DRPC.

The Treasury

The final arbiter was the Treasury. For a project such as Blue Streak, which could be considered as one of national priority, considerable delays were incurred as a result of Treasury refusal to release funds. The Spadeadam facilities were delayed for some six months as a consequence of Treasury reluctance, as this memo from the Ministry of Defence indicates.

> You will remember that in February the Minister of Supply wrote to you in connection with the project to develop a rocket testing site in Spadeadam and emphasised the necessity of settling as quickly as possible the fate of our medium range ballistic missile project. We agreed at the time that nothing should be done about this letter since we had not yet settled the problems raised by the Long Term Defence Review. The Ministry of Supply, however, are now being held up by Treasury refusal to agree any expenditure at Spadeadam until the Financial Secretary has seen your reply to the Minister of Supply's letter of 15th February [1956].[2]

Sir Frederick Brundrett, Chief Scientist at the Ministry of Defence and chairman of the DRPC, wrote:

> There is no doubt whatever that the political uncertainties stemming originally from the reports of the meeting at Chequers, and particularly the bitter hostility of the Exchequer and the Treasury to the project, have contributed to the difficulties, and in particular, specifically caused the work at Spadeadam to proceed at a speed less than the maximum that would have been possible had money been available.[3]

Again, an excerpt from a minute to the Minister of Defence in October 1957 reads:

> During most of 1956 we were defending the very existence of Blue Streak against savage attacks by the Treasury.[4]

But amidst the controversy the military case was being made that

> the conclusion from these arguments is that of all the weapons under consideration only the ballistic missile looks like having a reasonable chance of remaining comparatively invulnerable by 1970. What is more the firing sites for ballistic missiles will be difficult targets to destroy. It is clear, therefore, that unless we change our present policy of maintaining continuously in being an effective contribution of our own to the strategic deterrent, we must retain in the programme the ballistic missile.[5]

If, in 1957, Britain intended to maintain its deterrent, then it needed a ballistic missile, and Blue Streak was the only option, whatever the Treasury might have thought. Often delays meant that, in the long run, the whole project cost even more. Then came the financial crisis of 1958, when the entire Government

Treasury team resigned in protest at the size of public spending. £100 million had to be cut from Government expenditure, with the consequence that Macmillan wrote to the Minister of Defence and the Chancellor in December 1958: 'on Blue Streak we should take all steps to reduce expenditure which can be taken without giving any widespread impression that the approved programme is being abandoned or retarded. ... of the order of £1M.'[6]

It is difficult to see why the Treasury seems to have opposed the project so bitterly: other defence programmes such as the V bombers or the nuclear programme had been equally costly. It is impossible to judge how such economies affected the project, but it would not be unreasonable to say that the first flight of Blue Streak would have been put back by at least six to twelve months by the delays imposed whilst obtaining Treasury clearance. The point was also made more than once by Sandys that such economies would mean that as a consequence of the delays, the system would be late in service, and its useful service life concomitantly reduced.

Politicians

The strongest political personality that looms out of this story is Duncan Sandys, who made an early reputation for himself during the Second World War in the context of German guided weapons. He was an effective if abrasive Minister, being Minister of Supply in the early 1950s, then Minister of Housing and Local Government. In that context he was also responsible for setting up the Civic Trust. In 1957 he was appointed Minister of Defence by Macmillan, and was arguably the first to get a grip on the Ministry with its divergent Service interests. Before Sandys there had been a rapid turnover of Ministers, who did not have time to stamp their authority on their Ministry. Certainly, he retains a considerable notoriety among aircraft buffs for the 1957 Defence White Paper, with its unspoken 'no more manned aircraft' philosophy. Given the speed of the White Paper, he was probably implementing policy that had been already laid out by others, principally Sir Frederick Brundrett, Chief Government Scientist at the Ministry, and also Chairman of the influential DRPC. A further motive behind Sandys' appointment was to cut the cost of defence in general, and he was not a man to be easily deflected from his objectives. Certainly, he reduced defence expenditure to 7% of Gross Domestic Product (GDP) at which level it broadly remained for many years. Part of the increased dependence on nuclear weapons was to cut the cost of conventional defence.

Sandys started the ball rolling for Blue Streak whilst Minister of Supply, and he remained a vigorous proponent whilst at Defence. At the end of 1959, he was moved to a new Ministry, Aviation, which took over many of the functions of Supply. It has been asserted that Macmillan appointed him to this post to start the rationalisation of the aircraft industry. It also cleared the way for a new Minister of Defence, Harold Watkinson, to cancel Blue Streak. Watkinson's ministerial career was relatively short. It is quite likely that the fallout from the cancellation led to Sandys' sideways move to Colonial Secretary, although this too was a post that would require a man not afraid to take unpopular decisions. As Aviation Minister he was succeeded by Peter Thorneycroft, who had previously resigned from the Macmillan Government as Chancellor over the level of government spending, and it was Thorneycroft who began the Anglo-French talks which later led to ELDO. Thorneycroft later became Minister of Defence.

The Wilson Government, despite its rhetoric of the 'white heat of technological revolution' cannot be seen as a government that pushed British technology. Beset by economic problems, research and development (R & D) is an easy target for politicians looking for economies. Few in his Government were scientists or technologists: to make Frank Cousins, a trade union boss of the old school, Minister of Technology was no doubt politically astute, but can be seen as typical of the political cynicism with which Wilson operated. He was succeeded as Minister of Technology by Tony Benn, who seems to have had enthusiasm for but not a great deal of understanding of modern technology.

Crossman, another important Minister of the Wilson Government, writing in his diaries, bemoans time spent in cabinet discussing Black Arrow. Technology had no appeal for him either. But how was Britain going to survive in the latter part of the twentieth century without exploiting advanced technology?

The Wilson Government from 1964 onwards brought a number of economists in to evaluate programmes on a cost-benefit analysis basis (Wilson himself had been an economics don at Oxford). The problem was that research programmes were analysed for their economic benefits, and this is a difficult if not impossible task. One of the points of starting research programmes is that their outcome is not always predictable, since the object of research is to look at matters that are uncertain or unknown.

Surprisingly, the greatest number of files on Blue Streak in the Public Record Office relate to the Foreign Office. This is as a consequence of the attempt to convert Blue Streak into a European satellite launcher. Hence it is not surprising that the Foreign Office were firm advocates of the project irrespective of any technical or economic merit.

The US comes into the picture indirectly, since, with its vastly greater resources, it had covered most of the ground in space and rocket technology before the UK. The warheads that would equip Blue Steel and would have equipped Blue Streak were of American origin, as was a good deal of the technology that went into Blue Streak. Competing with the US in space research or satellite launching was also often thought by those in Government to be pointless, given the progress that had been made in America and the resources available to the American Government. The closeness of the US and UK defence, intelligence and research establishments also often meant that the UK concentrated on certain rather narrow areas (for example, re-entry research) so as to have useful information to trade with the US. (The US would never give information away for nothing, but it would exchange information on a fairly generous basis.)

Officials

Whilst the major policy issues were the province of the politicians, the day-to-day or month-to-month work was carried out by officials at the various Ministries. One of the most influential, by virtue of his post, was CGWL, or Controller of Guided Weapons and Electronics at the Ministry of Supply and its successors. For almost all the period, with a break of two years, Sir Steuart Mitchell held the post. From the Ministry papers he appears to be a sensible and capable administrator. Dr Robert Cockburn filled the break.

However, delving deeper into the Ministries, one drowns in a soup of alphabetic titles: in the RAF there was VCAS (Vice Chief of the Air Staff, who dealt with nuclear matters); DCAS, the Deputy Chief; DCAS (OR) Deputy Chief of Air Staff (Operational Requirements). There was DRAE (Director of the Royal Aircraft Establishment); DDRAE (his deputy); DAWRE (Director of the Atomic Weapons Research Establishment) and DDAWRE, his deputy. Then there are all the Ministry and Establishment departments with their heads: Guided Weapons, Space Department, and so on. Ministries have Private Secretaries (PS), Permanent Under Secretaries (PUS), and varieties of subordinate secretaries. It was part of their job to turn policy into hardware. But they were also responsible for the papers that went to Ministers, and, as a result, a good deal of the policy was made at a lower level than is often supposed.

Firms

De Havilland Propellers (later Hawker Siddeley Dynamics or HSD) was of course the largest contractor, building up to 18 flight models of Blue Streak (not all of which were completed) as well as several non-flight vehicles. Large test stands had also to be erected at Hatfield for proving purposes. Rolls Royce developed the prototype RZ 1 engines (copies of the American S3 engine) then designed and built the RZ 2.

De Havilland was also responsible for the Sprite and Super Sprite, designed to assist take-off for the likes of the Comet and the V bombers, and also the Spectre, used in the rocket interceptors and early test models of Blue Steel.

Armstrong Siddeley, which became Bristol Siddeley Engine (BSE) before being absorbed into Rolls Royce, was one of the first of the firms to be involved in rocket development, with the Snarler and Screamer motors. They were then chosen to develop the Gamma motor for Black Knight, the Stentor motor for Blue Steel and the later Gamma motors for Black Arrow. Their test site was at Anstey, near Coventry.

Napier was also involved in HTP work, producing the Scorpion, installed in Canberra reconnaissance aircraft, and a rocket pack intended for the Lightning fighter.

Many other firms were also involved as subcontractors, and in particular Sperry and Ferranti were responsible for inertial guidance platforms.

All these were mainstream aircraft manufacturers, and as such, their involvement in these projects is immediately obvious. What is less obvious, however, is the large part played by an otherwise rather obscure subcontractor and builder of somewhat indifferent flying boats: Saunders Roe (taken over by Westland in 1959, becoming the British Hovercraft Corporation in 1964, the Westland Aerospace in 1985, before being finally absorbed into GKN Aerospace).

Why Saunders Roe? Their previous history had been that of a small but enterprising firm, involved both in marine work and in aviation, and thus, not surprisingly, concentrating in the main on flying boats. It would be fair to say that many of the flying boat designs were rather indifferent. It would also be fair comment to say that later, from the 1950s onwards, throughout their existence as Saunders Roe and later in various Westland guises, they worked on idiosyncratic and often quite advanced projects that would reach prototype stage, but rarely ever reached production. A review of the projects they undertook reveals programmes with technological fascination, but which were often dead ends. These include:

- the SRA/1, a jet engined flying boat fighter. Three prototypes were built, the first of which flew on 16 July 1947.
- the Princess, a very large turbo prop passenger flying boat. Three prototypes were built, the first of which flew on 22 August 1952.

- the SR53, a mixed power plant (rocket/jet) supersonic interceptor. Two prototypes were built. The project had its inception in 1952, and the first flight was on 16 May 1957.
- the SR177, an extended version of the above. Prototypes were being built at the time of cancellation. Inception 1954, cancelled 1957.
- a design for the specification of F155, producing what would have been the very last word in rocket powered interceptors.
- a 'hydrofoil missile' for the Admiralty. This was a design for a large hydro-foil craft, powered by a jet engine driving a large wooden airscrew, under radio control, and carrying sonar and a torpedo. Design study 1957.
- the Black Knight research ballistic rocket. More than 25 built; 22 flown. Inception 1955, first flight 1958, last flight 1965.
- the design brochure for Black Prince (see Chapter 8) 1960.
- a design brochure for a liquid hydrogen stage for the Blue Streak satellite launcher (1961).
- the Black Arrow satellite launcher. Five vehicles built, four launched. Inception 1963, first flight 1969, last flight 1971.
- the SRN-1, Britain's first hovercraft. Indeed, the firm for some years was known as the British Hovercraft Corporation, developing and building all the British hovercraft.

This is not an exhaustive list. Ironically, all these projects fulfilled their requirements. If Saunders Roe were asked to produce a design, they did so, and it would be fair to say that the designs were exactly what was asked for. If that is the case, then it has to be asked whether the requirements were reasonable to begin with. Hindsight is very valuable, but it is pointless to castigate others for not foreseeing the future. However, a more polite way of rephrasing this would be to say that the projects investigated possibilities which might have had a fruitful outcome, and which were worth investigating for their potential.

In addition, the firm undertook a large number of design studies for other projects. Any firm of this sort will always be thinking of new designs, many of which will never see the light of day, but the Saunders Roe team produced an astonishing array of ideas. Again, most of these, like the ones listed above, are noted as much as anything for their eccentricity. Highest on such a list, second only to the hydrofoil missile, might come a study for a nuclear powered flying boat undertaken for the US Navy.

Money values

It is almost impossible to convert from 1950s and 1960s prices to current prices. One measure is the Retail Price Index (RPI). The RPI in 1960 was 12.6; in 2009 it was 218.0, an increase of more than seventeen fold. At a very rough estimate, multiply by twenty. Thus, Black Prince at £35 million could be obtained for the price of the Millennium Dome!

It can be argued that inflation with regard to defence projects has been higher. The cost of deploying Blue Streak was put at perhaps £600 million, or perhaps £20 billion in today's currency. On the other hand, the cost of replacing the present Trident system is put at somewhere around £80 billion over twenty years.

---

[1] R. Maudling. Memoirs. (1978). Sidgwick & Jackson Ltd.
[2] The National Archives (TNA): Public Record Office (PRO) DEFE 7/2245. Development of Blue Streak.
[3] TNA: PRO DEFE 13/193. Development of Blue Streak missile.
[4] TNA: PRO DEFE 7/2245. Development of Blue Streak.
[5] Ibid.
[6] TNA: PRO AVIA 92/24. Blue Streak: correspondence leading to the Cabinet decision to cancel as a weapons system.

Chapter 2

Rocket Motors

Rocket motors work by ejecting gases at high speed. From the physics point of view, the momentum given to the gases will be counteracted by an equal and opposite momentum given to the rocket. Rocket motors are designed to make this momentum change as large as possible.

A change in momentum implies a force, since force × time = change in momentum. The force is given by change in momentum per second (strictly speaking, rate of change of momentum). This is usually referred to as the rocket thrust, measured in newtons, or, in Britain in the 1950s and 1960s, in lb – a shorthand for pounds force.

Gases moving at high speed have kinetic energy, and in almost all rocket motors this kinetic energy comes from chemical energy. With solid fuel motors, the fuel and the oxidant are melted together and poured into a casing to cool and solidify. Almost all liquid fuel rockets need both a fuel and an oxidant. There are a few chemicals which can be used by themselves (referred to as monopropellants) – hydrazine (N_2H_4) and hydrogen peroxide (H_2O_2) are examples. They can be decomposed directly to gases (usually by means of a catalyst). The drawback is that they are not very energetic and tend to be used only for small control jets.

The most common rocket fuels are:
- hydrocarbons, referred to generically as 'kerosene' (some early British documents refer to 'kerosine'), usually as some form of jet fuel.
- hydrazine or some related compound (usually UDMH – Unsymmetrical DiMethyl Hydrazine or $(CH_3)_2N.NH_2$).
- liquid hydrogen.

Kerosene is cheap, easy to handle, not volatile and not poisonous. Hydrazine is easily storable, and it is mostly used in combination with dinitrogen tetroxide, N_2O_4. Both produce highly poisonous fumes, and dinitrogen tetroxide is also very corrosive. They ignite spontaneously on contact (i.e. they are hypergolic). The combination is often used in missiles which are left fuelled up on a long-term

27

basis, or in upper stages of satellite launchers, particularly when a restart capability is needed.

Liquid hydrogen is the most energetic and effective fuel, although it suffers from two major drawbacks: it boils at −253 °C or 20 K, and has an extremely low density of 71 kg/m^3 compared with 1,000 kg/m^3 for water. Low density implies large tank volume and, as a consequence, extra weight.

Kerosene was the usual fuel of choice in the UK, with either liquid oxygen or HTP as the oxidants. Although a good deal of research and development was done on liquid hydrogen, including test firing of liquid hydrogen chambers, sadly no rocket stages were built using liquid hydrogen.

Common oxidants are liquid oxygen, and as mentioned, dinitrogen tetroxide in combination with hydrazine. However, Britain was to make extensive use of another oxidant, hydrogen peroxide (H_2O_2), and the way it was used was and still is unique. Hydrogen peroxide was used in the form of High Test Peroxide or HTP, a solution with 85% of hydrogen peroxide and 15% water. Hydrogen peroxide can be decomposed to steam and oxygen at a high temperature using a catalyst – nickel gauze plated with silver, the silver being the catalyst. In this way, the HTP could be used as a monopropellant, but it was much more efficient to inject a fuel such as kerosene into the hot gases to be burnt in the oxygen produced in the decomposition. HTP was also thought to be safer and easier to handle than liquid oxygen. In 1952, the decision was taken to use only HTP motors for all liquid propellant rockets used on, or in, aircraft[1].

Specific Impulse

One measure of the effectiveness of a rocket motor or fuel combination is Specific Impulse or S.I. One way to define it is (thrust × burn time)/(total mass of fuel burnt). Another way is the thrust obtained from each lb (or kg) of fuel burned each second.

In modern units, S.I. would be quoted in Ns/kg – which turns out to be dimensionally equivalent to m/s. This is because S.I. can also shown to be the same as the exhaust velocity of the gases from the combustion chamber. Furthermore, the final velocity of a rocket (in the absence of gravity, air resistance and so on) can be shown to be:

$$v_{final} = v_{exhaust} \times \ln(\text{mass at start/mass at end})$$

Hence the greater the exhaust velocity, the greater the final velocity of the rocket (ignoring all the complications such as gravity, air resistance and so on).

The difference between the mass at the start and the mass at the end is effectively the mass of fuel (unless anything is jettisoned along the way!). The skill of the designer is to build a vehicle as structurally efficient as possible – all fuel and no structure is not possible, but the structural mass must be as small as possible. Blue Streak had an unusual tank structure, being a lightweight stainless steel 'balloon', which needed to be constantly pressurised to keep its structural integrity. Black Knight and Black Arrow were also structurally very efficient.

The equation above only applies in ideal circumstances. The effectiveness of a rocket motor is decreased in the atmosphere, since the thrust from a rocket engine derives from the pressure difference between the pressure inside the combustion chamber and the pressure outside. In a vacuum there is no outside pressure. Another way of looking at this is to say that the exhaust velocity is reduced by the air outside. Thus S.I. is sometimes quoted at sea level and sometimes in vacuum. Vacuum S.I. is typically 10–15% higher than sea level.

The HTP/kerosene combination has a relatively low S.I., around 210–220 at sea level. Oxygen/kerosene gives an S.I. of around 245 at sea level. Hydrogen/oxygen is the most effective combination of all, some motors reaching S.I.s of at least 400. One way of looking at this is to say that there is double the thrust for the same weight of fuel.

Thrust and Weight

The hot gases that exit from the rocket motor provide thrust. Rockets are (usually) launched vertically. The thrust must be bigger than the weight for the vehicle to start moving upwards. Usually, the ratio of thrust to weight is about 1.3, which means the initial acceleration is 0.3 times that due to gravity (the force driving the rocket upward is thrust minus weight). As fuel is burned off, the weight (and mass) decreases and the acceleration increases. The final velocity is governed by two factors: the speed of the exhaust gases (another way of expressing S.I.) and the ratio of the total weight of the rocket to the final, empty weight, as in the equation cited above. This latter figure becomes very important for satellite launchers, where the payload may be around only 1% of the initial weight. A small increase in weight can have a considerable impact on payload.

Thus if the Blue Streak motors could provide a total thrust of 300,000 lb, its maximum lift-off weight would be around 230,000 lb. If Blue Streak itself weighs 183,000 lb, then 47,000 lb is left for the upper stages and payload, only around 25%. A satellite launcher usually consists of three stages, and as each stage burns out, it is discarded. To make this staging as efficient as possible, the first stage should be around a half or two thirds of the total. Blue Streak was designed as a ballistic missile, and so was not as efficient a satellite launcher as it might have been.

The same problem applied to the early American launchers, but one way of overcoming the problem is to add strap-on solid fuel boosters to augment the thrust for the first part of the flight. These boosters might then provide extra thrust for the first 30 seconds or so of flight, whilst fuel from the main stage is being burned off and so reducing its weight.

Turbopumps or Pressure Feed

The pressure inside a combustion chamber can be very high – typically 500 psi or 33 bar. In the vacuum of space, a lower pressure can be used, but the efficiency of any rocket motor is reduced if used in the atmosphere, and one way of increasing the efficiency is by using as high a chamber pressure as possible. The question then is how to feed a large quantity of fuel into the chamber at such high pressure.

There are two options: a pump, or by pressurising the fuel tanks.

Pressurising the tanks had one big drawback: the tank walls had to be strong enough to withstand the pressure, which implies they are also going to be heavy.

The tanks can be pressurised from separate gas bottles, but, for large tanks at high pressure, that has a considerable weight penalty: the gas bottles themselves will be thick-walled and thus heavy. The alternative is a gas generator – two chemicals being mixed to produce large volumes of gas. The French stage of Europa used a gas generator; the German third stage was pressurised by helium in gas bottles. The great advantage of the system is that it is extremely simple and so there is little to go wrong.

A pump has to be driven by something – there needs to be a turbine which is normally driven by fuel from the main tanks. In HTP motors, the kerosene and HTP were well suited to the purpose; the RZ 2 motors in Blue Streak had a turbine which used an excess of kerosene – that is, it burned fuel rich – to keep the temperature down. This can be seen very clearly in Blue Streak launches: the turbines produce bright yellow flames as a result of the excess of carbon.

The great advantage of pump versus pressure is that with a pump, the tanks can be as thin-walled as structurally possible (Atlas and Blue Streak took this rather to extremes). Some small pressure is still needed in the tank for the pump to function, but it is relatively small. One drawback is the extra weight of the turbine and pumps. Another is that the system is relatively complex, and provides another opportunity for something to go wrong.

One of the major problems, particularly with regard to the higher thrust engines, was producing pumps powerful enough to cope with the quantity of propellant at the high pressures needed, as this chart shows:

	Flow rate (fuel + oxidant) (lb/second)	Combustion chamber pressure (lbf/in²)
Snarler	10	300
Screamer	24	600
Beta	14	320
Gamma 1	39	450
Delta 1	205	500
RZ 2	560	525

Cooling

Rocket chambers will, not surprisingly, get very hot, and ways have to be found to stop them getting too hot. One method is to use film cooling. At the top of the chamber is an injector, where the fuel and oxidant enter the chamber, and it usually resembles a shower head. It is designed so that the fuel and oxidant can mix and burn as quickly as possible. In film cooling, the injector feeds in only fuel at the very edge of the chamber, so that the sides of the chamber wall will be in contact with relatively cool unburnt gas. This usually means a slightly reduced efficiency, as the motor is running fuel rich. The next and most obvious method is to use the incoming fuel to cool the chamber walls before being injected into the chamber. This is called regenerative cooling. One way to do this is to have a double walled chamber, with the fuel flowing between the two walls. This tends to be rather heavy – the pressure in the chamber is obviously quite high, and so the walls need to be robust. This is illustrated in the de Havilland Spectre motor shown in Figure 6.

A refinement on this method was to make a chamber of thin nickel tubes brazed together. Reinforcing bands were usually attached around the tubes, but overall, the construction was much lighter. This technique was first used in the RZ 2 motor for Blue Streak and the chambers of the Stentor motor for the Blue Steel missile. Fuel or oxidant then flows through the tubes to cool the chamber.

Sometimes the expansion cone for upper stages is left as plain metal, and this is sometimes referred to as 'radiation cooling' – in other words, the metal is left to glow red hot and literally radiates its heat away into the vacuum of space.

Figure 5. A double walled chamber. In this de Havilland Spectre motor, HTP is used as the coolant.

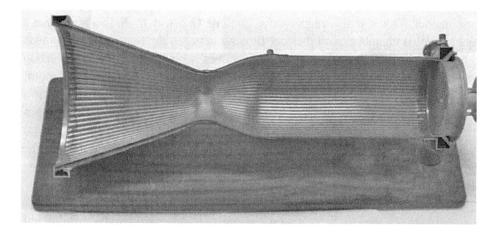

Figure 6. A sectioned example of a liquid hydrogen chamber made at RPE. It has been fabricated from a series of nickel tubes pressed into shape and brazed together. Fuel would flow through the tubes as a coolant.

Rocket Motors in the UK

Lubbock and his co-workers at the Fuel Oil Technical Laboratory in Fulham, London, were the first in the UK to work on liquid fuelled rocket motors. Their first design, called Lizzie, was fuelled by LOX (liquid oxygen) and petrol. It was a very simple device: the propellants were forced into the combustion chamber using high pressure nitrogen, so no pumps and other ancillary equipment were needed. It was intended for rocket assisted take-off in aircraft such as the Wellington. In 1946 it was the first liquid fuelled rocket motor to be test fired at the RPD Westcott, and was eventually developed to give thrusts of up to 2000 lb.

Lizzie was used to power the Liquid Oxygen Petrol/ Guided Aerial Projectile, or LOP/GAP, which was an early test vehicle, fired from

Figure 7. 'Lizzie' in a test shed at Westcott.

Aberporth in Wales, and later, the Rocket Test Vehicle 1 or RTV-1. Problems cooling the engine led to a change of fuel to a methanol/water mixture. As a consequence of work with Lizzie, it was realised that hydrocarbons did not act as good coolants for rocket engines, and that their flame temperatures are also relatively high, exacerbating the cooling problem. Kerosene was not considered again for rocket motors in Britain for some years.

Snarler and Screamer

In 1946 the Ministry of Supply asked Armstrong Siddeley Motors to develop a liquid fuel rocket motor with a thrust of 2,000 lb for use as a booster unit for fighters. Initial ideas were for a hydrocarbon/liquid oxygen motor, but after talks with the RAE, the fuel was changed to a mixture of 65% methanol and 35% water. At the time, it was thought that hydrocarbon fuels were not well suited for cooling the chamber. The first test run of the motor, now named Snarler, together with a fuel pump was achieved in February 1949.

Motors designed to be installed in manned aircraft need to be safe and reliable: by May 1950, the motor had achieved 71 minutes of full thrust in testing. A Snarler motor was then installed in the tail of the Hawker P.1040 prototype (forerunner of the Sea Hawk), the new aircraft now being designated the P.1072. Flight trials began in November 1950.

Snarler was capable of being throttled – not always easy to achieve with rocket motors – and had a maximum sea level thrust of 2200 lb. The S.I. of the motor was 195, with a combustion chamber pressure of 300 lb/sq in (2 0bar).

Work on Snarler's successor, Screamer, began in 1950. Initially intended to give 4,000 lb thrust, the specification given by the Ministry of Supply was changed to a more powerful motor of higher thrust. One of the main differences between Screamer and Snarler is that the Snarler pump was driven externally; Screamer would have its own gas generator to drive the turbines which would power the pumps.

In these early days there was very little design knowledge of gas generators, and it was decided to add water to the combustion mixture to reduce its temperature. Because water was being carried for the gas generator, it was decided to use its excellent cooling properties for the combustion chamber jacket, the heated water then being injected into the combustion chamber itself. Unusually, the combustion chamber had no throat, being a simple cylinder followed by the usual expansion cone.

By 1954, the complete motor was ready for testing, and by September, thrust ratings of 8,000 lb had been achieved. The motor was later installed in the underside of a Meteor for flight testing, but with the cancellation of the Avro 720 rocket interceptor in favour of the Saunders Roe SR53, and the decision to use only HTP in manned vehicles, Screamer was not developed further.

Delta

A pressure-fed liquid oxygen/petrol rocket motor was first tested in Britain as early as 1941. It could provide a thrust of 2,000 lb and was intended as an assisted take-off device. The motor was also used in the early LOP/GAP and the later RTV-1. Hydrocarbons have a high flame temperature and cooling proved to be a problem – the solution was found by changing from petrol to a water-methanol mixture as in Snarler. This led to a prejudice against the combination that persisted through the early 1950s.

In the early 1950s, RPE began taking an interest in larger rocket motors, and a design for a liquid oxygen/liquid ammonia motor was drawn up, notable mainly for its spherical combustion chamber[2]. Discussions were held with ICI at Teeside concerning the availability and supply of liquid ammonia, but visits to America by members of the technical staff at Westcott re-awakened their interest in kerosene as a fuel. The idea of lox/kerosene motors moved back up the agenda, and the ammonia design was dropped. These new designs were named Delta, following the Alpha/Beta/Gamma sequence.

Chamber	Geometry	Thrust (lb)
Delta 1	Spherical	50,000
Delta 2	Spherical	135,000
Delta 3	Cylindrical	185,000
Delta 5	Cylindrical	13,500
Delta 7	Cylindrical	12,500

Delta 3 is the most interesting in that it would have been the starting point for a ballistic missile. Indeed, reasonably detailed sketches were made for a design of such a missile.[3]

Once Rolls Royce had licenced the North American S3 design and developed it into the RZ 2, there was little point in continuing with the work on the larger Delta chambers, and work on them ceased in 1957. Firings of the smaller Delta 5 and 7 continued until 1966.

HTP

The Germans pioneered the use of hydrogen peroxide as a rocket fuel in the early 1940s, powering the Me163 rocket fighter, and the V2's turbine and fuel pump. British work was to take this much further. The key to a successful HTP motor is the choice of catalyst. When the HTP is passed over a suitable catalyst, it decomposes into steam and oxygen, and the decomposition is sufficiently energetic for the HTP to be used as a monopropellant. However, it is much more effective then to inject a fuel into the steam and oxygen. In British rocket motors this was always kerosene. The kerosene ignites spontaneously in the hot gases. Silver plated nickel gauze was used as the catalyst, and such catalyst packs could be easily inserted into the rocket chamber. The ratio of HTP to kerosene was around 8 : 1. Although the combination does not give a very high S.I. compared with many other fuel combinations, it has other advantages. Not being cryogenic, it can be left in the vehicle and does not need topping up. Nor does it need insulation as liquid hydrogen does: the insulation adds to the vehicle weight. Further, HTP is quite dense, 1375 kg/m³, as opposed to 80 kg/m³ for liquid hydrogen. This makes for a very much smaller volume and thus smaller tanks, again saving on vehicle weight. The later rockets developed by the UK using HTP technology were structurally very efficient.

Other engines were then developed using this combination: Spectre, Sprite, Scorpion, Stentor and Gamma. These were initially for aircraft use, although Stentor would be used in the Blue Steel stand-off missile, and Gamma would go on to power Black Knight and Black Arrow. Most of these were developed by commercial firms: Scorpion by Napiers; Sprite and Spectre by de Havilland; Stentor and later Gammas by Armstrong Siddeley, as they were then. Sprite and Super Sprite were designed to assist the take-off of large aircraft such as the V bombers and the Comet, but the increase in effectiveness of the jet engine meant that these units were obsolete before entering service in any major fashion. Scorpion and Spectre were intended for aircraft, to augment the jet engine. However, the HTP combination was to represent the principal British contribution to the rocket field.

The UK was to make hydrogen peroxide technology very much its own: no one before or since has made use of it on such a large scale. Early German and British work used compounds of manganese in one form or another to decompose the peroxide, often injected with the fuel, leading to a very messy exhaust. The secret lay in a metal gauze, through which the HTP was passed, and as it did so, decomposed to steam and oxygen at a temperature of around 500 °C. The gauze was made of silver coated nickel, and a catalyst pack was fitted at the top of the combustion chamber. Into these hot gases a fuel could be injected, and at that temperature they burned spontaneously, meaning there was no further ignition needed. This was very convenient, particularly in the rocket aircraft and the Blue Steel missile.

The largest HTP motor produced was the large chamber in the Stentor motor for Blue Steel, seen above, which had a thrust of around 24,000 lb at sea level. Although the small chamber would find use in Black Knight and Black Arrow, the large chamber was not developed further.

It has been argued that, in some respects, HTP was a technology in search of an application, and in some cases this was certainly true. The Sprite and Super Sprite were developed as rocket assisted take-off units for the Comet airliner and the Valiant bomber, but were far too sophisticated for simple RATO units, which were normally made from clusters of small solid fuel motors. The advantage of using several motors in clusters is that it was far less catastrophic if one failed. Having just two motors, one on either side, was much more hazardous, since the failure of one of the two would result in an off-centre thrust sufficient to make

Figure 8. A later Gamma chamber, as used on Blue Steel, the later Black Knights, and Black Arrow. The ring at the top of the motor was where the HTP entered the motor, which was made of thin tubes formed to the shape of the chamber and brazed together. The catalyst pack is shown on the lower right.

the aircraft lose control. Such an elaborate system, whereby the used motors would be jettisoned, parachuted back to the ground, then serviced for re-use, made very little sense.

The Scorpion was produced by Napier, and a twin-chambered version, the Double Scorpion, was fitted to Canberra bombers, enabling one of them to reach a new record altitude of 70,310 ft. They were to have been used for high altitude cloud sampling at the H bomb trials at Christmas Island (Operation Grapple), but

Figure 9. The Stentor motor developed for the Blue Steel stand-off missile.

the second Canberra was grounded during the crash investigations. There was also a proposal to fit it to the English Electric Lightning, but the Lightning's performance proved to be quite good enough without the rocket. Rocket assisted take-off and rocket interceptors very soon became obsolete; the main contribution of HTP motors was to Black Knight, Black Arrow and Blue Steel – and it is questionable whether HTP was the correct choice for Blue Steel. However, a new use was to be found for HTP motors – in ballistic rockets. The original Gamma 201 motor for Black Knight used four Gamma chambers, a double-walled chamber developed by RPE. This chamber was later replaced by the small chamber from the Stentor motor, which used the tube-walled construction. Equally importantly, the 301 allowed better adjustment of the kerosene/HTP mixture ratio, making the motor more efficient.

The Stentor small chamber was carried over into Black Arrow, where the first stage motor, the Gamma 8, had, not surprisingly, eight chambers, arranged as four pairs. The second stage of Black Arrow was powered by the Gamma 2, which had two chambers, but with an extended expansion cone, as it would be operating in the near vacuum of altitude. This gave it a higher thrust than the first stage chambers.

There is a final footnote to British HTP work. Bristol Siddeley (who became part of Rolls Royce in 1966) were given a contract by the Ministry of Aviation to develop a high performance HTP motor of 7,500 lb thrust, following on from suggestions made by the firm in 1963. The development programme ran from January 1964 to December 1966[4]. The chamber was designed to run at much higher pressures than usual – 1,000 psi – and a total of 118 firings were achieved, totalling 78 minutes. The thrust level of 7,500 lb was chosen deliberately so that the chamber could be used as a direct replacement for the existing Gamma chamber.

Unlike the existing Gamma chambers, the new chamber (named, for some inscrutable reason, Larch) was double-walled. The reason given for this was that 'HTP tends to decompose on the hot surfaces in the cooling tubes, producing insoluble gases which can occupy an unacceptable proportion of the restricted passage of one or more of the tubes and lead to burnout'.

Figure 10. A Gamma 201 motor for Black Knight being test fired at the Armstrong Siddeley test site, Ansty.

The higher chamber pressures also gave an improved S.I.:

	Standard Gamma	Larch
Sea level SI	217	226
Vacuum SI	251	269

The new chamber (Figure 11), would also have been slightly lighter.

Replacing the existing Gamma chamber in Black Arrow with the new improved version meant that the vehicle could be stretched. As a consequence, the payload could be increased from 232 lb in polar orbit to 375 lb.

Despite the time and money that had been spent on the development, it was not taken further. When RAE did decide to uprate Black Arrow, it went for the solid fuel strap-on booster option. The Gamma motors of Black Arrow were to be the last HTP motors to be developed, but HTP motors did put Britain's only satellite into orbit, and it is fitting that a British developed technology was used to do so.

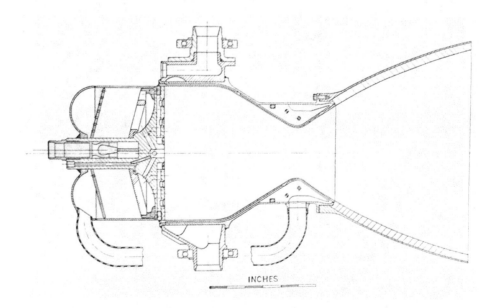

Figure 11. The experimental high pressure 'Larch' chamber.

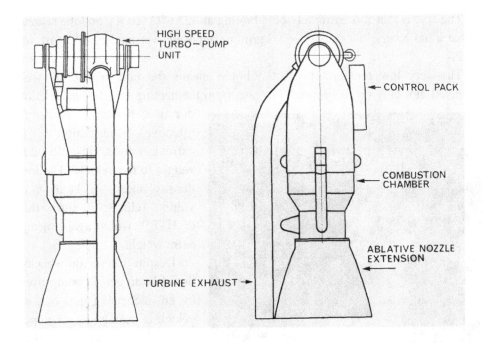

Figure 12. 'The Larch' HTP/kerosine test chamber.

Hydrogen

Liquid hydrogen is usually regarded as the most effective fuel for rockets. (In this section, it may be assumed that liquid oxygen is the oxidant. Fluorine is better theoretically, but is very hazardous environmentally, if from no other point of view.) This is because it has a very high exhaust velocity, or looking at it another way, a very high S.I. Thus the HTP/kerosene combination used in Black Knight and Black Arrow has an exhaust velocity in vacuum of around 2,500 m/s, whereas a well-designed liquid hydrogen motor can achieve exhaust velocities of around 4,400 m/s.

Using the rocket equation $v_{final} = v_{exhaust} \times \ln(\text{mass ratio})$, a liquid hydrogen stage of the same mass ratio would achieve a final velocity around 75% greater. On the other hand, the structural penalties of using liquid hydrogen means that the mass ratio would be significantly lower than an equivalent HTP stage. There are complications to liquid hydrogen vehicle design.

The first is that it is extremely cold boiling at −253 °C (20 K), and the second is that it has a very low density – 70 kg/m^3 as opposed to around 1,300 kg/m^3 for HTP.

The very low temperature of the liquid means the tank has to be well insulated, not only on the ground, but also from the heating effect of air friction during launch. Although effective insulation is extremely light, this still adds weight to the vehicle. The low density means a large tank volume (almost 20 times that of HTP!), which again means extra weight.

Despite these drawbacks, liquid hydrogen is being used in an increasing number of vehicles, usually as an upper stage. The Ariane 5 central core uses liquid hydrogen, although it has the two large strap on solid fuel boosters to lift it to altitude. The Ariane 5 ECA (Evolution Cryotechnique type A) core has a burn time of 650 seconds.

RPE began work on hydrogen chambers in the late 1950s. At that time, they had no facilities for storing or producing liquid hydrogen, but instead used gaseous hydrogen pre-cooled by liquid nitrogen. A number of fully working chambers were built and fired at Westcott (see Figure 14).

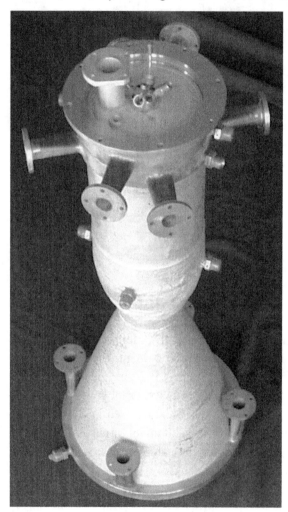

Figure 13. A hydrogen/oxygen test chamber built at RPE Westcott.

The larger chambers were capable of around 4,000 lb thrust: it would have been relatively easy to scale them up to, say, 8,000 lb, which would be well-suited to upper stages for Blue Streak, Black Knight, or Black Arrow. Such stages would have increased payloads very considerably. Whilst developing the chambers would not have been difficult, building a liquid hydrogen stage would have needed a considerable amount of development work and thus cost.

Based on this work, a variety of designs for launchers using Black Knight as the first stage were drawn up[5]. Some were pressure fed, others used turbopumps. Sketches of the designs can be seen below.

Calculations were carried out for a variety of configurations. Four different first stages were considered and three different second stages. A Cuckoo solid fuel motor was taken as third stage (calculations were also carried out for a two-stage version, without the Cuckoo motor, but only two of the combinations were able to put any payload into orbit at all). Payloads could no doubt be increased somewhat by a purpose-built third stage.

Figure 14. A hydrogen/oxygen chamber being test fired at the RPE, Westcott.

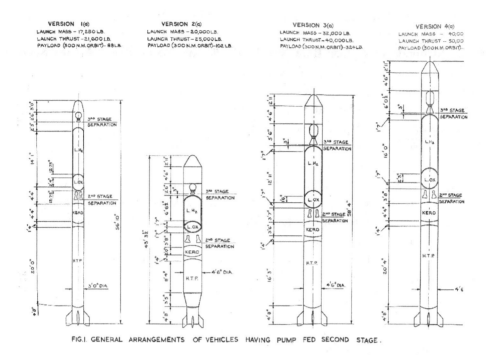

FIG.I. GENERAL ARRANGEMENTS OF VEHICLES HAVING PUMP FED SECOND STAGE.

Figure 15. Various proposals for satellite launchers using Black Knight as the first stage and a liquid hydrogen/oxygen second stage.

Version 1 is the standard Black Knight, with a tank diameter of 3 ft and a sea-level thrust of 21,600 lb. Versions 2, 3 and 4 have a tank diameter of 4 ft 6 inches and sea-level thrusts of 25,000, 40,000 and 50,000 lb respectively. Version 3 would have a six chamber motor, and version 4 an eight chamber motor (effectively the first stage of Black Arrow). The lift-off weight was derived by assuming a thrust : mass ratio of 1.25.

Three variants of the second stage engine and propellant feed systems were examined:

(a) An engine having four chambers w ith turbo-pump feed of the propellants.

(b) An engine having four chambers with pressurised tank feed.

(c) A single chambered engine with pressurised tank feed.

The estimated payloads for each variant was calculated as being:

Version	Launch mass	a	b	c
1	17,280 lb	88 lb	18 lb	56 lb
2	20,000 lb	102 lb	56 lb	76 lb
3	32,000 lb	324 lb	169 lb	248 lb
4	40,000 lb	377 lb	187 lb	289 lb

Versions 1 and 2 are really non-starters. Versions 3 and 4 are, on the face of it, fairly promising. However, the first stage of version 4 is in effect Black Arrow. Developing Black Arrow, where the intention was to keep the cost down by using as much Black Knight technology as possible, stretched the budget. Developing a liquid hydrogen stage, which would have been technically challenging, would have been much more expensive, and, as can be seen, payloads were not very significant. Some improvement could have been achieved with solid fuel strap-on Raven boosters, but not enough to make the design worthwhile.

Another major proposal was for a liquid hydrogen third stage for the Blue Streak satellite launcher and the Anglo-French launcher proposal. Although the Saunders Roe brochure for Black Prince is sometimes taken as the 'definitive' version of the Blue Streak launcher, there was, in reality, no such thing. Black Prince shows an HTP third stage, but the RAE realised that a liquid hydrogen stage could increase the payload considerably, and in this period, it was looking at 6- or 12-hour orbits for communications satellites. It is interesting to see the emphasis that this stage is given in the initial brochure for the Anglo-French launcher.

Two quite comprehensive studies were carried out: one by the RPE and one by Saunders Roe. Both go into considerable detail, including detailed analysis of the thermal cladding that would be needed for the liquid hydrogen tank.

The RPE produced the report for its design in April 1961.[6] One of the more unusual features of the report is that it seems to be the only one written in this period (other than some ELDO reports) which uses entirely metric units. This leads to some slightly awkward conversions. For example, the diameter of the fuel tanks is 1.37 m... or 54 inches! This was obviously designed as a third stage for a Blue Streak/Black Knight combination. Indeed, the RAE had calculated the optimum mass for a liquid hydrogen third stage for the Blue Streak launcher to be 2,270 kg, and the stage was designed around this weight, although later calculations showed the optimum mass as 3,630 kg.

There would be four motors in the stage, each of which was intended to produce a thrust of 9 kN (2,000 lb) with a chamber pressure of 50 N/cm² (5 bar or 75 psi). One design being considered was what might be described as self-pressurising: a pressure-fed system, with the gases being used to pressurise the tanks coming from the fuel itself via a heat exchanger. The tank pressures could be relatively low given the low chamber pressure – 80 N/cm² (8 bar or 90 psi) was the value being considered. This is quite an elegant solution, dispensing with the weight and complexity of a turbopump, yet avoiding the weight penalties of thicker tank walls and heavy gas bottles. The only drawback is that with the relatively low thrust, the burn time will be quite prolonged, which means carrying the unburned fuel up the Earth's gravitational potential well as the vehicle gains in height.

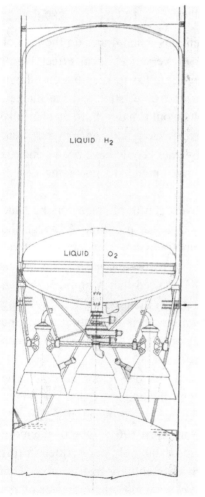

The specification for the BSSLV third stage investigated by Saunders Roe required a motor which had:

(a) A thrust of between 3000lb and 4000lb (in vacuo) lasting for about 15 to 20 minutes.

(b) A thrust of between 40lb and 60lb (in vacuo) lasting for about 2½ to 3½ hours.

Communication satellites need to be in as high an orbit as possible, and the new vehicle could have put an appreciable payload in an orbit around 8,000 miles high. The usual method of doing this is the apogee motor, as discussed before. Bristol Siddeley came up with a design for a motor which used a motor with two large chambers and two small chambers. The large chambers would take the vehicle up to orbital height, but the small chambers would then be used to raise the orbit, with a

Figure 16. The RPE design for a liquid hydrogen/oxygen third stage to be used as part of a Blue Streak launcher.

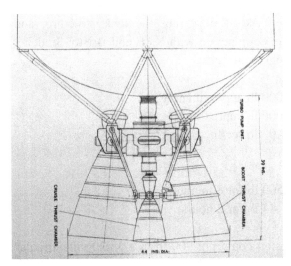

Figure 17. The BS 600 proposal with two large pump fed chambers and two small pressure fed chambers.

burn time of two or three hours.

To power such low thrust chambers with a pump was impractical. Pressurising the tanks usually meant carrying large and heavy gas bottles. Instead, the proposal was to use a heat exchanger to produce 'hot' (relative in this context) hydrogen gas. The gas could then be used to pressurise the tanks (a further heat exchanger would be needed for the liquid oxygen tank).

Rocket chambers are usually at quite high pressures – perhaps 40 times atmospheric pressure. At sea level, the escaping gases are opposed by atmospheric pressure, and higher chamber pressures make the motor more efficient. In the vacuum of space this does not apply. Chambers can be run at quite low pressures, and it was suggested in this case that a chamber pressure of only one atmosphere might well be feasible. This avoids the complication of pumps and the weight of gas bottles.

On the other hand, pumps are needed for the earlier boost phase, and unless they are discarded (which they were not), the vehicle is carrying unnecessary weight during the long cruise phase.

Bristol Siddeley had not done any work on liquid hydrogen motors up to now, and this proposal was marked in the Ministry of Supply file with a hand written comment:

Downey thinks we would be nuts to bring yet another firm into the space business.

Downey thinks we would be nuts to bring yet another firm into the space business.[7]

Downey was one of the senior officials in the Ministry of Aviation – the criticism is slightly unfair since Bristol Siddeley were already producing the Gamma motors for Black Knight.

Saunders Roe were given the task of designing the tank structure, and produced a substantial brochure[8]. In conjunction with RAE and Bristol Siddeley Engines, the parameters for the design were set:

All Up Weight:	7,000 lb approximately
Propellants:	5,000 lb approximately
S.I.:	400 lb.sec/lb

Two thrust phases:

(1) Boost:	3,500 lb for 8 minutes
(2) Cruise:	44 lb for 2 hours

It was estimated that such a design could put 900 lb in a 5,000 mile circular orbit or 600 lb into a 9,000 mile orbit.

In many ways this is an interesting idea, but on closer inspection has as many drawbacks as advantages. During the cruise phase, the large chambers and their plumbing are still attached to the vehicle, but as dead weight. Jettisoning them would make the proposal much more efficient but would not be easy. Secondly, a 'slow burn' is less efficient from a different point of view: as the vehicle climbs the gravity well as it moves further from the Earth, then it is taking unused fuel with it. From a gravitational potential energy point of view, it is better to expend the fuel in one big burst at the start of the orbit transfer – this does the opposite!

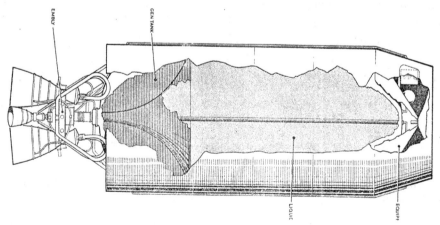

Figure 18. Saunders Roe's proposal for a liquid hydrogen/oxygen stage.

In the end, of course, all of this became moot. The third stage of Europa was to be developed by Germany, and the design chosen was very different. As a consequence of the ELDO B proposals put forward by the French in 1964, contracts were awarded for research into liquid hydrogen motors. Rolls Royce was one of the firms involved, and building on the work done at the RPE, began the design and testing of a motor called the RZ 20. The contract was shared with the French firm of SEPR (Société d'Etudes pour la Propulsion par Réaction). Rolls Royce was to produce the thrust chamber part of the motor, with SEPR providing the turbopump assembly.

Val Cleaver, Chief Engineer of the Rocket department at Rolls Royce, wrote to the Ministry of Technology asking whether he could build the test stand at Spadeadam, which was a Ministry establishment, but being run as an agency by Rolls Royce.

JEP Dunning, Director of the RPE, protested that the test site should be built at Westcott, which already had extensive facilities for testing and firing liquid hydrogen chambers, although none of them had been as powerful as the proposed Rolls Royce chamber, seen below being test fired at Spadeadam (Figure 19). Cleaver pointed out that Rolls Royce were building the facility as a private venture. As he put it in a letter to the Ministry of Technology:

> It was not possible for ELDO to commit any money for the necessary test facilities for these chambers. Therefore (and not without some difficulty, as you can probably imagine) I persuaded our Main Board to sanction the cost of a modest test cell for the purpose, as a P.V. Extension to the Components Test Area at Spadeadam.[9]

He went on to say:

> We are most anxious to have the test facility at Spadeadam, for two very definite reasons:-

> (a) Because if any larger operational programmes for hydrogen rocketry ever arise in the UK, it will be inevitable that their testing should be done at Spadeadam. It is highly desirable, therefore, to begin as we mean to go on, and start gaining early experience there as soon as possible.

> (b) Because the team at Spadeadam desperately need some injection of new work, to raise morale and inspire some confidence in the future of the establishment (I am sure I do not need to emphasise to you the problems of this sort we have had since 1959, with the 1960 military cancellation and all ELDO's subsequent ups and downs.)

Figure 19. Rolls Royce RZ 20 hydrogen motor being test fired at Spadeadam.

Given that it was a private venture, the costs were substantial: £54,000 for the construction costs, £15,000 for two liquid hydrogen road trailers, and £5,000 for the chamber itself, making a total of £74,000 (a contingency figure of approximately 10% was added to the estimate to bring it up to £81,000).

In the end, two firings, each of ten seconds duration, were achieved before the programme ran out of money. The total cost of the programme was £250,000.

Ion Motors

Ion motors are an extremely promising technology, and have remained mainly at the 'extremely promising' stage for the past 40 years. Instead of converting chemical energy into kinetic, as in conventional motors, it converts electrical energy into kinetic. This is done by ionising atoms, then accelerating them through a very large voltage. The main problem is that the only source of electrical energy in space (other than a nuclear reactor) comes from sunlight via solar panels. The amount of energy is not great, and thus the thrust available is small.

The RAE began investigating ion motors as early as 1963. Initial thoughts were for an attitude control and station-keeping capability. Another possibility emerged as a requirement for a high energy upper stage for the Black Arrow launcher, utilising the spiral orbit-raising principle from an initial low altitude parking orbit. Interestingly, at that time ELDO was also considering the augmentation of the payload capability of its launch vehicle by the same means.

Figure 20. The T1 ion motor developed at the RAE.

The initial designs used mercury: an ion accelerating potential of about 1.5 kV was chosen, giving a S.I. of close to 3,000 s. The success of the initial tests with the T1 thruster resulted in the design of a new device, the T2, for which a 10 cm beam diameter was selected to provide a thrust of 10 mN with a beam accelerating potential of 2 kV.[10]

10 mN thrust might be adequate for attitude control, but not for orbital adjustment. Further development has continued, with the main change being the substitution of xenon gas for mercury as the fuel, but the further story is rather beyond the scope of this book.

As a measure of the work being done by British firms on rocketry, the following figures were given in 1961 for the total expenditure to date (i.e., effectively, since the war):

	£ in millions	dates
Napier:		
Double Scorpion	1.486	1955–1959
de Havilland:		
Sprite & Super Sprite	0.881	1947–1961
Spectre	5.576	1953–1961
Research	0.254	1954–
Bristol Siddeley Engines Ltd:		
Snarler	0.226	1946–1953
Screamer	1.222	1946–1953

Stentor	3.40	1956–
PR.37/2 (Jindivik)	0.029	1960–
Gamma	1.60	1956–
Research	0.39	1955–
Rolls Royce:		
Blue Streak	5.379	1954–
Supply of HTP	3.50	1946–

These costings do not include the work done at RPE Westcott, which was considerable. The series of motors leading up to Gamma (Alpha and Beta) were developed there in the late 1940s and early 1950s. Rolls Royce used Westcott's facilities for early RZ 2 work, and RPE also had its own on-going liquid hydrogen work. In addition, Westcott was a major centre for the development of solid fuel motors.

However, by 1968 the picture had changed radically. Napier no longer existed, de Havilland were doing no more work on rocket motors. Bristol Siddeley and Rolls Royce had been amalgamated. There was just the one firm, and work was shrinking. Val Cleaver, in charge of the rocketry work at Rolls Royce, wrote to the Ministry of Technology to ask how he could keep his team together:

> In this atmosphere, it is hard to maintain staff morale, or to retain the good people. Many of our best men have gone out of rocket work over the years (apart from a few who have left our projects, only to emigrate to the States), and we have been able to justify the recruitment of only a mere handful of bright youngsters in recent years.

The Director General (Engineering) at the Ministry of Technology then wrote to CGWL to explain:

> My object in encouraging Cleaver to write to you was the conviction that the presently foreseen programme of liquid rocketry development implies the winding up of Rolls Royce's R&D activity in this field by the early seventies, and the facts need to be faced now, both by Mintech and by the Rolls Royce management.

He then went on to give the following figures for future planned expenditure, based on no further commitments to ELDO, one Black Arrow firing a year, and a limited programme on packaged propellants:

Year:	1967	1968	1969	1970	1971
Expenditure: (in £ million)	1.656	1.900	1.117	0.667	0.437

It is not clear whether this includes work on the RZ 20 liquid hydrogen motor, which was being carried out under an ELDO contract (and part of the funding had come from Rolls Royce itself).

Indeed, one of Cleaver's complaints was that it was all obsolescent technology. The RZ 2 had been designed in 1955, 13 years earlier, and although since refined, there was nothing further to be done with it. Similarly with the HTP work: all that was left was derivatives of the Gamma motor, which had started life again in the mid-1950s as the small chamber from the Stentor engine. The liquid hydrogen work was being run on a shoe string.

In the event, work effectively finished in 1971, with the demise of Black Arrow and of Europa. Since then there has been no significant rocketry work done in the UK.

Solid Fuel Motors

In principle, solid fuel motors are very simple. A tube is filled with the fuel/oxidant mixture, which is then ignited – but as always, there is rather more to it than that. Early motors used simple cordite, a mixture of nitroglycerine and gun cotton, and were end burning – that is, the cordite at the end of the tube is ignited, and the cordite burns upwards towards the other end. Cordite was replaced by propellants based on ammonium perchlorate (NH_4ClO_4) and ammonium picrate ($C_6H_2(NO_2)_3O.NH_4$) with small amounts of other material added.

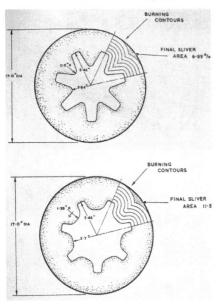

Figure 21. Cross sections through two solid fuel motors.

A British innovation was that of centre burning. An empty cylinder runs the length of the tube. The igniter is at the top, and when initiated, the fuel burns from the centre out to the edges. One obvious problem is that the surface area increases as the burning spreads out, and one way to overcome this is to have a star-shaped cut out (see Figure 21).

From the military point of view, solid fuel missiles are vastly preferable to liquid fuelled ones. The solid fuel tube has to be very strong to withstand the high pressures and temperatures inside, thus making it very robust when it comes to handling. Liquid fuelled missiles, however, have very thin tank walls, and in any accident there is the potential to spill a good deal of rather nasty liquid. With solid fuel motors, it is a question of point and fire; liquid fuelled missiles need a good deal of careful setting up.

Solid fuel motors have other advantages: by varying the geometry or the combustion mixture, motors can be made that give very large thrusts for very short periods of time, or smaller thrusts for a longer period. The Gosling boosters for the Bloodhound missile accelerated the vehicle to over Mach 2 in three seconds. The thrust is not uniform, as the graph below shows[11]. In particular, there tends to be a long tail off as the last slivers burn away (see Figure 23). For these reasons, the thrust and burning time given in reports are only approximations.

Solid fuel motors have two disadvantages in a satellite launcher: they tend not be very energetic (have a low S.I.) and have a poor mass ratio (mass full/mass empty). S.I. is related to the exhaust velocity of the gas (in modern units, they are the same), and final velocity of a rocket stage is given by:

$$V_f = V_e \times \ln(\text{mass ratio})$$

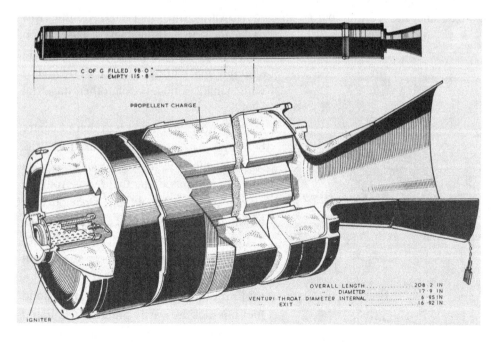

Figure 22. Rook solid fuel motor.

Thus the first Black Knight rocket, BK01, had an all up weight at launch of 13,072 lb, and 1,424 lb when empty. Hence its mass ratio was (13,072lb/1,424lb) = 9.18. With an S.I. of around 220, its final velocity in the absence of any other forces would be (220 × 9.8) × ln(9.18) = 4,800 m/s. Performing the same calculation on the Cuckoo II motor, used as the second stage on later Black Knight vehicles, gives 3,750 m/s – quite a significant difference.

There are two obvious ways of improving performance: increasing the S.I. of the fuel, and making the case lighter. Hence later solid fuel motors became more efficient. The solid fuel boosters either side of the Shuttle have an S.I. of 242 at sea level (268 in vacuum). There is also another way to improve performance, which is simply to build them bigger. The mass ratio improves with size since the amount of material for the case is proportional to the radius of the tube, whereas the amount of fuel inside is proportional to the square of the radius.

Most British solid fuel motors were relatively small. The largest was the Stonechat, with a diameter of 36 inches. The Stonechat formed the basis of the Falstaff vehicle, which was used to test components of the Chevaline system. (Chevaline was a Polaris upgrade programme whereby one of the three re-entry vehicles and its warhead was removed to make way for an elaborate system of decoys.) Even so, its total impulse was only 1,700,000 lb.s as against Black Knight's 2,300,000 lb.s.

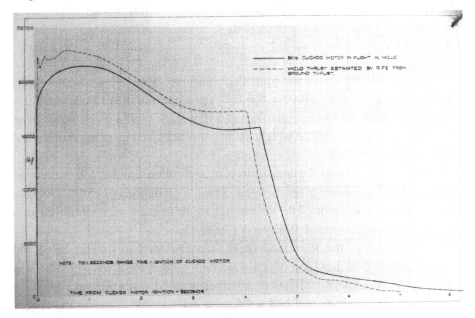

Figure 23. Thrust/time curve for a Cuckoo motor, showing the tailing off of the thrust near the end.

It is interesting to compare Stonechat to the Algol 1 motor (first flown in 1960), which was used as the first stage of the original Polaris missile and also as the first stage of the Scout satellite launcher.

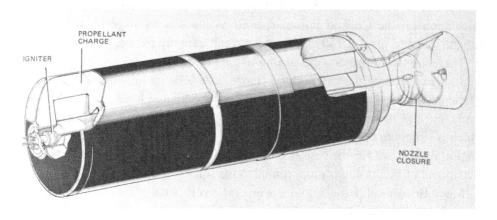

Figure 24. The Stonechat 36-inch solid fuel motor.

	Stonechat:	Algol 1
Weight:	10,300 lb	23,600 lb
Diameter:	36 inches	40 inches
Thrust:	32,000 lbf	106,000 lbf
Burn time:	53 seconds	40 seconds
Sea level S.I.:	212	214

In terms of S.I., the two look equivalent, and the mass ratios compare quite favourably, being (23,600/4,100) = 5.8 for Algol and (10,300/1,800) = 5.7 for Stonechat. The later A3 Polaris missile had a first stage with a much better mass ratio: (24,400/2,790) = 8.7. The weight saving was achieved by using a fibreglass casing.

Several 17-inch motors derived from a motor called Smoky Joe, so named from the plume it produced when burning. These include the Albatross, Cuckoo, Goldfinch, Raven and Rook. The Raven and Rook motors were employed in a variety of different roles in the 1950s and 1960s.

The motor tube of the Rook and the Raven consisted of two wrapped and welded cylinders 90 inches long which were butt welded together. The tube was made of steel of thickness 12 SWG (0.104 inches or 2.64 mm). Head ends were welded to the tube: the top end had a threaded opening for allowing the charge former to be centralised during propellant pressing and allowing excess propellant to 'bleed' off. Later, the igniter would be fitted in the opening.

These two motors formed the backbone of the various solid fuel vehicles used for a variety of research purposes, with several hundred motors being fired. The Raven formed the basis of the Skylark vehicle.

Below is a table listing a few of the motors developed at RPE. This is taken from a manual of solid fuel motors which listed data for a total of 73 different types of motor[12].

Motor	Thrust (lb)	Burn Time (seconds)	S.I.	Weight (lb)	Length (inches)	Diameter (inches)
Cuckoo I	18,200	4.1	204	524	51.7	17.2
Cuckoo II	8,200	10	213	500	51.8	17.2
Raven VI	15,000	30	191	2,540	206	17.2
Smoky Joe	2,900	39	171	925	123	17.2
Stonechat	32,000	53	212	10,300	216	36.3
Waxwing	3,500	55	*282	761	49.7	28

* in vacuum

These data are taken from an index of solid fuel motors developed at RPE Westcott in the mid-1960s. The table shows only a small selection – 73 motors were listed in all. These rockets were used for a variety of different purposes:

Cuckoo I:	Extra boost for first stage of Skylark.
Cuckoo II:	Black Knight re-entry tests.
Raven VI:	Skylark.
Siskin II:	Black Arrow – stage separation and to settle propellants in tanks.
Smoky Joe:	Red Shoes, which became the Thunderbird SAM.
Stonechat:	Falstaff vehicle for testing of Chevaline components.
Waxwing:	Third stage of Black Arrow.

The most famous solid fuel rocket produced in Britain was Skylark, which had a remarkably successful career. First launched in 1957, from Woomera, its final launch took place from Esrange, Sweden, on 2 May 2005. In all, there have been 441 launches, from sites in Europe, Australia, and South America.

The design first dates to 1955, when initial work was carried out by the RAE and the RPE. The first vehicles were ready less than two years later, and sent for testing to Woomera during the International Geophysical Year.

During the 1960s Skylark evolved into an excellent platform for space astronomy, with its ability to point at the Sun, Moon, or a star. It was used to obtain the first good quality X-ray images of the solar corona. Within the UK national programme, the frequency of Skylark launches peaked at 20 in 1965 (from Woomera), with 198 flights between 1957 and 1978.

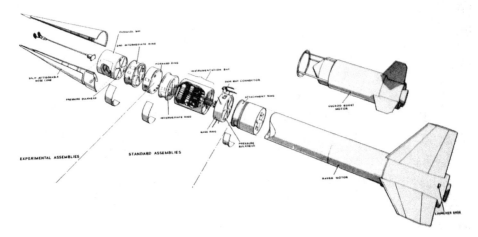

Figure 25. The Skylark sounding rocket.

Skylark began as a simple one stage vehicle, with three fins and a relatively long burn time of 30 seconds, using the Raven motor. This was to keep the accelerations within reasonable values. A series of different Raven motors were produced, each with a different filling as requirements changed. As a consequence of the low acceleration, a tower was needed to guide the rocket for the first few seconds. This was simple in construction and used components from Bailey bridges!

An extra boost stage was added to improve performance. Initially, this was the Cuckoo motor (so named, apparently, because its function was to kick the Raven out of its nest). Later versions used the Goldfinch motor in place of the Cuckoo.

The following description of how Skylark was used by the space science community was written by Professor Mike Cruise[13], who has had a long and distinguished career in space science.

Many of the senior space scientists around the world were trained in space instrument design, data analysis and space project management on projects using the Skylark sounding rocket as the space platform. In the nineteen sixties and seventies over two hundred Skylarks were launched from sites in Norway, Sardinia, Australia and South America offering five minutes of observing time above 100 km and substantial payload carrying capacity. Many of the Skylark flights delivered data which ended up in Doctoral Theses, launching the careers of the students involved. A PhD gained by this route involved science, engineering, travel and exposure to many different professional cultures ...

The scientific instrument was constructed on a circular bulkhead of magnesium alloy which was previously delivered from BAE as part of the Skylark 'Meccano kit' ... The design of the Skylark provided great flexibility for the experimenter. Holes could be cut in the cylindrical bays or in the circular bulkheads provided the design was approved by BAE at Filton. The strength of the vehicle was in the magnesium alloy cylindrical skin. Generally four or five cylindrical bays would be mounted on top of one another containing the parachute, batteries, the control and telemetry systems and then the attitude control system if one were employed. Usually the experiment bay was mounted at the top under a conical nose cone which split longitudinally in two sections after reaching altitude.

A few hours prior to launch, the stack of Skylark bays with the nosecone at the top and the parachute bay at the bottom was mounted on a small trolley and taken by road to the Skylark launcher. The vehicle was rail launched – that is, there were three parallel rails mounted vertically in the launch tower and metal shoes were fitted at various positions along the length of the vehicle to engage with these rails. The fins extended outside of the rails, in the azimuthal spaces between them ... The launcher tower was about 50 metres tall and the whole launching assembly could be tilted to angles of about 15 degrees from the vertical to adjust the trajectory for winds.

It was necessary to make calculations of the ballistic winds at various heights to predict the trajectory as the vehicle was only powered for 35 seconds of the ascent phase and had no guidance system.

Balloons were launched and tracked by radar for several hours beforehand to provide this data on the winds up to 10 or 15 kilometres altitude. In addition, there were various instrumentation checks and the firing of sighter rockets to check that all the radars and kine-theodolites were functioning correctly before a firing took place ...

Figure 26. A Skylark launch from Woomera.

Normally the experimenters watched the launch proceedings from the block house, EC2, a concrete building below ground level, built into the edge of the concrete launch apron. All the control connections to the vehicle came to EC2 and there were telemetry receivers to check data from the instrumentation and the experiment. In a separate room in EC2, an Australian military technician did the actual firing by starting an automatic sequencer two minutes before launch. This counted down and issued the firing pulse to the detonator in the booster motor at the pre-programmed time. Up to two seconds before launch the launch could be stopped using a line attached via a small snatch connector to the side of the instrumentation bay. Several people in EC2 had 'Stop Action' buttons which could abort the launch via this route. The snatch connector was left in place as the launch proceeded and the wires literally snatched from the side of the vehicle as it departed.

What did the Skylark programme produce in the way of benefits to the UK and the students concerned? Some very new science in most cases. Studies of the ionosphere, the middle atmosphere, X-ray sources, UV spectra of stars and, towards the end of the programme, some Earth observation data – all were progressed by Skylark experiments and contributed to the early development of space science. The Skylark engineering design was conservative to say the least, and most of the experiments were far in advance of the instrumentation that supported them. Mechanical switches were still being used to multiplex the telemetry while semiconductor storage was being employed to capture science data in the experiment. This conservatism was a lost opportunity for UK space companies who, given a freer hand, might have built more advanced equipment with consequent spin-off for the emerging satellite telecommunications industry. Undoubtedly the restraining hand of RAE Farnborough was at work in this respect.

The parachute failures dented the effectiveness of the whole programme and were a factor in letting the US pull ahead in many scientific fields. As the payloads became heavier and longer, the parachute design remained the same and success rates suffered. It must be recorded that, by the middle of the seventies, sounding rockets were losing their place to satellite borne equipment. Why spend three years building rocket borne equipment to gather five minutes of data when you could spend five years building satellite borne equipment that would deliver three years of data? The economics were against investing in new rocket technologies. The range at Woomera was extremely effective in the late sixties and the BAE team did their very best within the hardware limitations to ensure the experimenters gained the data they wanted.

The big contribution of the programme was the opportunity for young scientists and engineers to experience a space project from beginning to end within a PhD duration of three or so years. Vicarious benefits included seeing a snapshot of the whole British colonial experience in the space of a few days journey across the world and the opportunity to test oneself in management terms against time, technology and resource constraints. The nostalgia felt by those who experience a Skylark PhD is fuelled by the current lack of any replacement for the horribly realistic management training it provided.[14]

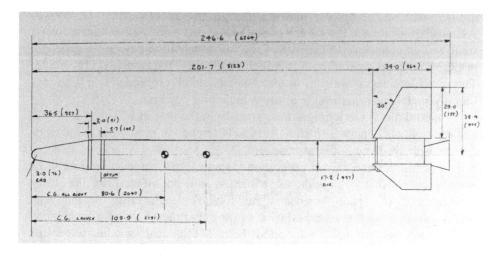

Figure 27. The lay out of a typical test vehicle for solid fuel motors – in the case, a Rook motor. (Dimensions are in inches and mm.)

---

[1] TNA: PRO DSIR 23/28964. A review of the use of liquid propellant rockets in the UK, past and present influences and some considerations for the future (RPE TM 233). Paper by JEP Dunning, Director, Rocket Propulsion Establishment, prepared for the Third LPIA Meeting, November 1961.

[2] TNA: PRO AVIA 48/30. Delta.

[3] The Delta Project – Early Lox/Kerosene Engines in the UK. J. Harlow, presented at the 46th International Astronautical Congress, 1995.

[4] Rolls Royce Rocket Research Report No. 85. A High Performance Combustion Chamber using 85% HTP/Kerosine.

[5] TNA: PRO DSIR 23/30814. Orbital capabilities of Black Knight with a hydrogen-fuelled second stage (RAE TN Space 30).

[6] TNA: PRO DSIR 23/28768. The use of liquid hydrogen in the third stage of a satellite vehicle. AB Bailey, RG Cruddace and BWA Ricketson, April 1961.

[7] TNA: PRO AVIA 65/1568. Use of liquid propulsion in rocket engine development.

[8] Saunders Roe Technical publication S.P.510.

[9] TNA: PRO AVIA 92/232 May 1967. Correspondence between Cleaver and the CGWL.

[10] TNA: PRO AVIA 6/24109. Ion Engine Development at the Royal Aircraft Establishment, Farnborough. Technical Report 71102. BP Day, DG Fearn, GE Burton, May 1971.

[11] TNA: PRO AVIA 6/19255. Proving trial of the Black Knight vehicle for re-entry physics experiments.

[12] TNA: PRO AVIA 68/111. Index to solid propellant rocket motors.

[13] Professor Cruise started his research career at the Mullard Space Science Laboratory of University College London, gained his PhD in 1973, was appointed to a lectureship at

UCL in 1979 and became Deputy Director of MSSL in 1985. In 1986 he moved to the Rutherford Appleton Laboratory and became the Associate Director for Space Science in 1993. Moving to the University of Birmingham in 1995, he was appointed Professor of Astrophysics and Space Research and in 1997 became Head of the School of Physics and Astronomy and subsequently Pro Vice Chancellor for Research and Knowledge Transfer.
[14] This section was published in an expanded version in issue 5 of the journal *Prospero*, published by the British Oral History Project.

Chapter 3

Rocket Interceptors

An Overview

In 1945 the RAF and USAF had the world's most powerful strategic bomber fleets, yet they were on the point of becoming obsolete, and the factor that was driving them obsolete was the jet fighter. The increase in performance that the jet engine gave to interceptors rendered the likes of the Lancaster and its derivatives hopelessly vulnerable. If airborne radar and guided weapons are added to the armoury of the fighter, the balance tilts even further away from the bomber.

One answer, of course, was to build jet powered bombers. The Air Ministry had been aware of this for some time, and before the Second World War had ended, had issued the Operational Requirements that would lead eventually to the V bombers, which were, together with guided missiles and the development of atomic weapons, a major part of the post-war defence programme. Similarly, the Americans, while having pushed their propeller driven designs as far as feasible, were also busy designing jet bombers in the 1940s and 1950s, culminating in the B52, still in service.

In post-war Europe, the strategic focus for the Western Allies switched very rapidly from Germany to Soviet Russia. The Soviet Air Force was also a formidable fighting machine, although it had evolved along lines more tactical than strategic. It had, on the drawing board, some impressive interceptor aircraft such as the MiG 15.

The first rocket powered interceptor of all was the German Me 163, which was used in the latter stages of the war against the high-flying daylight bombing raids by American B17s. It was small and simple, using a wheeled trolley for take-off and a skid for landing. Its endurance was extremely limited, but it had, by the standards of the time, a phenomenal rate of climb. However, despite its impressive performance, it had very few 'kills' credited to it – one source gives a total of nine.[1]

But the Me 163 obviously impressed the British Air Staff, and proposals for a very similar aircraft began to emerge in the late 1940s. The designs being considered were for a very similar aircraft: a rocket motor with no other means of propulsion, a simple skid for a dead stick (i.e., unpowered) landing, and a battery of unguided 0.5 inch rockets. It was intended for point defence, for airfields and the like. With its limited endurance, it was not suitable for much else. Such an aircraft would have been able to carry enough fuel for only three or four minutes powered flight. In effect, it was almost a manned guided missile, and the unpowered landing technique would not have made it popular with pilots.

In 1945, the whole strategic equation had been rewritten with the advent of the atomic bomb. There was no great urgency for the rocket interceptor in the immediate war years, since the Russians had built a formidable tactical air force, but had almost nothing in the way of heavy bombers. In addition, at that time it was thought that the Russians would not have atomic weapons until the mid-1950s. In the event, the first Russian fission bomb was exploded in 1949.

A further difficulty to the problem of interception was that any jet atomic bomber would be flying very high, very fast. Up until the 1960s, the bomber's best defence had always been height. The higher the aircraft, the more difficult it is to detect, the more difficult it is to hit with conventional anti-aircraft shells, and the more difficult it is to intercept. For interceptor fighters, the choice was either to loiter at high altitudes, which, given their limited endurance, was not usually a feasible option, or to reach these heights as quickly as possible. In the 1940s, the performance of the jet engine was not sufficient to do this. The problem was to get an interceptor to that height quickly enough and with a sufficient speed differential to be able to manoeuvre into a position in order to be able to attack. It was further realised that such an attack would probably be made by guided weapons of some form – either heat seeking, using infra-red sensors, or radar controlled.

There was a fundamental problem with an aircraft as small as the proposed rocket interceptor, in that it would have been able to fly only in daylight and reasonably good weather, and this problem would plague all the designs until the later P177 and F155 designs. It is curious, given these limitations, that there was so much interest in the design. When Churchill was returned to government in 1951, he took a personal interest in the project, asking Lindemann, his scientific advisor and *éminence grise*, to look further into the idea. But the RAF strategic offensive had been entirely night based, and the RAF had rarely encountered the Me 163, and knew of it mainly by reputation. Similarly the German offensive against the UK had been mainly night-based after the early attacks in 1940. It was only the Americans, with high-flying well-armed Flying Fortresses who attacked during the day. So why were the RAF so interested in a fighter that could be used only in daylight? One answer, of course, is the defence of the airfields where the V bombers would be based – effectively point defence.

But despite this, the Air Ministry issued OR 301. The main points of the designs requested were that they should be relatively simple and would use rockets for the main propulsion. However, quick calculations would show that the endurance of such an aircraft is extremely limited. Let us do some order of magnitude calculations.

Given a rocket motor with an S.I. of 200 seconds and mean thrust of 4,000 lb (the Spectre was rated at up to 8,000 lb thrust, but could be throttled) then the fuel consumption is around 20 lb per second. Given that the aircraft might carry 6,000 lb of fuel, this gives a powered flight time of 300 seconds or 5 minutes! This is not long in which to take off, intercept and shoot down an incoming aircraft at an altitude of almost 10 miles.

There are other problems too: high-speed, supersonic aircraft make very poor gliders! If the pilot's interception takes him too far from his base, then he will be forced to eject. Similarly, every landing will have to be one chance only. Landing such an aircraft unpowered would be a pilot's nightmare. It soon became obvious that an auxiliary turbojet would have to be fitted. This extended the post-interception phase and enabled the pilot to 'go round again' if there was a problem on landing.

But there can be other criticisms of the basic concept. The OR stated 'in order to facilitate ease and speed of production, the aircraft and its equipment are to be as simple as possible.' This, however, was a mistake. Although it is very tempting to go for a simple design on these grounds, any such design would have some fatal flaws. The first is that there was no inbuilt air-to-air radar, which would have been no novelty in 1952, and the lack of it would be a severe handicap for high-flying interceptor aircraft. It can also be argued that, owing to the limited nature of the OR, obsolescence was inevitable. The aircraft would be restricted to ground control and daylight interception. Would ground control be readily available in a nuclear war scenario?

Again, to quote from the OR:

> Current day interceptor projects are expected to be adequate in performance to match the enemy threat in normal circumstances, but may be unable to destroy enemy aircraft carrying out special operations at exceptional heights.

An aircraft to fulfil this requirement must have an outstanding ceiling and altitude performance. So far as is known at present, the characteristics can only be provided by rocket propulsion, and, although aware of the probable operating limitations of this method, the Air Staff consider that the promise of tactical advantage more than outweighs other considerations.

It is surprising in other ways that the OR was put in this way. As mentioned, Bomber Command throughout the Second World War carried out the vast majority of its raids at night. It is unlikely that the Russians would attack by day, knowing how vulnerable such an operation would be with the advance warning that would be given as the bombers crossed the width of Europe. So OR 301 was in danger of becoming a requirement for an interceptor without a target.

But another key phrase is, of course, 'special operations at high altitudes'. This was an oblique way of referring to the nuclear armed bomber, and there is one crucial difference between a conventional and a nuclear bomber. With conventional bombing, it is accepted that most of the bombers will get through the defences. Indeed, during the Bomber Command offensive, the German defences would be congratulating themselves if they inflicted 10% losses on a night's raid. In nuclear terms, this is completely reversed. Even 90% losses on a bomber fleet could mean devastation wreaked by the remaining 10%. This was the philosophy behind the rocket interceptor.

In any event, designs were sought from all the major aircraft firms – Blackburn, Westland, Fairey Aviation, Saunders Roe and Bristol, among others. Saunders Roe were not originally on the list, and given their previous work, this is not surprising. However, they had gained experience of modern aircraft with the SRA1, a jet-propelled flying boat fighter. Bizarre though this concept might seem (it was intended for the Pacific war against Japan), it had two axial flow turbo jets, and, given the limitations on the design posed by its aquatic role, had a very respectable performance.

These designs were passed through to RAE to 'score' them on a complicated points system. The two that fared best were the Saunders Roe P154 and the Avro 720. The basic difference between the Avro design and the others is that Avro chose liquid oxygen and kerosene as fuels, as opposed to HTP/kerosene. The Gamma and the de Havilland Spectre rocket motors were the HTP choices. HTP was undoubtedly safer in a crash, although any rocket aircraft was inherently dangerous, owing to the explosive nature of fuel and oxidant.

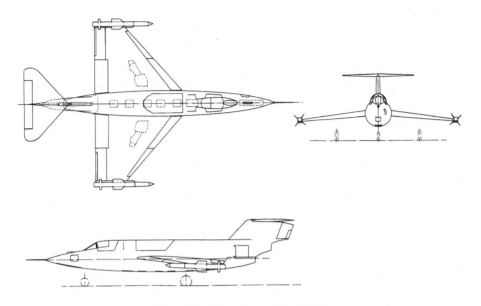

Figure 28. Plan view of the SR53.

But the limitations of these designs became obvious. Saunders Roe then came up with the suggestion that the aircraft should carry an auxiliary jet engine and have proper landing gear. The point of the jet was to supplement the rocket, and then to provide a limited cruise facility, followed by a return to base. The jet engine was of relatively low thrust compared with the rocket, but had high endurance. The Spectre was of 8,000 lb thrust; the Viper jet engine of 1,850 lb. This suggestion was also under consideration by the Ministry, and so Saunders Roe produced modified designs. The SR53 design then emerged from the various proposals.

Avro and Saunders Roe were instructed to build three prototypes each, before the first of many defence economy axes fell. The projects were put on hold. Eventually the Avro prototype, though nearly complete, was to be dropped. Saunders Roe was asked to build two prototypes of the F138D/SR53 (the first designation was the Ministry code for the project, the second was Saunders Roe's).

Saunders Roe pressed on with further designs since the SR53 was felt to be too limited. Saunders Roe proposed the P177, with a much more powerful jet engine, and limited Air Interception capabilities, in other words, a radar set mounted in the nose. Both the RAF and Navy were impressed with this design, and for once, the two Services were in full agreement over a project. The P177 was given the go-ahead and Saunders Roe were asked to produce an initial 27 aircrafts. The two prototype SR53s were proceeded with so as to give experience with the concept.

The Air Staff went further and issued another requirement for a rocket assisted interceptor, F155, with an even more demanding specification. A number of proposals were put forward, with the 'winner' being a development of the Fairey Delta.

A variety of factors led to the cancellations of the P177 and the F155. The main reason, although not the commonly accepted reason, was a change in defence policy. At that time, the decisions about future defence projects and related policy were taken on the basis of reports by the DRPC, the current chairman being Sir Frederick Brundrett. The work that had been done over the past ten years on guided weapons, or surface to air missiles, was reaching fruition in the form of the Bloodhound missile.

Bloodhound was a remarkably successful missile, with a range of over 50 miles, being deployed in British service between 1958 and 1991. It was also deployed by Australia, Singapore, Sweden and Switzerland. Not only could it do the same job as the rocket assisted interceptors, it could do it a good deal cheaper. It costs a good deal less to maintain a squadron of missiles sitting on an airfield for ten years than it does to maintain and fly an equivalent squadron of manned aircraft.

There was also another reason for dropping the rocket assisted fighter: conventional jets with afterburners gave a performance not far short of the rocket. The English Electric P1, which went into the service as the Lightning, had a performance nearly as good as the P177. It too suffered from endurance problems!

Almost coincidental with this change of policy came a change of Prime Minister, when Eden resigned to be replaced by Macmillan. Macmillan wanted defence economies, and with that in mind, appointed Sandys as Minister of Defence. Very soon after the appointment came the 1957 Defence White Paper; indeed, so soon that most of the policy must have been established prior to Sandys. The 1957 White Paper became famous for three points: the abolition of National Service, considerable cuts in Defence spending, and cancellation of various aircraft projects in favour of missiles. On closer examination it is difficult to see how many other projects other than the rocket interceptors were cancelled, but it produced a considerable psychological shock to the British aircraft industry. There was a strong sense that there would be 'no more manned aircraft' for the RAF.

Figure 29. A somewhat fanciful artist's impression of the SR177 in Luftwaffe service.

So despite a bitter rear-guard struggle fought by the Minister of Supply, Aubrey Jones, the P177 was cancelled. The Admiralty in particular pressed Sandys hard, and forced him eventually to admit that although the Navy still needed the aircraft for its carriers, the Defence Budget could not afford it. Both Saunders Roe and the Ministry of Supply tried hard to sell the aircraft overseas: there was Luftwaffe interest, and Saunders Roe prepared brochures for the Australian and Swedish Governments. However, Brundrett was against even this idea, arguing that we were trying to sell an aircraft which was obsolete, and that the Germans would be better off buying missiles from the UK. At the end of 1957, the Germans decided not to buy the aircraft; instead, they bought the Lockheed F-104 Starfighter, which became notorious in later years for its accident rate. Lockheed also had rather persuasive selling tactics unlikely to be matched by a small firm on the Isle of Wight!

The Lightning interceptor remained after the 1957 White Paper: later marks had almost the same capability as the P177 (and the same weakness in terms of endurance). It also showed the usefulness of a manned aircraft in the many interceptions carried out along the north and east coasts of the UK against long-range Russian aircraft probing British air defences.

Both the Lightning and the P177 fitted the specification for which they were designed more than adequately: the problem was not with the aircraft but with the specification and the changes in both technology and policy as the Ministry of Supply lumbered through its slow development procedures.

The apotheosis of the concept of the rocket interceptor was a design submitted by Saunders Roe for the Air Staff requirement F155. Saunders Roe's design brochure was very impressive, leading to a behemoth of an aircraft with two jet engines with reheat and four rocket motors. This was an interceptor capable of taking on anything. It was also immensely huge, rivalling in scale even the TSR 2. Indeed, its size was to be its downfall. The original specification had been for an aircraft to carry two infra-red guided missiles and two radar guided missiles. Issue 2 specified one type or the other, with an option to switch. But whereas the other firms submitted their modified design, Saunders Roe stuck to their leviathan, and it was promptly discarded by the Ministry on grounds of size and expense. Fairey and Armstrong Whitworth were the chief contenders, with machines half the size, but again the project was never completed, falling foul of the same change in defence policy.

So the end result of all the work was the two flying prototypes of the SR53, and these were never to be anything other than research machines. But more than anything else, the concept had proved fatally susceptible to 'mission creep' over a period of ten years, from an extremely simple, almost crude, initial concept, to a highly sophisticated final series of designs. There is a saying, attributed to Voltaire, that the best is the enemy of the good ('Le mieux est l'ennemi du bien'). If the RAF had wanted a good point defence high-flying interceptor, it could have had such a machine with 50 or so SR53s by the late 1950s. The concept of the rocket interceptor never had a chance to prove itself, partly because the window of opportunity in the technologies available was relatively narrow. This window was never fully utilised by the often slow progress of Operational Requirements through the Air Ministry, Ministry of Supply, the budget limitations and the desire to go one better as each design became finalised.

Rocket motors did have their drawbacks in the form of relatively limited operating time and use of exotic and expensive fuels. Handling such fuel would have produced difficulties when servicing the aircraft, and even though HTP is reckoned to be relatively benign as far as rocket fuels go, it is still hazardous to handle.

Figure 30. The SR53 in flight, but powered only by the small Viper jet engine.

The other factors leading to cancellations were the enforced defence economies, the constant improvement in jet engines, and the development of guided weapons. The role of interception was to be taken by the Lightning aircraft, which although it too had an impressive rate of climb to altitude, was also limited by range, at least, in the earlier marks. But it was the advent of guided missiles, principally Bloodhound, deployed along the East Coast V bomber bases and at RAF bases elsewhere, which finally killed off the rocket interceptors.

Designs in Detail

The F138/SR53

As mentioned, the original Operational Requirement in July 1951 did not specify a jet engine; the aircraft was to be purely rocket propelled. A meeting of the OR Committee at this time said that 'it would be a very local defence weapon' and so 'The possibility of fitting a small turbine to assist the landing was then discussed but ruled out on the score of weight'. The armament was to be a battery of 2-inch air-to-air rockets. As an indication of the number of aircraft firms that there were at that time, tenders were invited from Bristol, de Havilland, Fairey, AVRoe, and Short Brothers. Copies for information were sent to Armstrong Whitworth, Blackburn, Boulton Paul, English Electric, Gloster, Percival, Saunders Roe, Supermarine, Westland, Folland, Handley Page and Scottish Aviation!

In the minutes of the F.124 tender Design Conference in July 1952, it was announced that 'General agreement was reached that only the Saunders Roe, AVRoe and Bristol A design remained in the competition ...' and that 'Saunders Roe had submitted very good designs on two previous occasions and he felt that their design team were so good that it would be a mistake for it to be disbanded as would be the case unless the firm received a contract soon.'

There are two curious points about this: the fact that it had taken a year to evaluate the designs, and the comment on the Saunders Roe design team: contracts were often awarded for seemingly obscure reasons. The recommendation was made that three prototypes be ordered each from Avro and from Saunders Roe.

However, the huge increase in the defence budget in 1950 could not be sustained, and economy again became the watchword. This meant that three prototypes could not be afforded; despite RAF preference for the AVRoe design, the Ministry of Supply decided to press on with the Saunders Roe design, but with only two prototypes.

But there is an interesting comment from the OR committee sometime later, in June 1953:

> ... the changes in requirement that have been brought in from time to time have moved the design some way from the basic conception of a simple rocket aircraft – and there is some danger, in my opinion, that the final weapon will be less effective than it might be ...

Indeed, in a sense this remark could be said to be the essence of the whole story.

However, work progressed rapidly with the SR53: in October 1952 there was a structures meeting between Saunders Roe and the RAE, and a preliminary mock up meeting in Cowes in September 1953.

The final delay in the completion of the first aircraft, XD 145, was delivery of the Spectre I motor: this could not be delivered to Cowes before mid-December 1955. The motor had earlier been installed in Canberra for flight trials. By mid-June 1956 the aircraft was completed, then dissembled for transport to the Aeroplane and Armament Establishment at Boscombe Down. Here it was put together again, and the first rocket engine firing was on 16 January 1957. The first flight took place on 16 May. The second aircraft, XD 151, was first flown on 18 December 1957.

The SR53 was intended to be a lead in for the P177, and the cancellation of the P177 meant that it had lost its purpose. However, it was felt that in many ways that the SR53 was a unique aircraft that could be used for aerodynamic research, rather as the X planes in America. Accordingly, proposals were put forward for enhancing the performance[2]. It was estimated that the Spectre 1 rocket motor in its then state of 7,000 lb thrust, and S.I. of 190s gave a maximum velocity of Mach 1.8 at 60,000 ft and a maximum height of 76,000 ft. Thus a meeting in May gave some options for further development:

(i)   Spectre 5 engine 95,000 ft or M1.8;

(ii) Twin Spectre 125,000 ft or M2.3;

(iii) Spectre 5 and airlaunch at 40,000 ft and M0.8 gives 115,000 ft or M2.9.

The Spectre 5 was an improved version of the rocket motor; the Twin Spectre, as its name suggests, stacked two such motors together.

Figure 31 shows various scenarios for improving the performance of the SR53, and the results that might be obtained.[3] It must be said, however, that some parts of this scenario do look a little optimistic. A further problem was that the SR53 as designed was not that well suited for the task, and most of this kind of research had already been done by 1958. As it stood, there were constraints on the airframe, being all aluminium. Kinetic heating meant that certain key parts of the airframe would have to be replaced by stainless steel, and work was done on this by the design team at Osborne House on the Isle of Wight.

Several scenarios were sketched out: with no armament, more fuel could be carried. The jet engine could be removed, and with a more powerful rocket engine and still more fuel, the flight envelope could be extended further. Also suggested was the use of solid fuel Mayfly rockets to assist take-off, and, in the *pièce de résistance*, air launching from a Valiant was proposed. All this would have made XD154 Britain's answer to the X-15! The Valiant could take the aircraft up to 40,000 ft at a speed of Mach 0.8. Unlike the American X-15, which was underslung, the proposal was to mount the SR53 on top of the Valiant.

In the meantime, flight trials went ahead with XD145 flying supersonic for the first time at around 45,000 ft on its 31st flight, in May 1958. But disaster overtook the programme in June when XD151 crashed during an aborted take-off on its 12th flight. A long and thorough investigation followed. Here is an excerpt from the report[4].

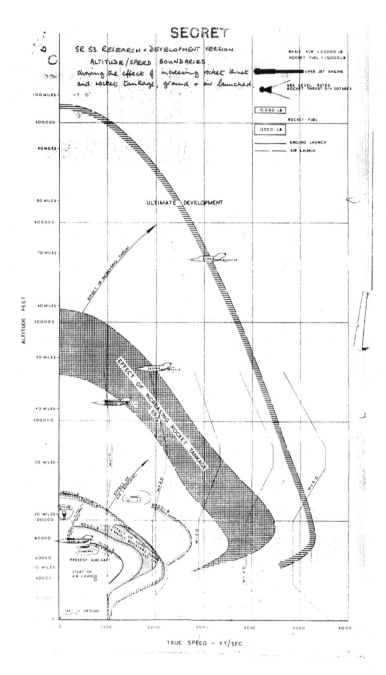

Figure 31. The SR53 as a research aircraft – although some of the ideas do seem on the optimistic side for such an aircraft.

The aircraft taxied out at 1200, and the Spectre was started at 1203. Approximately 5 seconds elapsed before the engine went 'hot' but this is understood to be normal. The aircraft lined up on the runway and after cockpit checks were completed, and 10° flap selected the aircraft commenced to take off. The aircraft accelerated normally and the nose wheel was raised. About 30 seconds after the pilot had reported 'hot', he was heard to call 'Panic Stations' and then a moment later, 'Come and get me will you'. The anti-spin parachute was seen to stream, but the aircraft ran off the end of the runway. Upon impact with a runway marker light pole, a chain fence with concrete posts and finally a large marker light the aircraft broke up and caught fire. The pilot was killed.

An analysis of the film shows that the take-off was abandoned at a very critical stage when the aircraft was half way down the runway and on the point of becoming airborne. The rocket is shown to have ceased to run hot i.e. no flame, at this point. It could not be determined whether or not the rocket was deliberately throttled back by the pilot. There is no sign of the airbrakes having been opened and the aircraft left the runway at an estimated speed of about 145 knots.

The aircraft was on its take-off run, just becoming airborne, when the rocket motor abruptly cut. Whether the pilot took this action was never established, like so many other details of the accident. The aircraft braked hard, but overran the runway. It might still have survived but for a wing catching an obstruction, a runway lamp post. As the aircraft disintegrated, the rocket fuel ignited in a fierce conflagration. No cause for the accident was ever established, and there was no evidence of pilot error. Saunders Roe's chief test pilot, John Booth, was killed in the accident.

With the investigation producing no clear result, the flight test programme continued, and XD145 made a total of 56 flights, or 22 hours flying time. Peter Lamb, Booth's successor, described the SR53 as 'an extremely docile and exceedingly pleasant aircraft to fly', which, given the kick the rocket engine must have produced, says a lot about the aircraft. It reached a maximum speed of Mach 1.33, not an exceptional speed, altitudes of up to 55,000 ft, but certainly lived up to expectations with a climb rate of 29,000 ft per minute.

By the time the Ministry had decided to go ahead with the possible research project, Saunders Roe had been taken over by Westland. Westland's policy was to drop fixed wing aircraft development to concentrate on helicopters (and Saunders Roe would later become, for a time, the British National Hovercraft Company). Saunders Roe's Chief Designer, Maurice Brennan, responsible for all the fixed wing designs, had moved to Hawker Siddeley. The Ministry talked to Hawker Siddeley, but concluded that

... it was not clear for some time whether Westland would be willing to take on this work using the existing Saunders Roe team for the purpose, but they eventually decided to concentrate their activities on helicopter work and decline all fixed wing business. We had no alternative but reluctantly to accept their decision. However, Saunders Roe's Chief Designer had by this time joined Hawker Siddeley and asked the latter to consider taking the job on. Having examined the matter with them, we reached the conclusion that we could not obtain by this means the programme of work that we wanted within the amount we had set aside for it ...

The programme was finally closed in July 1960.

XD154 was set aside at RPE Westcott, and fortunately has been preserved. It is now in the Aerospace Museum at Cosford, with many other famous prototypes including the TSR 2 and Bristol 188.

Facts and figures

The SR53 was a relatively small aircraft, only weighing 7,400 lb empty, but 18,400 lb fully fuelled. It was an extremely elegant aircraft, its only drawback in looks being a slightly tubbiness in the fuselage as a consequence of the large amount of fuel that needed to be carried. The wing span without missiles was 25 ft 1 inch, and it was 45 ft long.

The P177

This project grew from the realisation of the limitations of the SR53. It would have been a larger aircraft, with a much more powerful jet engine: the balance of jet and rocket would be almost equal. In addition to the de Havilland Spectre motor, capable of 8,000 lb thrust, it would also have a de Havilland Gyron Junior turbojet, also of 8,000 lb thrust. In addition, it was given a limited Air Interception capability, which extended the range of weather conditions and night time operations. Coincidentally, the Navy was looking for a high-flying supersonic interceptor at the same time, and both Services, rather unusually, concluded that the same aircraft would suit them both. Proposals were sent to the Ministry of Supply in March 1954 and a design contract was placed with the company in May 1955. With remarkably little fuss, development of the P177 began in autumn 1955, and work proceeded steadily throughout 1956.

A memorandum by the First Lord of the Admiralty in April 1956 set out the Navy's case for the aircraft[5]:

The P.177 is a single seat high altitude fighter capable of operating at a sustained speed of M = 1.4 in the 40,000 to 60,000 ft height band; its ceiling will be in the region of 75,000 ft and it will be capable of reaching M = 2 for short periods. It will be armed with 2 Blue Jay Mark 3 air to air Guided Missiles, with unguided rockets as an alternative armament. It will be equipped with AI [AI = airborne interception] and will have a limited capacity for night operations ...

The P.177 was selected by the Naval Staff because:-

(i)   It is one of the very few British aircraft likely to be better than anything American when it comes into service.

(ii) It is the only aircraft required by both the Navy and the RAF.

(iii) Its engine, radar and weapons are already being developed for other air-craft, so that the development costs should be comparatively low.

(iv) It is smaller and cheaper than the N.113 [Supermarine Scimitar] and its early substitution for that aircraft will save production expenditure.

However, two factors now intervened to threaten the project. One was a change in defence policy with regard to manned aircraft, and the second was a need to reduce the UK's defence spending. With the arrival of Duncan Sandys at the Ministry of Defence, and the resultant 1957 Defence White paper, the RAF version was cancelled forthwith. As the White Paper put it:

Work will proceed on the development of a ground-to-air missile defence system, which will in due course replace the manned aircraft of Fighter Command. In view of the good progress already made, the Government have come to the conclusion that the R.A.F are unlikely to have a requirement for fighter aircraft types more advanced than the supersonic P1, and work on such projects will stop.

Given the extent to which the RAF was threatened, the P177 was relatively small fry, and the P1 interceptor, or Lightning as it became known in service, gave many years of service in a very similar role.

However, the Navy was still very insistent that they needed the P177, and at the same time, Aubrey Jones, who was Minister of Supply, wanted to keep the project going since there was some German interest in the aircraft. A prolonged Whitehall battle then ensued. Sandys told the Navy quite firmly that the aircraft, irrespective of merit or the Navy's need, could not be afforded. In the same way he told Jones that if the Ministry of Supply wanted to keep the project under way then it would have to go on his budget, not Sandys'.

All this wrangling cannot have helped the progress at Saunders Roe, or the prospect of trying to sell it to the German Air Force. After all, if the RAF version had been cancelled and then the Naval version, this cannot have inspired confidence in the project. In addition, not everyone wanted to sell it to the Germans. Sir Frederick Brundrett, Chief Scientist at the Ministry of Aviation is on record as saying that the German requirement.

…was for a high-altitude quick climbing defensive weapon. There was no doubt that they would be best satisfied with a guided weapon and in his view the right course was to interest them in our own Stage 1 and Stage 1 and a half guided weapon developments rather than the P.177.

The Germans did not share his view, however, and in December 1957 the project was finally cancelled.

Figure 32. The P177 as it might have appeared in naval service.

Would the P177 have lived up to expectations? There is a strong probability it would have done: its smaller sibling, the SR53, did all that was expected of it. The RAF was left with the Lightning, the Navy with nothing. It could be argued that the P177 would have been too specialised for the Navy: it would not have been able to maintain a continuous Combat Air Patrol as its endurance was too limited, although it was intended to fit a probe for air-to-air refuelling. But that was not its function: it was intended to by-pass the patrolling function by its ability to reach high fast targets quickly. The concept was never put to the test, but by the late 1950s it was becoming obvious that jets such as the Lightning with re-heat on the engines could do the job as effectively as rockets.

The Lightning, however, went on to prove its usefulness against the Soviet aircraft that during the 1960s and 1970s attempted to probe British airspace. A manned aircraft does have some advantages over missiles, as the Falklands demonstrated. In reality, both are needed since they are complementary.

The Germans went on to buy Lockheed F-104 Starfighters, as did many other NATO countries. Saunders Roe was a minnow by comparison with Lockheed, and whether as many P177s would have fallen out of the sky as F-104s is also an interesting question. But the most fascinating image is of the P177 in Luftwaffe colours!

The P177 was a good deal less elegant in appearance than the SR53, the rather bulbous fuselage and pointed nose above the air intake detracting from its appearance. It was twice the weight at 14,500 lb empty, and 28,000 lb at extended load. The span was 27 ft, and it was 50 ft 6 inches long. It was within about six months of its first flight[6] when the project finally cancelled in December 1957.

A history of the SR53 and P177 projects written by the Air Ministry can be found in Appendix A.

---

[1] WJ Boyne, *Clash of Wings*. New York: Simon & Schuster, 1994.
[2] TNA: PRO AVIA 65/546. SR53 Finance and Policy. The details of the proposed improvements are outlined in a series of papers in this file.
[3] TNA: PRO AVIA 65/287. F138D Saunders Roe rocket interceptor aircraft for RAF.
[4] TNA: PRO BT 233/40.
[5] TNA: PRO AVIA 65/649. Saro rocket interceptor: RN aircraft NA 47.
[6] Ibid.

Chapter 4

Blue Steel

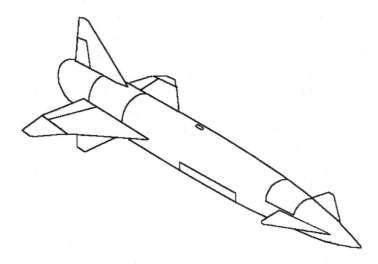

Figure 33. Perspective view of the Blue Steel stand-off missile.

If Britain had built the V bombers as strategic bombers which were capable of launching a nuclear attack, then it was logical to think that the countries threatened would take care to defend their cities against such an attack. In the Second World War, this had been done by means of night fighters equipped with radar, searchlights and anti-aircraft guns. But a new weapon was appearing on the scene: the guided missile, such as the Bristol Bloodhound, which would be deployed from 1958 onwards. Bloodhound was radar controlled, used a ramjet engine and had a range of up to 50 miles. There were even proposals at one stage to equip them with nuclear weapons to increase their destructive power. But threat implies counter-threat, and the Air Staff was working on the assumption that from about 1960 onwards the V bombers would be unable to penetrate the

Moscow air defences – in other words, that Moscow would be protected by guided weapons similar to Bloodhound.

The next question was then how to deliver the Bomb from the time the defences became too formidable for the V bombers to penetrate. In the early 1950s, ballistic missiles were not yet an option. Britain had invested heavily in the V bomber force, so that any ideas to prolong their active life would be very welcome. Hence the idea of a 'flying bomb' evolved. Initial ideas in the late 1940s had centred on a system called Blue Boar, which was a television guided glide bomb. This proved too limiting, since cloud and bad weather could obscure the television picture, whilst the radio link between missile and aircraft could easily be jammed. Instead, thoughts turned to a longer range powered device. This was not a flying bomb in the V1 sense, whose guidance had been extremely limited, but one that would be able to deliver its payload with considerable accuracy. Nor would it be a cruise type missile, since the technology for long-range guidance, terrain following radar or satellite navigation also did not exist at that time. Instead, inertial guidance would be used, which could be accurate enough at relatively short ranges. The missile would be released from the aircraft, immediately climb to a considerable height, cruise at high speed – around Mach 2 or so – then dive down onto the target.

The theory was all very well, but in 1954 reality was something else. Britain was still working on its first fusion bomb, so it was difficult to estimate what payload size and weight would have to be carried. The next problem was the inertial navigation. As the US was discovering with Snark, Matador and Regulus, navigation over considerable distances was a problem. This was one of the reasons why a range of 100 nautical miles was chosen for Blue Steel. The problem was made more difficult since it would be launched from a moving aircraft whose own position might be uncertain. A further problem with cruise type missiles is their relative vulnerability to enemy defences unless they fly at a very low level, which was, again, not possible in 1954. Hence OR 1132[1] specified a speed of Mach 2.5 at 70,000 ft or higher for the vehicle, although in 1954 supersonic speeds were an area still fraught with unknowns, and supersonic wind testing was still very difficult. Yet another problem was that at these speeds the skin of the vehicle would start heating up as a result of friction with the air. Aluminium airframes would not be suitable at high Mach numbers (this is one of the reasons why Concorde and similar aircraft are limited to around Mach 2.2) and the only real alternative was stainless steel, which was difficult to work with. This was more unknown territory.

All of this was summed up by a memo from the Ministry of Supply, appropriately enough on 5 November 1954, by saying:

Present estimates are that medium range GW defences will make it excessively dangerous for the V bombers to fly over, or within about 50 miles of the target in 1960... The requirement is therefore for a flying bomb which will have its maximum use between 1960 and 1965... It is expected that a fusion warhead will be available by 1960 and it seems generally agreed that the bomb should be designed to carry this warhead.[2]

[GW = Guided Weapons. In this context, the reference is to surface-to-air missiles.]

The time period is significant – it is assumed that by 1960, the V bombers will no longer be able to attack the target directly. The next assumption is that by 1965 some other form of delivery will have taken over, although this is still too early for the Air Staff to be thinking specifically of ballistic missiles.

The early files note that in some respects the project is almost equivalent to building a fighter aircraft. It would have no instruments and so on, but certainly was aerodynamically novel, and with an autopilot connected to the inertial navigator in place of a human pilot. As a further minute of November 1954 put it:

This bomb is indubitably much simpler than a fighter aircraft in the range of equipment that has to be provided. On the other hand it is a big step to go from the present speed range to Mach Numbers of 2 and above and again in comparison with the development of a manned fighter, it is to be expected that the production of the many vehicles required for firing trials will lengthen the development time.

That was certainly an accurate forecast!

Hence in September 1954 a letter was sent to the V bomber firms with data for a possible missile[3]. At this point the design had not been thought through in any detail, and there were up to six different possible configurations under consideration, some with ramjets and solid boosters, others with liquid fuelled rocket motors. The V bomber firms had then to start thinking how they were going to carry the missile on the aircraft.

Vickers, who made the Valiant, were the initial favourites for the contract, which eventually, despite misgivings, went to Avro. The extant papers of the Ministry of Supply give no real reason at all for the choice. Avro had no guided weapons expertise, and had to set up a special department for the purpose, headed by RH Francis, who had previously worked at RAE. His approach was the more measured one of a government department rather than the more urgent and commercial approach of a firm engaged on an urgent defence project.

The Air Staff had specified a missile to be available by 1960, and by about 1965 the missile would effectively have become obsolete since its range was only 100 miles. By this date, it was expected that the Russians would have developed defence in depth, so that a combination of missiles and interceptor aircraft would mean that the bombers would not be able to reach their release points unscathed. Hence its service life would be short. It need not be a particularly sophisticated design, but it did need to be in service on time. Given the performance limits, light alloy would have sufficed for the structure. Propulsion could have been by turbojet, ramjet or rocket motor.

Unfortunately the design produced by Avro was a good deal more sophisticated than was necessary. The airframe was to be built from stainless steel, a difficult material to work with, particularly when it came to bending it in two planes at the same time. A rocket motor was chosen for propulsion, and this would have a considerable impact on the serviceability and availability of the missile when it entered into service.

Blue Steel was not a ballistic missile, but instead was controlled like an aircraft. It was intended for release from the bomber at around 40,000 ft. The aircraft would be travelling at around Mach 0.7 or so. After release, it dived down, and after 4 seconds, at around 32,000 ft, its motors lit up. From there it climbed to around 59,000 ft, then increased speed to Mach 2.3. After that, the missile began a cruise/climb, using only the small chamber of the rocket engine, to over 70,000 ft. When it ran out of fuel or arrived at the target, the motor cut, and the missile dived down to its target.

The engine designed for Blue Steel was designated the PR9[4], which became known as the Stentor, built by Armstrong Siddeley (later to become Bristol Siddeley). Early test versions used the de Havilland Double Spectre engine until the Stentor became available. The Stentor had two combustion chambers, one of which was of fixed thrust of 25,500 lb at 45,000 ft, the other was to have a variable thrust of between 1,000 lbf and 6,200 lbf at 45,000 ft, and was capable of being throttled. Burning HTP and kerosene, it produced an S.I. of around 220. The large chamber was intended for the boost phase to high altitude; the small chamber for the cruise phase thereafter. The motor turned out to be reliable and effective; so much so that when reports of the failures of the early rounds stated that the rocket engine had failed, the chairman of Armstrong Siddeley wrote a sharp letter pointing out that this was not the case: the engine had been starved of fuel as a result of sloshing in the tanks. He did not want his company associated with the poor reputation the missile had at that time.

The Stentor chamber was the first HTP motor to be made from tubes braised together and formed into shape, rather than the double-walled chambers such as the Gamma or Spectre. The small chamber would, in the course of time, go on to power Black Knight and Black Arrow.

Avro tested the aerodynamics on a 1/8th scale model, moving up to a 2/5th scale model. These were tested by using solid fuel boosters to launch them at the RAE range at Aberporth in Wales, then variants made of light alloys rather than stainless steel were tested. For full scale missiles, the range at Woomera was brought into use, which entailed further delay as specially equipped Valiants had to be prepared (although the Valiant was not to carry the operational weapon, they were used in early trials). There was an attempt to speed up the testing by doing trials at both Aberporth and Woomera. This turned out to be a mistake. Two aircraft had been converted for trials purposes, and so one was based in the UK and the other in Australia. The problem was that they would frequently go unserviceable. If two aircraft had been available in Woomera, then the other could have been used as a back-up.

The problems were many and varied. The Auxiliary Power Unit, made by de Havilland, was particularly troublesome. The unit used HTP, which was catalytically decomposed, and the gases produced then drove a turbine that in turn drove the generator. Sloshing of the fuel caused problems as the missile went through some vigorous manoeuvring as it climbed, levelled off, then went into its final dive.

The weapon was supposed to have been ready to enter service by 1960. In April of that year, the first three rounds of the final version of the missile had been fired at Woomera and each had failed to follow the launch programme. The minutes of the second meeting of the Blue Steel Management board talk about Avro's 'dismal record at Aberporth'[5] when talking of the delays in flight testing. The first round approximating to the final version was not successfully flown until 1962.

The problems led to the RAE being called in to assess Avro's performance, and a series of Study Groups were set up. In December 1960 comments were made to the effect that the design of the missile was sound enough, but that 'the standard of engineering is poor in a number of respects, with far too little emphasis on reliability studies'.[6]

In addition to work being done on the missile, a considerable amount of work had to be done on the V bombers to prepare them for the missile. The large size of the missile meant that it was carried semi recessed into the fuselage. One of the problems with the Victor bomber was that its ground clearance was quite small – there is a story (quite possibly apocryphal) that the only way to get the missile under the bomb bay was to deflate the tyres of the trolley carrying it, so as to give sufficient clearance.

Figure 34. A Blue Steel missile being serviced in a hangar.

Blue Steel came into service at a time when Britain's deterrent policy was changing rapidly. It had been originally intended to extend the operational life of the V bombers until Blue Streak was deployed in the mid-1960s. When Blue Streak was abandoned in favour of Skybolt, the need for Blue Steel was far less acute, since the Air Staff hoped to have the first squadrons of V bombers equipped with Skybolt deployed by 1964. To the consternation of the British Government and the Air Staff, Skybolt was cancelled at the end of 1962. Some rather tense negotiations led to the Nassau Agreement, whereby the United States agreed to supply Britain with Polaris missiles. The drawback was that it would take some years to build the submarines and equip them with the missiles, and so Blue Steel would have to soldier on long past its 'sell by' date.

The Treasury was adamant that there would be no more money for Blue Steel and the V Force. The only way that the bombers could now penetrate Russian airspace was to fly as low as possible: a role in direct contradiction to the design of both aircraft and missile, which would now be launched at an altitude of 1,000 ft.[7] This would halve its range, making the system even less viable.

This was not the only problem. The one advantage of choosing a rocket motor was that the missile did not need an air intake, and as a result its radar cross section was very much less – in modern jargon, it was 'stealthier'. The use of HTP gave rise to a whole host of other problems. The first and unexpected problem was that aircraft de-icing fluids exploded when coming into contact with HTP, which necessitated a rapid change in procedure for de-icing during winter. The time needed to prepare a missile was of the order of seven hours. One option was to keep some missiles partly prepared, which might mean filling with HTP. The HTP had at some stage to be drained out of the missile, and the tanks were then flushed out with water and dried. Only one drying unit was provided for each station, so that after a full-scale exercise or sudden emergency, it could take as long as two weeks for the station to recover its normal peacetime preparedness!

The use of HTP also meant obtaining safety clearances from the Ordnance Board and Nuclear Weapons Safety Committee. Worries about safety meant that the Nuclear Weapons Safety Committee withheld the authority to fuel the missile on the aircraft when the warhead was fitted, the authority to fit the thermal batteries to readiness missiles and the authority to fly the aircraft with the warhead fitted to the missile to a dispersal base. Thus for a considerable part of its service life, Blue Steel could have fuel, or a warhead, but not both.

Blue Steel was only intended as an interim measure, and the Air Staff issued two further Operational Requirements, OR 1149[8] and OR 1159[9]. These were for air-launched missiles, but ones with much greater ranges. The extra range effectively ruled out a rocket powered missile: the only way such ranges could be achieved was with an air-breathing missile using a turbojet or ramjet, and as such are beyond the scope of this book.

Various proposals were made for extending the range of Blue Steel, but such proposals were effectively just tinkering at the edges. This was summed up in a memo from the Deputy Chief of Air Staff to Watkinson, the then Defence Minister:

> In considering Blue Steel and any possible developments of it we must take note of some pretty unpalatable facts. We first started thinking about this weapon in 1952. The OR was accepted by the Ministry of Supply in 1954 for an in service date of 1960 and events have I think proved that had this date been met the weapon would have had a useful and viable life. An in service date of 1963 for a weapon with a range of only 100 miles is however a different matter.
>
> In our submission to the Treasury in 1955 the total R&D was estimated at £12.5M. This is now estimated to be £55M...

In all the circumstances I cannot see that we would be justified in pouring more money into the development of a weapon which has made unsatisfactory progress to date and which remains dangerously close to being a non-viable weapon at the time of its introduction into service.

There was also the fear among the Air Staff that if Avro were given the go-ahead to start work on any of the improved versions then there was a good chance that Mark I might be delayed even further.

Blue Steel eventually came into service with the RAF in a limited capacity at the end of 1962, although it was not until April 1964 that clearance was given for a filled and fuelled operational Blue Steel to be used with the Vulcan Quick Reaction Alert. However, its reliability was not good: in 1963 the RAF estimated that the chances of a powered missile being fit for powered launch at the target was around 40%, and of these, only about 75% would reach their targets. The Victor squadrons were operational until 1968, when they were retired from the Blue Steel role. The Vulcans continued with the missile until 1970, when it was withdrawn from service.

Figure 35. The Stentor rocket motor as fitted to Blue Steel.

It is difficult to avoid the conclusion that HTP/kerosene was a poor choice of propellants for Blue Steel, and that Avro, completely inexperienced in missile development, was a poor choice by the Ministry of Supply for its development. Both these factors extended an already critical development time and led to many operational difficulties. The delays in the programme, the cost overruns and the difficulties involved in handling the missile all conspired to reduce even further the reputation of the British aircraft industry. Blue Steel would be replaced by Polaris, then by Trident – both American-built. It is now half a century from Blue Steel, and Blue Steel was the last offensive British missile to be developed and deployed.

From the narrower perspective of British rocketry, the most useful contribution of Blue Steel was the two thrust chambers, which were the most successful part of the project. The smaller chamber went on to become the chamber that powered the later Gamma engines of Black Knight and Black Arrow. The larger chamber was suggested for various vehicles, but despite its apparent usefulness, was never exploited further.

---

[1] TNA: PRO AVIA 65/91. Propelled air-to-surface missiles for 'V' class bombers: application to Vulcan aircraft.
[2] TNA: PRO AVIA 65/1687. Blue Steel. Development programme.
[3] TNA: PRO AVIA 65/91. Propelled air-to-surface missiles for 'V' class bombers: application to Vulcan aircraft.
[4] TNA: PRO AVIA 65/1289 Blue Steel. Main propulsion unit.
[5] TNA: PRO AVIA 65/1467 Blue Steel. Management Board.
[6] TNA: PRO AVIA 65/1468 Blue Steel. RAE Study Group.
[7] Ibid.
[8] TNA: PRO AIR 20/12143. Proposals for OR 1149.
[9] TNA: PRO AVIA 65/1697. Stand off bomb, OR 1159: research and development.

Chapter 5

Blue Streak – The Origins

The story of Blue Streak divides into two phases, phases which are very sharply divided from each other: the military and the civil. The civil phase was an afterthought, a by-product from the military. Blue Streak was cancelled as a military weapon in 1960, and its life as a civil project began with the intention of creating a satellite launcher from what had been intended as a Medium Range Ballistic Missile (MRBM). But after 14 years and hundreds of millions of pounds of expenditure (much to the despair of the Treasury) neither programme yielded useful results – but that was not the rocket's fault. Technically, the design was excellent. Almost every launch was flawless. But it spent most of its life in search of a role.

The original intent behind Blue Streak was to produce a guided weapon capable of carrying a megaton range warhead to the strategically important parts of the USSR. Design work began in 1955 and the final result was a technological snapshot of rocketry progress circa 1957. In principle and design it was very close to its American and Russian counterparts, and very much their equivalent. But to realise this, we have to look briefly at the history of guided weapons.

The V2 was notoriously inaccurate, and given in addition its limited payload (1 tonne of high explosive), it was not an effective military weapon. Following the war, both the Americans and the Russians pressed on with improved designs, based around the V2 concept. The British devoted relatively little effort to large rockets at this stage, and such work was, in the main, theoretical, although three captured V2s were fired off in Cuxhaven, Germany (Operation Backfire) in what was effectively a familiarisation exercise. However, the post-war rocketry effort in the three countries began to solve some of the three major problems of guidance, accuracy and range.

Most of the work done in the UK in the early 1950s simply consisted of studies of possibilities. The technologies of the time were changing fast, and the problems were firstly to choose which would be the most fruitful, and secondly, to try and estimate how far the technologies could be usefully developed. One of the first significant British developments in this field of long-range delivery systems was a report commissioned from the English Electric Company Ltd.[1] Work on the report, entitled 'Long Range Project', by LH Bedford, started in March 1952, and the completed report was delivered in July 1953. The first consideration was that of range and accuracy. The report took a range for the weapon of 2,000 nautical miles with circular error probability (c.e.p.) of the order of 1,500 ft as the target to be aimed for (c.e.p. refers to the probability of 50% of the missiles landing that distance from the aiming point). Three types of missiles were considered: the ramjet, the glide rocket and the ballistic rocket. The guidance system for all of these was taken to be integrating accelerometers using gyroscopes. The times of flight were calculated as 16 minutes for the ballistic rocket, 25 minutes for the glide rocket and 50 minutes for the ramjet. The significance of this was that the accuracy of the system decreased as time of flight increased. Thus the ballistic rocket was to be preferred on two counts: that of accuracy and that of invulnerability. Intercepting a warhead from a ballistic missile is a task that even today is extremely difficult. Indeed, if attempted operationally on a system with even rudimentary decoys, it is well-nigh impossible.

The two major problems of the ballistic rocket were identified firstly as obtaining a sufficiently high S.I. from the rocket motors, and secondly the problems of re-entry into the atmosphere. In neither case were practical solutions offered, nor was there any attempt to suggest design features such as the fuel to be used. The report was still theoretical and speculative.

The RAE was conducting its own studies into the same areas as the English Electric Report, and it is interesting to note the estimated all-up weights of the three options[2]:

Range:	Ram Jet	Winged Rocket	Ballistic Rocket
500	24,000	22,000	35,000
1,500	34,000	56,000	105,000
2,500	48,000	135,000	210,000

(Range is given in nautical miles; weights in thousands of pounds)

This has important implications for the design of any ballistic missile: a mass of 210,000 lb implies a lift-off thrust of at least 250,000 lb. It was noted that the ballistic missile would be much heavier than the others, and that if it were to be chosen it would be because of its much greater chance of survival against enemy defences. The calculations were done assuming liquid oxygen and kerosene as propellants, and give a remarkably accurate prediction for the weight of what would become Blue Streak.

As a consequence of these deliberations, the RPD at Westcott pressed ahead with design studies on larger motors. But even at the end of 1954 no formal summary had been made of the RPD's overall policy on the ballistic missile, as it was felt the position was constantly changing. At the same time, research was continued on a series of rocket engines that went under the generic name of Delta. Despite RPD's earlier dislike of kerosene as a rocket fuel, these were lox/kerosene fuelled, and mainly for research purposes. There was no firm design for a missile at this stage, although speculative drawings of how to fit several motors together for such a missile were made.

Commercial firms were also interested in the concept: during a meeting between the RAE, RPD and Rolls Royce, the company stated that:

> In their own studies they had assumed a warhead of 10,000 lb and minimum range of 2,500 miles and had produced a preliminary design. This design was a single stage missile with an all-up weight of 250,000 lb and empty weight 18,000 lb giving a mass ratio (empty/all up) of 0.072. Thirty-three chambers of 10,000 lb thrust each were used.

Apart from the number of chambers, again this sounds very much like Blue Streak.

At around this time the ramjet and the glide rocket drop out of consideration. The glide rocket was never a very serious candidate. The ballistic missile with separating warhead and self-contained guidance system has the great advantage that it was, and still is, almost invulnerable to defensive counter measures. A missile is also much better equipped to carry decoys; in other words, dummy re-entry vehicles or devices that would look the same as a re-entry vehicle to enemy radar. Decoys have been an on-going area of research up to the present day.

The ramjet is not as vulnerable to guided missiles as the manned aircraft, but does not begin to compare with a free falling re-entry vehicle at velocities of several thousand ft per second. Height is also a factor here: today's turbofan subsonic cruise missile is designed to fly as low as possible since terrain following radar and accurate guidance have subsequently been developed to make this possible. Supersonic ramjets would be high flying and more vulnerable to defensive missiles.

In 1954 the Sandys/Wilson agreement was signed between the UK and the US, whereby the two countries agreed to collaborate on long range missiles; the British concentrating on medium range weapons whilst the Americans would aim for intercontinental ranges. However, to produce an effective military weapon, there were several important problems to be solved other than simply building a big enough rocket. For the missile to fulfil its function, all the systems had to work together. Loss of just one would render the weapon impotent.

The first of these problems was that of guidance. Radio/radar guidance during the launch phase was possible. However, such external guidance could easily be jammed or destroyed, particularly during a nuclear attack. The answer lay in internal inertial guidance, using gyroscopes and accelerometers to determine the vehicle's heading and speed accurately. The American Atlas missile used a form of radio guidance, but all other missiles carried inertial guidance. A form of radio guidance for Blue Streak was also developed for a time before being abandoned as too easy to jam and too easy to destroy, and also because there was an economy drive on! There were many obstacles to an accurate inertial guidance system in the 1950s, before the advent of transistors and when electronics depended on power-hungry thermionic valves for amplifiers. Suitable gyroscopes were difficult to manufacture, and eventually, partway through the project, gyros had to be bought from the American firm Kearfott. To give some idea of the accuracy which was needed over a range of 2,500 nautical miles, it was stated early in the programme that there was

> a requirement for a 50% circular error no greater than 8,000 feet at all ranges. If this requirement should result in undue delay in the introduction of the missile into service the Air Ministry will be prepared to accept a 50% circular error of no greater than 3 miles at all ranges in the first instance.[3]
> [8,000 ft is around one and a half miles.]

A novel feature of these designs for ballistic missiles was that at the moment the engines cut, some two to three minutes after launch, the warhead and its re-entry vehicle would separate from the empty rocket shell and travel along a ballistic path outside the atmosphere towards its target. After a flight time of some tens of minutes, the re-entry vehicle would descend on its target at very high speed – perhaps as much as 15,000 miles per hour. It had to re-enter the atmosphere at this speed, which led to the other unknown of the time: what would happen to such a vehicle? Would it survive re-entry or would it burn up like a meteor? In parallel with Blue Streak, the Black Knight programme was set up to investigate the problem. Guidance and re-entry were the two major imponderables, for which a good deal of work had to be done in parallel to the main project. But what of the rocket itself?

After the Sandys/Wilson agreement had been signed, the US and the UK set out to design missiles which were complementary to each other. Initially, the Americans were to produce the long-range Atlas missile, the British the medium-range Blue Streak. A team of Americans visited the UK in April 1955 to discuss progress. Sir Steuart Mitchell (CGWL) described the British plans, which involved the RAE as the principal designer for the first two years, control being passed to the firms in the second year. The American team was not impressed by this idea. The VCAS (Vice Chief of Air Staff) noted that he had been told in private that

> They felt themselves that unless we give it to industry with a free hand it might delay the project greatly. They voiced the opinion at the meeting that the British Technical Civil Service was of a much higher calibre than the American, but that a scheme such as that proposed by Mitchell would just never work in the US.

Whether this was the case is difficult to judge, but certainly, progress in 1956 seems to have been slow. On the other hand, uncertainties as to the warhead, as we shall see, contributed to the delays.

It is also noticeable that the many technical reports that came out of the RAE at this time were also speculative and academic. Typically, a half dozen different design solutions would be carefully evaluated in these reports, but they did not lead to a direct practical design in the way that an aircraft design team might work. British aircraft designers of the time would start with quite detailed sketches, which would be refined up to the final solution. A commercial firm was also under pressure to produce a prototype as soon as possible in a way that the RAE was not.

At the same time, a British team from RAE had visited America, and produced a report defining the problems more clearly. As a result of this, the Air Staff felt sufficiently confident as to issue an Operational Requirement (OR 1139) for the missile in 1955, which stated a requirement to deliver a megaton range warhead over a distance of up to 2,500 miles. An OR is one thing, a design is another. Throughout 1955 and 1956, whilst work started on Black Knight and on other aspects of the programme, arguments went back and forward as to the details of the design. The crucial point, from which all else flowed, was: should it have one motor or two? The motor under consideration, an American design, had a thrust of 135,000 lb. This implied a missile weight of no more than 100,000 lb with only one motor. A two motor design could be double that mass. The critical factor, and an unknown factor, was the payload, and the payload was, of course, a megaton warhead and its re-entry vehicle.

With only one motor and a warhead weight of 2,000 lb, the maximum range that could be expected was only 1,900 miles – not enough. There was a further snag: Britain did not have a thermonuclear warhead weighing only 2,000 lb. Indeed, Britain did not have a thermonuclear warhead at all – in 1955, the design of such a warhead was only just beginning. It would not be until late in 1957 that such a device would be tested successfully. The only warhead that might have done the job weighed 4,500 lb – far too much. Some lateral thinking would be needed.

The Warhead

The saga of the warhead is a story in itself. The urgency with which it was pursued is slightly puzzling in hindsight since it was not intended to deploy Blue Streak until the mid-1960s. The UK had taken the decision to develop thermonuclear devices (also known as fusion weapons or more popularly, the H bomb) in 1954, and the Atomic Weapons and Research Establishment (AWRE) at Aldermaston did not as yet have a working design. Given all the other development work that needed to be done, a decision on a warhead could have been deferred until a fusion weapon had been developed. There was a degree of risk in this: there was the possibility that no warhead could have been developed within the weight limit of one ton. On the other hand, the Blue Streak design was the largest possible for a single stage vehicle with only two motors. Designing anything larger would have led to a very unwieldy weapon.

What Britain did have was a relatively low yield lightweight fission device – Red Beard – and a design for a much more powerful warhead which incorporated fusion principles, but was not what would now be called a thermonuclear device. This was Green Bamboo – never tested, since it soon became obsolete. The snag was that this device had a weight in the region of 4,500 lb, but in the absence of anything else, it was Green Bamboo which was specified in OR 1142, entitled 'Warhead for a Medium Range Ballistic Missile'.

Red Beard, the lightweight fission weapon, had been ruled out since its yield was only 10–20 kilotons (kT), and given the predicted accuracy of Blue Streak, this was thought to be inadequate. Green Bamboo was too heavy. Thus William Penney, then Director of Aldermaston, was asked if a lighter warhead of similar yield could be built. He replied to say this might be possible, but that 'the figures of 1,800 lb weight and 30" diameter quoted for an unboosted fission bomb with a yield of about 1 megaton were purely estimates at this stage and could not be guaranteed.'[4] After further study, he decided that 'on current knowledge I could not guarantee to make a satisfactory warhead within the weight specified'. A weight of around 2,200 lb was more probable, and it would use around twice as much fissile material as Green Bamboo. It would be a pure fission device, but in the absence of any alternative, this warhead, now codenamed Orange Herald, was chosen, and, punctiliously, the wording of the OR was changed from 'thermonuclear' to 'megaton'. (Note: there is some ambiguity here. 'Megaton range' meant around a megaton yield, and 600 kT would qualify since 600 kT = 0.6 MT, which can be rounded up to 1 MT.)

Given that the lightest warhead that AWRE could guarantee would weigh 2,200 lb, this decided the issue of one motor or two for Blue Streak: if it was to reach the range required it would need two motors.

A memo concerning Orange Herald sums up the position with some rather interesting comments along the way:

Orange Herald – OR1142
Your loose minute of 31 October asks what is the present status of the official requirement for Orange Herald, and whether the Minister knows of the project. I propose to answer in some detail for I think that it will be useful to record the history of the project.

2. A Draft Operational Requirement No. OR1139 (May 1955) for the Ballistic Missile was discussed at an Air Ministry Operational Requirements Committee meeting on 9th June, 1955. ... In discussion of the range of the missile a warhead weight corresponding to Green Bamboo was assumed, and it seemed unlikely that the range which Air Staff desired could be met by a single-motor missile ...

4. In the meantime, however, CGWL had been discussing with AWRE the penalties implicated on the missile design by the weight of Green Bamboo, and DAWRE [Director AWRE, William Penney] had said that he could develop a megaton warhead of about half the weight, although requiring more fissile material ...

5. This smaller megaton warhead was considered by AWRE to be an important project, and in mid August I was asked by AWRE to arrange for an MOS [Ministry of Supply] code word to be registered for it, for AWRE use and to facilitate inter-departmental discussion. One copy of my minute notifying the choice of a code word for "A megaton warhead for the Medium Range Ballistic Missile" went to PS/Minister [the Minister's Private Secretary] but, of course, without any description of the warhead

7. In discussion on 14th October, DDAWRE [Deputy Director AWRE] made it clear that AWRE consider it important to go ahead with the lighter warhead, that they are in fact doing so (for the present on their own initiative) and that they are considering the possibility of a trial in 1957. AWRE's line is thus quite clear ...

9. I have also consulted the CGWL side, in the person of DGGW [Director General Guided Weapons]. He tells me that they are not yet in a position to put up the missile requirement OR1139 for the Minister's approval prior to acceptance: they still consider the project to be in the design study stage, and a meeting will probably be held in December (1955) to make a final assessment. It seems almost certain that a twin-motor design will be agreed, but whether the missile has one or two motors, they will still want the lighter warhead ... I pointed out that Orange Herald has at the moment the status of a private venture on the part of AWRE, and that it seemed to me desirable that an OR should be issued to cover it. So far as DGGW is aware, the Minister has not been given any data from the CGWL side on the characteristics of

Orange Herald.

10. Thus the answers to your questions are:

(a) The present status of Orange Herald is that of a private venture by AWRE, for the present Air Staff Requirement for a warhead for the OR1139 missile refers specifically to Green Bamboo. DGGW will inform Air Staff that the missile needs a lighter warhead; this should result in a revised OR calling in effect for Orange Herald, and we can then clear with AWRE the warhead parts of this OR in readiness for the eventual submission of OR1139/1142 to the Minister by CGWL and CAW [Controller Atomic Warfare].

(b) There is no evidence that the Minister knows anything of the project beyond the code word and its definition.

[Sgd] D. CAMERON DAW Plans. (10th November 1955)

That last line is worth repeating: 'there is no evidence that the Minister knows anything of the project beyond the code word and its definition'. Decisions involving the expenditure of millions of pounds were authorised not by the politician in charge of the ministry, but by the senior officials running the projects – and Orange Herald probably cost close on ten million pounds (as much as Black Knight or Black Arrow) to design, build and test. Indeed, it is rare that ministers were involved in anything other than the basic decisions. Mention is made of the OR being subject to final ministerial approval. It would be unfair to say that this comprised a 'rubber stamp', but on the other hand, how was a professional politician able to judge whether the OR made sense or not? Ministers would take decisions on matters such as whether a ballistic missile was needed, and the rest was up to the permanent officials. Ministers do take the final responsibility – thus when Blue Streak was ultimately cancelled as a weapon, ministerial reputations were at stake. Indeed, it might be said that Duncan Sandys' defence of Blue Streak while Minister of Defence and Minister of Aviation, even though it was his job to do so, probably hindered his future career. This memo does throw some light on the degree to which decisions involving very large sums of money are taken quite independently of the minister who is nominally in charge – and yet, in almost any system of government, this is the case.

Orange Herald was, in many ways, an extremely unsatisfactory device. It required a very great deal of weapons grade U235 – one Air Ministry paper gives a figure of 120 kilograms, although other figures have been quoted. Whether the UK had the facility to produce 60 such warheads is a very interesting question. Another memo noted that 'Orange Herald is one of the rounds to be tested at Grapple next year. The cost of the material may be £2½ million.' If those costs were carried through to the final deployment, then the warhead cost would be 60 × 2½ = £150 million!

Operation Grapple was a series of tests of atomic devices based at Christmas Island in the Pacific. Three devices were scheduled for the first series, held in mid-1957, and Orange Herald was the second. The first and third were attempts at fusion devices, which were not wholly successful, but showed 'proof of principle'. Grapple had been described by the British Government as 'H bomb' tests, and the press reported the Orange Herald test as an 'H bomb success'. Orange Herald was not an 'H bomb', but rather a large fission device, although one can understand why the Government did not draw attention to this. This has led some later commentators to suggest the deception was deliberate – that if the fusion tests had not worked then the Government could always claim success via the Orange Herald test, whose yield was around 720 kT.

Although as we will see, Orange Herald was never deployed in its original function as Blue Streak warhead, it did come into service in a rather indirect way. The fusion tests later in 1957 were successful, but there was still a long way to go between the testing of the 'physics package', as the part that goes bang is sometimes called, and deployment of a fully serviceable weapon with all its handling and safety devices. Mainly for political reasons, a high yield warhead was required for the V bombers so that the Government could say the RAF had a 'megaton capability'. The answer was a reduced yield (400 kT) version of Orange Herald, codenamed Green Grass.

Initially, the warhead was mounted in the casing of the original British atom bomb, Blue Danube, and the resultant weapon named Violet Club. Owing to the large amount of fissile material (U235) in the warhead, there were some very considerable operational limitations. Any mishandling of the device which caused it to become damaged might result in sufficient fissile material coming into contact so as to cause a low level explosion. One safety device used to prevent this happening was several hundred ball bearings, which filled a void inside the warhead and physically kept the fissile material apart. If Violet Club had ever been flown operationally (which is unlikely), the ball bearings would have had to be removed before take-off. A later version was produced in a different casing (Yellow Sun Mk 1) which allowed the ball bearing to be jettisoned in flight! The number of Green Grass warheads produced was quite limited, partly due to the amount of fissile material needed, and, given the constraints with which the weapon had to be handled, the RAF was quite glad when it was replaced by a fusion warhead.

The testing of Orange Herald meant that RAE could now move on with the design of a re-entry vehicle for Blue Streak, but to do so meant that AWRE had to supply details of the warhead in terms of masses and dimensions. Little was forthcoming, until a slightly indignant memo in May 1958 from the Deputy Director at RAE noted that

> ... it was learned that no work was in progress on Orange Herald at AWRE, nor was there any intention of doing any. Newley suggested that our work, based on Orange Herald, should be stopped, and that AWRE would offer instead a two-stage warhead of similar weight ... Orange Herald had very doubtful in-flight safety, and is highly vulnerable to R effects, and the new proposal is welcome in that it would be greatly superior in both these respects. Nevertheless, it seems to have emerged in a most casual fashion.[5]

The 'R effects' mentioned refer to a perceived vulnerability of warheads to neutron irradiation. Orange Herald, with its large amount of U235, would have been very vulnerable to such effects.

The two-stage device referred to would be some variation on the fusion weapons that had been tested at Grapple. These went under the generic code name of 'Granite' devices (e.g. Green Granite, Purple Granite etc.). Hence the RAE, who was responsible for the re-entry vehicle, dropped their work on Orange Herald, and waited in anticipation for details of the size, shape and weight of the new Granite devices.

There then followed one of the greater ironies of the British nuclear weapons programme. Nuclear co-operation with the United States had ceased soon after the end of the war, and Britain had gone it alone in the development of firstly a fission device and now a fusion device. In many ways, the achievement was quite remarkable, as Aldermaston, with its relatively small budget and limited resources, produced a working fusion design less than three years after work had begun. As a result of Aldermaston's success, the Americans agreed to resume nuclear co-operation. The UK was given the designs for two nuclear warheads, one of them being the Mark 28, approximate yield 1.1 MT. The advantage to the UK was that the design had been fully engineered as a weapon rather than just as an experimental device, and so the Granite designs, painstakingly developed by Aldermaston, were dropped in favour of the Mark 28. The anglicised version of the Mark 28 would go into service as Red Snow.

The major problem to continuing with 'weaponising' a Granite design was that further nuclear testing would have been required, and public and world opinion was turning very much against atmospheric nuclear tests. Britain did not have an underground testing facility, and so the advantages of adopting the American design were that much greater. Hence if Blue Streak had been deployed operationally, then the warhead would have been the 1.1 MT Red Snow.

Whatever the warhead, it seemed that its weight would not come under a ton, which imposed its own restraints on the design. Thus Air Vice Marshall Satterly wrote in July 1955:

> My views are that if we go for the single motor missile we shall always be in trouble over weight and range, and will find ourselves in the early 60's [*sic*] still striving to catch up. Let us be bold and go for the twin motor and exploit any future saving in weight in the warhead, or anywhere else, by increasing the range. Let us then review the position in a years [*sic*] time, when we can put much more reliance on the small warhead and when we are due to consider parallel development of a second missile.[6]

Sir Steuart Mitchell, CGWL, 'reluctantly agreed that two motors would probably be necessary to guarantee a range of 1500 nautical miles'. Dr William Cook, Deputy Director at Aldermaston, said that AWRE 'had not realised how significant was the weight of the small warhead in reaching this decision.' Reluctantly, given the uncertainties in the payload, it was realised that two motors would be needed to achieve the necessary range. A single motor missile would have been almost identical to the American Thor missile, but the heavier UK warhead coupled with the already limited range of Thor (1,500 miles with the US warhead) ruled out the use of Thor by the UK. This was summarised in a paper by Sir Frederick Brundrett, Chief Scientist at the Ministry of Defence, in June 1956:

> The co-operation between the Americans and ourselves on this missile development is extremely good except that which is limited by United States laws governing the passage of information on atomic weapons. We have, in fact, considerably more knowledge on this subject than we are supposed to have and it is vitally important that the Americans are not made aware that we have the information that follows.
>
> The warhead for Thor is being designed to a weight of atomic core of 1,500lb, but the weight of the warhead itself must include the metal sheathing designed to act as a heat sink. The total weight including this sheathing will be 2,600lb if the sheathing is of steel and 3,100lb if the sheathing is of copper. The comparable figures for our own design are 2,250lb, 3,600lb and 4,500lb...
>
> What this means, however, is that if an arrangement could be made for the Americans to provide vehicles to which we could fit our heads, which is a technical possibility, the range of the American vehicle with our head would be reduced to something of the order of 1,100 miles ...[7]

There is no doubt that the single motor design would have been simpler, cheaper, and would have taken less time to develop, but, in a sense, the Air Ministry had painted themselves into a corner. The missile had to be as big as it was to achieve a range of 2,000 miles with a megaton warhead. But no one appears either to have queried this requirement or even decided what the missile was for. Was it a deterrent? Was it actually intended as a weapon that could be used to win a war? Did it need a range of 2,000 miles? The Strath Report, also concluded in 1955, looked at the effect on the UK of just five megaton warheads, and was deeply pessimistic about the result. And if the intention was to 'win' a nuclear war with the USSR, how much destruction would have been necessary to annihilate it or, at least, force it to surrender?

The requirement could have been relaxed either in terms of range or warhead – and it does seem odd that no one ever contemplated the possibility of warheads becoming lighter. And given the range as 2,000 miles, why did it have to be a megaton warhead? Would 200 or 400 kT have been sufficient? And could Aldermaston have designed such a warhead within the new weight constraints? Further, the payload weight was pushed up by inadequate knowledge of the re-entry head. It might have been worth putting more research into re-entry before finalising the design. It is possible that a single motor design might have had a better chance of ultimate deployment, being smaller, cheaper and quicker into service. But the assumptions on which the criteria were based never seem to have been questioned in depth.

A Solid Fuel Design?

A report from Westcott dated December 1956 considered the 'Application of Solid Propellant Motors to Medium Range Ballistic Missiles'[8]. Its summary states that

> The studies are based chiefly on the studies of motors with plastic propellant charges of maximum length 25 ft and maximum diameter 3 ft 6 in. These maximum dimensions are considered feasible with radial burning plastic propellant charges ... and are within the pressing limits of facilities already planned and requested ... For a missile carrying a 4,000lb warhead, fitted with clustered motor units, the ranges calculated for single stage and two stage propulsion are respectively up to 1,300 miles and up to 2,500 miles.

Not surprisingly, given the lack of experience with solid fuel motors of such size, the report is somewhat lacking in precise detail, but instead takes various arrangements of mo tors and makes an estimate (or guess) at the range obtainable from each one.

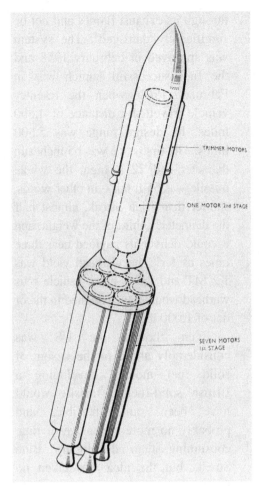

TRIMMER MOTORS

ONE MOTOR 2nd STAGE

SEVEN MOTORS
1st STAGE

Figure 36. Proposed solid fuel missile.

The individual motors shown in the sketches are also very generic: other than being 3 ft 6 inches diameter and 29 ft 2 inches long, there is very little information about them. Quite why these particular dimensions have been chosen is not obvious.

It is clear that the option of using solid fuel motors was not taken very seriously – there is no mention of them at all in policy papers, and it is quite possible that the study was undertaken so as to be seen to have covered all possibilities. It does not appear from the report that there had been wide consultation with those who were actually producing solid fuel motors – the limits imposed on the dimensions seem to have been rather arbitrary. Certainly there is no discussion of the degree of practicality of building motors as large as these or larger.

The payload used in the calculations is given as 4,000 lb – given the later reduction in the weight of the payload it might have been worth revisiting some of these ideas. Unfortunately the idea of a liquid fuel missile had become too firmly entrenched by then – which is, in many ways, a pity. For comparison, let us look at the American solid fuel Minuteman missile.

The US Air Force began looking at the possibility of solid fuel motors in August 1957, in response to the Navy's Polaris missile. The task was given to Colonel Edward Hall, who calculated that 'the ICBM version of Weapon System Q [i.e., Minuteman] would be a three-stage, solid-fuel missile approximately 65 feet long, weighing approximately 65,000 pounds, and developing approximately 100,000–120,000 pounds of thrust at launch'. The missile would be stored vertically in underground silos and 'would accelerate so quickly that it could fly

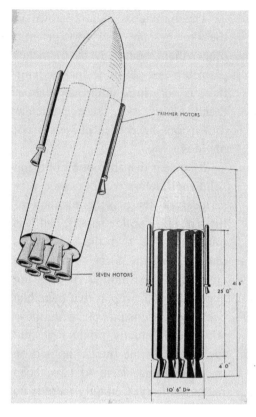

TRIMMER MOTORS

SEVEN MOTORS

41' 6"

25' 0"

4' 0"

10' 6" Dia

Figure 37. A distinctly unconvincing attempt at a solid fuel design, with seven rocket motors.

through its exhaust flames and not be significantly damaged'. The system was approved in February 1958 and the first successful launch was in February 1961, when the re-entry vehicle travelled a distance of 4,600 miles. Its design range was 5,500 miles. The first stage was 65 inches in diameter and 22 ft high; the whole missile was 55 ft tall – in other words, shorter than Blue Streak, almost half the diameter, a third of the weight, and it could deliver its payload near three times as far! The warhead yield was 1.2 MT and the re-entry vehicle plus warhead would have weighed in the or der of 1,000 lb.

Even though the US was considerably ahead in the design of solid fuel motors, developing a British solid-fuelled missile would have been quite feasible, and probably no more expensive or time consuming than developing Blue Streak, but the idea was taken no further.

The outlines of the design were now beginning to emerge: liquid fuelled, two motors, all up weight approaching 200,000 lb. It took some time for a more detailed design to emerge, however. Thus Joe Lyons of the RAE wrote in February 1956:

> It had been agreed in principle that it would be a thin steel missile with propulsion at rear and the warhead at front. Titanium had been considered for the skin but was not promising. A cylindrical structure of about 10 ft diameter and length of about 60–70 ft was generally agreed. It was probable that fins would be fitted but this was not completely certain yet.[9]

Even the use of the NAA motors was still to be debated. A note from Serby, DG/GW (Director General/Guided Weapons at the Ministry of Supply) in March 1956 reads:

Should the missile be designed as a single-stage weapon using 2 × 135,000lb NAA motors since the AUW (*All Up Weight*) could be reduced and the requirement for thrust control could be eliminated if a number of smaller motors could be used?[10]

The thrust control issue arose from the use of large rocket motors: towards the end of the flight, when almost all the fuel was consumed, accelerations became unacceptably high. Thus there was a proposal to throttle back the motors: not an easy task.

The firms detailed to do the work had been decided back in 1955.

It is proposed that Messrs de Havilland should be responsible for the airframe and general weapon co-ordination, Rolls Royce for the rocket motor and fuel system design, Sperry for the internal inertial 'guidance' and autopilot, Marconi for the ground radar launching system.

Whilst relationships between the firms and the Ministry were usually good, this was not always the case with de Havilland, particularly in the early days. There were considerable cost overruns at a time of financial stringency, and at one stage the Ministry went as far as sending in Cooper Brothers, a firm of accountants from the City, to check the costs and management. And with reference to talks with Rolls Royce in 1958, the Ministry noted that

they share the view with everybody else that de Havilland can be extremely difficult and very unsatisfactory, but have no complaints to make over their immediate contacts in this particular connection. Indeed, at the working technical levels, they have a very high opinion of the de Havilland staff, but, here again, they fully share the general view about de Havilland top level people.[11]

To be fair, we are not given de Havilland's views on Rolls Royce!

The debate as to the missile structure had been effectively settled by April 1957, when Wing Commander Bonser of the Ministry of Supply noted that:

A list of equipment required for the building of the 'Blue Streak' airframe has been submitted by the De Havilland Aircraft Division ...

The equipment is required to reproduce that used by Convair for the production of the same type of pressurised structure for an American Ballistic missile. This type of structure is unique to Ballistic missiles and consists of a series of rings in stainless steel and seam welded. These rings are then welded together and fitted with stainless steel domes to form the main tanks for the liquid oxygen and kerosene. The resulting structure is of such strength that it must be kept under pressure in order to retain shape.

This very light structure and the method of production has been developed by Convairs over a very long period (5–10 years) and to save time is to be copied by De Havilland. So important is this feature of the 'Blue Streak' programme that it has been decided that the British missile shall have the same diameter as the American one. This means that the tools, jigs and fixtures can be reproduced with the minimum loss of time – a most important feature as the first structure is required by mid-1957.

It might be thought that work could now go ahead on Blue Streak without any further problems, but with Blue Streak that was never the case. There was constant opposition to the project throughout its life within Whitehall. This surfaces most clearly in the Treasury, but other ministries such as the Admiralty, were also against the project, as we shall see. Indeed, even the Permanent Secretary at the Ministry of Supply, the Ministry whose job it was to develop Blue Streak, was against the project. Sir Roger Makins of the Treasury, one of the 'Great and the Good' of the 1950s and 1960s, reported a conversation thus:

> Sir Cyril Musgrave, of the Ministry of Supply, came to see me on 14th November [1956], to talk about the Medium Range Ballistic Missile. His primary objective was to talk about Spadeadam, and when I told him the Chancellor had made a decision, the main point of his visit was lost. However, he did say that the Ministry of Supply was having great difficulty in holding De Havillands at arm's length, particularly now that the American Government had approved the contract with Convair.
>
> I explained that the Chancellor had felt it desirable to hold up his approval of this transaction until he had an opportunity of considering the future of the M.R.B.M. in relation to the rest of the air weapons programme. On this, I believed that the Ministry of Defence were on the point of producing a paper. I would certainly do what I could to accelerate both its appearance and consideration. Sir Cyril Musgrave turned out to be a bitter opponent of the M.R.B.M. and a passionate advocate of the supersonic bomber [the Avro 730, cancelled in 1957]. He evidently relished locking horns with the Ministry of Defence on this subject.[12]

The transaction being referred to was the licencing by de Havilland of the technique for building the tanks – the decision had been taken to use the same construction method as the Atlas missile, with its 'balloon' stainless steel tanks. The passage about Musgrave is, on the face of it, extraordinary – the Ministry of Supply was simply a procurement ministry, and was not supposed to decide military policy. It demonstrates how blurred the lines can become at times.

The Chancellor certainly did hold up his approval. That conversation was in November 1956, the proposal had been made and put to the Americans; the Americans had agreed, but still the Treasury held out. The proposal reached the Chancellor himself ('Rab' Butler) on 4 July 1957 – eight months later. The memo began:

> This is a proposal that de Havillands should buy from the American Company Convair some 'knowhow' for the development and production of a British intermediate range ballistic missile (Blue Streak). This knowhow will cost $700,000.

There is no doubt that if the Blue Streak project were finally agreed there would be no question of not approving this purchase. But although Ministers have taken decisions which go a long way towards the final decision to go ahead with Blue Streak, that final decision has not yet been taken ...[13]

Butler's response was scrawled beneath in pencil:

No action. Anything could happen in this field in the next 6 weeks. America might offer us the knowhow. Russia might agree to a halt in atomic tests. Everyone might agree that we should not make more fissile material. We might decide not to make a British missile.[14]

This is misleading in so many ways that it is difficult to know where to begin. A halt in atomic tests would not make the slightest difference in the military need for a missile, nor would the amount of fissile material. The Americans might have given the UK the 'know-how' free (unlikely, and that avenue had probably been explored already), but not all the $700,000 was just for 'know-how' – it included specialised welding equipment for the tank sections.

Figure 38. Blue Streak's tanks – made of very thin stainless steel, they had to be kept pressurised to maintain their structural integrity. They were made from lengths of stainless steel rolled around into a cylinder and welded. The 48 stringers on the kerosene tank can also be seen quite clearly.

In correspondence which took place last summer, the Financial Secretary agreed that work on the MRBM should go on but asked that expenditure and commitments should be kept down to the minimum essential until the United States Government had replied to an approach regarding the sharing of information on this and other defence R & D subjects. As far as I can gather the prospects of obtaining substantial US Government help in this field are not at all encouraging. Further, they are least encouraging in the spheres of atomic weapons, of which the M.R.B.M. is, of course, one. It is not necessary here to discuss the rights and wrongs of this state of affairs as between the US Government and her most important ally; but it is worth considering what courses of action are open to us:

(i) we can drop the whole MRBM project. This would mean either that we ceased to contribute actively ourselves to the strategic deterrent or that we did so only during the lifetime, now relatively restricted, of the bomber.
(ii) we can proceed as at present, buying (with the US government's permission) what American information we can, but in the main relying on our own brains and effort (but knowing we are far behind both the Americans and the Russians in the ballistic missile field).
(iii) we can try to regain as much lost ground as possible, by pressing the Americans, by every means within our power, to let us have the information, or the weapons, or both, that we require.[15]

This is an extraordinary memo. First, it recognises the dilemma that would face Whitehall for the next four years: no missile, no deterrent. In practice, the Treasury would have been more than happy to abandon the deterrent – in the mid-1960s, it thought it had succeeded. (The Foreign Office and the Ministry of Defence were described by the Treasury at one stage as the 'last two remaining retentionist [*sic*] departments'. It was the politicians of the Wilson Government that wanted to keep British nuclear weapons.) The other extraordinary feature is the way the Americans are regarded as some kind of fairy godmother. There were no 'rights or wrongs' in this case: there had been some controversy with regard to nuclear information in the 1940s, but that certainly did not apply to missiles. The word 'sponging' comes to mind on reading memos such as these.

Part of the uncertainty with regard to the MRBM was due to the uncertainties in British defence policy, and Suez had a part to play in this, as Sir Cyril Musgrave noted in November 1956:

I believe, however, that Suez has once more put the Policy Review into the background and it becomes necessary to decide immediately whether we should authorise de Havillands to sign the agreement or whether we should reveal by our continued refusal that the future of the project is in doubt. This means revealing the matter to the Americans.[16]

The outcome of the Suez debacle was a further rethink in defence policy under the new Prime Minister, Harold Macmillan, who appointed Duncan Sandys as the new Minister of Defence with increased powers. Part of Sandys' policy rested on missiles and nuclear weapons, which should have made Blue Streak more secure – although, paradoxically, this proved not to be the case.

The licencing of the motor proved to be much more straightforward. Rocketdyne had been set up by North American Aviation (NAA) soon after the war to build rocket motors. There were links between NAA and Rolls Royce dating back to the Second World War, when NAA had developed the Mustang fighter. The Mustang had originally been powered by an Allison engine, which was replaced by the Packard V-1650 – a variant of the famous Rolls Royce Merlin engine. Lord Hives of Rolls Royce and NAA President 'Dutch' Kindelberger were thus old friends, and the agreement for Rolls Royce to licence the Rocketdyne S-3 rocket motor was relatively informal (Rolls Royce had difficulty locating the contract in the early 1960s when ELDO was being formed; Val Cleaver, the chief rocket engineer at Rolls Royce, said that Hives and Kindelberger had probably signed the deal 'on the shake of a hand'). The agreement provided 'for the exchange of Technical Information on Rocket engines over a period of 10 years on payment by Rolls Royce to NAA of a capitol sum of $500,000 and an annual payment of $100 000.'[17]

Rolls Royce initially copied the S3 design and then refined and anglicised it, so that the motor could be built with purely British components. The S3 was being developed for the American Thor and Jupiter missiles, having evolved from the original V2 design via the Navaho missile. This motor burned kerosene and liquid oxygen, standard for the time, but a combination that might, in retrospect, have appeared out of date by 1960, although this is still a matter of controversy. A copy of the design, designated the RZ 1, was built by Rolls Royce and tested at Westcott. From this, the anglicised design, the RZ 2, evolved.

The Missile Design

The rocket structure, like Ancient Gaul, could be thought of in three parts: the engine bay at the bottom, the main tank structure containing all the fuel in the missile, and the ancillary equipment, guidance and payload at the top. The engine bay, containing two RZ 2 motors, was 9 ft in diameter (so designed for transport by air), but the elegance of the final shape of the missile was rather spoiled by two panniers either side containing nitrogen to pressurise the kerosene tank. The liquid oxygen tank could be pressurised by oxygen gas derived from the liquid via heat exchangers. So in June 1957, de Havilland stated that the propellant tanks, constructed of 0.019 inch thick stainless steel, remained unaltered. External stringers on the rear (kerosene) tank would permit the weight of the head to be supported without pressuring the rear tank. This would in turn allow the kerosene to be drained from the missile in the event of a failure occurring on the launcher.[18]

The upper tank had to be kept pressurised at all times to prevent the structure collapsing under its own weight. These 48 stringers also helped to give Blue Streak its distinctive appearance. Inside the fuel tanks were various baffles to prevent the sloshing of fuel, but missiles such as Blue Streak are not much more than gigantic thin-walled tanks.

For Atlas, skin gauges varied throughout the structure, being tailored to meet local stresses. The heaviest skin gauge was forty thousandths of an inch. By comparison, the skin gauge for Blue Streak was nineteen thousandths, but the lower section, the kerosene tank, was re-inforced with stringers. Blue Streak was simpler in being a pure cylinder, whereas the Atlas tanks tapered at the top. The most probable cause of the failure of such a structure in compression is what is known as Euler buckling – the process that occurs when you step onto an empty soft drinks can. But there were other reasons for the reinforcements.

A structure such as Blue Streak or Atlas is also very vulnerable to sideways bending forces, particularly when transmitting large loads vertically. These can originate from sideways gusts of winds, and also from the act of swivelling the rocket motors off centre for control purposes. Indeed, the two motors were to be inclined inwards slightly so that their thrust lines passed through the centre of gravity of the missile. Another problem to which liquid fuel rockets are prone is 'sloshing' which occurs when the liquid sloshes from side to side in the tanks as the vehicle rocks. Although it is often said that Blue Streak performed impeccably for ELDO in the 1960s and 1970s, this is not quite true. Sloshing of the fuel towards the very end of the first flight, F1, on 5 June 1964, overcame the control system and caused the missile to tumble uncontrollably.

The most important parameter for a ballistic rocket using no aerodynamic lift forces is the engine thrust. Two of the RZ 2 motors (see Figure 39) gave a thrust of 270,000 lb. Given that the smallest practicable initial acceleration is 0.3 g (and there is a good case to make this bigger in a missile) then the lift-off weight is of the order of 200,000 lb. Some of this, perhaps 4,000 lb, is payload. The rest is divided between fuel and structure, so that structure plus fuel amounts to 196,000 lb. Given 10% as structure, as an arbitrary figure, then this gives fuel weight as around 175,000 lb. Given the densities of the fuels, their volumes can be calculated. Given a diameter for the rocket – say 10 ft – then the length of the tanks can be estimated. Using these 'back of the envelope' calculations, then the outline of Blue Streak is quite easily arrived at. For comparison, the F1 vehicle with a dummy load of a ton, had a lift off mass of 205,000 lb, 190,000 lb of which was fuel. Detailed design is, of course, another matter.

Some of the design details were more obvious than others – for example, the tanks needed pressurising. For the oxygen tank this was simple enough: a small amount of the liquid can be vapourised in a heat exchanger and piped up to the tank. Pressurising the kerosene tank with oxygen gas would not have been a good idea: instead, nitrogen gas was used, being stored in spherical bottles in panniers either side of the engine bay.

Whilst the tank section was to be built and tested at de Havilland's site at Hatfield, testing the rocket motors was another matter. A purpose-built site would be needed for engine development and also for static firing of the complete vehicle. Not only would this be extremely noisy, it was potentially quite hazardous given the amount of combustible fuel contained within Blue Streak's tanks. The site chosen was Spadeadam on the moors near Carlisle.

Figure 39. The Rolls Royce RZ 2 rocket motors that powered Blue Streak.

The Treasury was of course concerned with the cost: an estimate of £10.2 million for the construction of the Spadeadam site in April 1956 had become £12.3 million by October (and the final cost would be much higher). There is an interesting comment in a slightly later memo:

> If a decision were taken to stop work on the MRBM ... there would be a saving of some £70m. or more over the next ten years.[19]

If only the total cost had come to a mere £70 million! The decision to go ahead with Blue Streak was not yet firm at this time, and a further memo noted:

> ... it is probable that in the Policy Review a choice may have to be made between the supersonic bomber and the MRBM as research and development projects. The cost of R. and D. for the supersonic bomber would be about £70 million over the next 10 years – roughly the same figures as those for the MRBM, but the costs of producing and maintaining an appropriate number of supersonic bombers ... would probably be higher than the costs of an appropriate number of MRBMs.[20]

Spadeadam was split into five areas: the Administration area; the liquid oxygen factory, which was owned and run by the British Oxygen Company; the Component Test Area situated at Rushy Knowe; the Engine Test Area at Prior Lancy; and the Rocket Test Area situated at Greymare Hill.

The site is described in a Ministry of Aviation paper of November 1961 (the English is slightly eccentric at times):

> The Spadeadam Rocket Establishment was built by the Ministry of Works on behalf of the Ministry of Aviation for the purpose of developing and the static testing of the British Ballistic Missile 'Blue Streak'.
>
> The Establishment is situated on the Cumberland Fells about twenty miles North-east of Carlisle and covers an area of approximately 8,000 acres. It comprises five main areas, three of which are test areas for the static testing of the complete missile, propulsion units and of the rocket engine component parts respectively.
>
> As a safety measure these areas are separated by distances of up to one and three-quarter miles. This dispersion has required the construction of six miles of road connecting the 'Areas' on the Establishment.
>
> MISSILE TEST AREA
>
> This Test Area comprises two missile stands each with a traversing servicing tower on which the missiles are statically tested including the firing of the propulsion units.
>
> By means of the gantry incorporated in the servicing tower, the missile is erected into the vertical firing position on the concrete emplacement situated at the end of a 300-ft concrete causeway.
>
> Built into the emplacement is a steel flame deflector weighing nearly seventy tons for deflecting to the horizontal plane the jets of the rocket motors.

The large quantity of water required to maintain the temperature of the flame deflectors at a safe temperature level is pumped to the test stands via 36" diameter pipelines supplied from a one-million gallon reservoir situated adjacent to the Missile Test Area.

The necessary liquid propellants and high pressure nitrogen gas used for pressurising are stored in this area.

The Missile tests are instrumentated and controlled remotely from a central block-house situated approximately 1000-ft. from the test stands (both of which stands will be evacuated when firing is taking place on either stand) built of reinforced concrete. The tests may be observed from the Control Block-house by means of periscopes and closed-circuit television. In addition to the recording of test data on magnetic tape, film records of the tests are made by cine-cameras situated at strategic points around the test stand.

The main Instrumentation System comprises 19 Control Consoles, 4 Checkout Consoles, (46 Chart-type Recorders) with a capacity of 285 channels and three types of magnetic tape recorders with a total of 32 Information Channels. The Control Centre and each test Stand are connected by over 3,500 wires.

ENGINE TEST AREA

This area, in which the individual propulsion units are test fired, consists of three engine test stands... spaced 250-ft apart. A fourth test stand is partially complete.

Each stand consists of a massive concrete and steel structure in which the liquid propellant rocket engines are mounted to fire vertically downwards into a water cooled flame deflector which deflects the flame into the horizontal plane. The propellants used are Kerosine for the fuel and liquid oxygen for the oxidant. The early engine produced and evaluated by R-R Ltd. developed 135,000-lb thrust.

Figure 40. The picture above shows a flight model Blue Streak (note the painted spiral) on a test stand at Spadeadam. The vehicle would be assembled, filled with fuel, and static fired before being shipped out to Woomera in Australia for launch.

The quantity of water used for flame deflector cooling, storage facilities and transfer systems are similar to those provided in the Missile Test Area.

At a distance of 600-ft from the nearest Engine Test Stand is the Control Block-house constructed of 2-ft thick reinforced concrete. This building is equipped with 130 chart-type recording instruments, four 24-channel oscillographs and, when fully equipped, eight control consoles for the remote control of the test equipment and the rocket engine during test. A large number of chart recording instruments are needed to obtain the maximum amount of technical data during the short duration of the test.

An underground concrete duct, 7-ft square and 1,100-ft in length, inter-connects the test stands with the control room for the routing of approximately 8,000 instrumentation and control cables.

The test firings are also recorded by cine-cameras from various locations around the test stands, the cameras being controlled remotely from the control room. These filmed records in addition to the other test records are processed and analysed in the Establishment.

LIQUID OXYGEN MANUFACTURING PLANT

To provide for the large consumption of liquid oxygen, a manufacturing plant has been constructed at Spadeadam. The plant is capable of producing a total of approx. 100-tons of liquid oxygen per 24-day in addition to liquid and gaseous nitrogen. The liquid oxygen is transported by road tankers from the plant to the storage tanks in the test areas, the nitrogen gas is pumped by pipeline to high pressure reservoirs in each of the areas.[21]

Figure 41. Rocket test stands at Spadeadam.

Spadeadam was built by the Ministry of Works under the supervision of the Ministry of Supply. It was one of the few areas of the whole MRBM programme over which the Treasury had direct control. (It was probably also the most expensive – the estimate of the cost at the time of cancellation was around £24 million). They kept spending under a tight rein, an example of this being when PFG Twinn of the Ministry of Supply wrote to the Treasury asking to be able to spend up to £10,000 on particular items without seeking direct Treasury authorisation. Reluctantly, the Treasury agreed.

Spadeadam was obviously extremely remote, which is why it was chosen in the first place. This did lead to transport difficulties, particularly when showing visitors round. Twinn requested authorisation for some transport in the form of:

| 1 Ford Consul | Black with Heater | Saloon | 4-6 seats | £556 |
| 1 Ford Prefect | ditto | ditto | 4 seats | £424[22] |

To which the Treasury replied:

> We approve the purchase of the Ford Consul, but would be grateful if you would substitute a Ford Popular for the proposed Ford Prefect. Hundreds of Populars are in use in Government service, and we would rather keep to this cheaper 4-seater model.

It really is quite extraordinary that the ministries should be debating the relative merits of a Ford Prefect or a Ford Popular. One cannot quite see the same problem arising at Cape Canaveral.

In July 1959, an official from the Treasury, Mr JA Marshall, set off to inspect Spadeadam in the company of Mr William Downey from the Ministry of Supply. 'The visit was enjoyable (in spite of almost continual rain) and instructive,'[23] he remarked.

In general, he approved: 'The administrative block is a modest and simple construction as far as I saw it, and there certainly appears to be nothing lavish here.'

But later he was to observe one example of:

> ... what I would regard as some extravagance.
> This is in the housing of the lines which run from the control and recording centre to each of the four stands. It has been put wholly underground ... in a tunnel some 400 yards along ...

Other cables had been put in an over ground duct. Mr Marshall enquired why this had not been done here. He was told: '... had it not been so placed, it would have obscured the view of those in the control centre'.

It also had to be carried under a road, but Mr Armstrong was not entirely convinced it had been necessary to put the entire length of cable underground.

Despite this particular example of 'extravagance', Mr Marshall seemed satisfied with his visit. He was 'well impressed by all I saw', which included the control rooms, full of:

electronic instrumentation provided to record every possible facet of the test for subsequent analysis. To the layman this produces a hopelessly bewildering mass of knobs, buttons, recording graphs, lights etc'

He goes on:

Mr. Downey had the courage to ask the schoolboy question: "Which is the button which actually sets it going?". The scientist who was showing us round was utterly at a loss for about a quarter of a minute!

There were other problems. 'Commander Williams, the senior Rolls Royce representative, who is the general manager of the place, had a go at me on two minor things'.

This related to the housing which had been built for those working at Spadeadam.

He said he thought the Treasury were unduly mean in not allowing them to build a higher proportion of garages ... a number of people who have no garages ... do have cars and keep them standing in the street ... Since the streets are fairly narrow – the whole estate is on very modest lines – this could be a serious nuisance ...

And also:

He also thought we underestimated the difficulties they faced when we insisted on a certain proportion of the houses having two bedrooms instead of allowing them to be all three, or in a small number of cases, four-bedroomed.

How much was the Treasury saving? Mr Marshall noted that '... the additional capital cost of the three bedroomed house is only £70 ...', but on the other hand, 'we get an extra 5s. a week rent back from it'.

5s. means 5 shillings, or 25p in decimal money. The extra room would have paid for itself in just over five years.

If Blue Streak had gone ahead in its military guise, then a considerable number (no figure has been found, but a reasonable estimate might be between 10 and 20) of experimental and proving vehicles would be needed, as well as the 60 or so production missiles, starting in 1960 and ending by about 1966. Somewhat optimistically, the development schedule had called for the first missile to be fired from Woomera around mid-1960. As it was, the first vehicle was not fired until 1964, and there were a total of 11 firings by 1971 (in practice, more were built and tested, as there were some basic, non-flight, development vehicles, and also some which were used to check at the sites at Woomera and Kourou).

Thus after the cancellation, the tempo of work slowed considerably, and only one of the missile test stands, C3, was completed. There were also inevitable redundancies and a considerable drop in morale. In May 1967, Val Cleaver, Chief Engineer of the Rocket Department at Rolls Royce, wrote of the need 'to raise morale and inspire some confidence in the future of the establishment'[24], but apart from the test site for the liquid hydrogen RZ 20 motor, no further development work was carried out at Spadeadam.

Figure 42. RZ 1 chamber on P site at RPE Westcott.

Spadeadam would take time to build, so whilst it was being constructed, Rolls Royce used the P stand at Westcott for preliminary testing of the Rocketdyne derived RZ 1. Most of the runs were of very short duration – usually just a few seconds. Progress was delayed by a spillage of liquid oxygen onto the steel girders, which caused them to crack. One of the longest test runs was also the last, on an open day at which the Press was present. A motor was fired for 20 seconds, and on being checked after the firing, it was found that the gear box between the turbines and the pumps had disintegrated. A fraction of a second more, and the firing might have become very spectacular.

So from mid-1957, testing and development proceeded apace. At Hatfield in Hertfordshire, large structures were built to house the testing of early, non-flight vehicles. These were for checking the strength of the vehicle structure and for such tasks as determining whether half a ton of fuel could be pumped from the tanks each second. Engine testing was carried out separately by Rolls Royce, first at a test site at Westcott and then at the purpose-built facility at Spadeadam in Cumbria. Here, in addition to engine development, assembled vehicles could be static fired. Once tested, they would then be taken apart for transport to Woomera.

Transport proved to be a difficulty. The large size of the tanks meant that very few aircraft would be suitable, and those which were, such as the Bristol Freighter, had limited range. This meant a great many countries would be overflown on the route from the UK to the Antipodes, and, given the nature of the cargo, political difficulties were foreseen. The problem was solved when the RAAF bought Lockheed Hercules aircraft: one of these could be leased from Australia for the purpose. And given the extra range of the Hercules, the overflying problem could be reduced by going westward round the world, over Canada and the US. As it would turn out, the tank of the first missile was at Los Angeles when the cancellation was announced, and it was returned to the UK. All subsequent civil vehicles were transported by road and sea.

Another of the major problems in the early development was that of guidance. The contract had been given to Sperry, but they were struggling: '... accuracies being demanded from the inertial equipment for this project are extremely high and very marginal. It does seem that these accuracies are just about obtainable ...'[25] in January 1956. There is a despairing cry from the Director of Air Navigation in

Figure 43. Cutaway view of Blue Streak.

October 1956: 'There does not yet exist, I believe, anywhere in the world a gyroscope suitable for Blue Streak.'[26] This was not quite true: American gyroscopes were very much more advanced, and Ferranti was eventually given the job of adapting Kearfott gyroscopes for the guidance system. Although these were probably not the best the Americans had under development, they were perhaps the best they were prepared to make available to the UK. And in November 1956: 'It will thus be seen that, though the Blue Steel situation is parlous, the Blue Streak position is even more desperate'. A working design was finally produced, although in the end it was never used: the guidance system was cancelled at the same time as the missile. A satellite launcher needed a much less complicated system, and scrapping the Blue Streak inertial guidance could save money.

As the 1960s progressed, inertial navigation was to become much improved: a system designed for the TSR 2 aircraft was adapted for Black Arrow. Blue Steel was left to soldier on with a much earlier design, which, with its valves, was very power hungry. Again, the penalty was being paid for being early in the field.

But the next, and most controversial, point about Blue Streak was to be its proposed means of deployment. To have the missile sited on the ground in the open, as the Thor missiles were, was pointless. A pre-emptive strike by relatively few Russian missiles would have destroyed every site before a missile could be fired in retaliation. As all sides in the Cold War realised, land-based missiles would have to be sited in 'invulnerable' silos, although the word 'silo' was not in usage in the UK at that time. Indeed, the two phrases used became code phrases, showing which side of the controversy you were on. 'Underground launcher' was the phrase used by those in

Figure 44. An early test vehicle being lifted up into the stand at the de Havilland factory at Hatfield.

favour of the project, 'fixed sites' if you were against. 'Fixed sites' were perceived as being vulnerable. Airfields, somehow, were not perceived as being vulnerable – although they were as 'fixed' as any missile launcher.

---

[1] TNA: PRO AIR 2/13206. ATOMIC (Code B, 12): Long range surface to surface weapons.

[2] TNA: PRO AVIA 48/7. Rocket Research Panel: minutes of meetings.

[3] TNA: PRO DEFE 7/2245. Development of Blue Streak. The Operational Requirement for a Medium Range Ballistic Missile system. Note by DCAS 9 August 1955.

[4] TNA: PRO AVIA 65/1193. Warhead for a medium range missile: Air Staff requirement OR 1142 Orange Herald (DAW plans action). This file is the source for all the quotes in this section.

[5] Ibid.

[6] TNA: PRO AIR 2/13745. Warhead for medium range ballistic missile Blue Streak (OR 1139 and 1142).

[7] TNA: PRO AIR 20/10299. Ballistic missiles: minutes of Joint US/UK Medium Range Ballistic Missile Advisory Committee and related papers.

[8] TNA: PRO AVIA 6/17210. Application of solid propellant motors to medium range ballistic missiles. Technical Note RPD 156. WR Maxwell, December 1956.

[9] TNA: PRO AVIA 54/2141. Blue Streak development: A1 Coordination Panel; minutes of meetings.

[10] TNA: PRO AVIA 54/2146. Blue Streak development: Joint US/UK Advisory Committee; technical correspondence.

[11] TNA: PRO DEFE 7/2245. Development of Blue Streak.

[12] TNA: PRO T 225/1775. Guided weapons development contracts: Blue Streak and Black Knight.

[13] Ibid.

[14] Ibid.

[15] TNA: PRO T 225/1150. 'Blue Streak' medium range ballistic missile: Spadeadam test site. Memo by DR Serpell, 18 October 1956.

[16] TNA: PRO T 225/1775. Guided weapons development contracts: Blue Streak and Black Knight.

[17] TNA: PRO AVIA 65/714. Spadeadam engine testing: Rolls Royce Ltd.

[18] TNA: PRO AIR 20/10299. Ballistic missiles: minutes of Joint US/UK Medium Range Ballistic Missile Advisory Committee and related papers.

[19] TNA: PRO T 225/1150. 'Blue Streak' medium range ballistic missile: Spadeadam test site. Memo by DR Serpell, 30 October 1956.

[20] Ibid. Memo by DR Serpell, 3 November 1956.

[21] TNA: PRO AVIA 65/715. Spadeadam engine testing: Rolls Royce Ltd.

[22] TNA: PRO T 225/1150. 'Blue Streak' medium range ballistic missile: Spadeadam test site. PGF Twinn to JA Marshall 17 March 1958.

[23] TNA: PRO T 225/1424. 'Blue Streak' medium range ballistic missile: test site, Spadeadam. JA Marshall 10 July 1959.

[24] TNA: PRO AVIA 92/232. Spadeadam: liquid hydrogen-liquid oxygen thrust chamber test facility. Cleaver to CGWL.

[25] TNA: PRO AVIA 54/2135. Blue Streak development: inertia guidance.

[26] Ibid.

Chapter 6

Blue Streak – the 'Underground Launcher'

There was one feature in which Blue Streak was ahead of its time, and this feature, by a process of tortuous logic, was to provide the means by which its cancellation was achieved.

One of the defining images of the Cold War was that of the missile silo: the sight of a missile erupting from a hole in the ground conjured up images of mushroom clouds, radioactive fall-out, megadeaths. One of the less known facts about missile silos is that much of the initial research on launching missiles from underground was carried out near a small village in Buckinghamshire – Westcott.

The first American missiles to be deployed – Thor, Atlas, Jupiter – were all deployed on surface sites, with the only protection for the missile being protection from the elements – wind and rain. These missiles were also very large and very fragile. A sniper a mile away could put a bullet through the tanks and disable the missile permanently. They would be damaged beyond use by the blast from an atomic explosive even though it might have been miles away. In the jargon of the day, they were hopelessly vulnerable to a pre-emptive strike. Later versions of the Atlas missiles were kept in hardened shelters – hardened in this context meaning strengthened against attack – although they still had to be removed from the shelters to be erected, fuelled and fired.

The slightly later Titan I missile was housed in a 'lift and fire' silo: the missile was prepared for firing in an underground tube, then lifted to the surface for the moment of firing. This reduced the window of vulnerability to a few minutes, but the vulnerability still existed. The only way to make sure the missile could not be destroyed before launching would be the 'fire in the hole' method.

The Minuteman solid fuel missiles that America was in the process of deploying were far more robust than the relatively fragile liquid fuelled rockets. They also had a much greater initial acceleration, meaning they would clear the silo quicker, and so providing a hole in the ground for them was relatively straightforward. Blue Streak, on the other hand, would take several seconds to clear the silo, and there were two major problems to be investigated. The first was whether the acoustic energy produced from the rocket motors would be sufficient to damage the thin tank walls, the second was the problem of gas flow. How could the exhaust gases (and Blue Streak burned nearly half a ton of fuel a second) be deflected away from the rocket? And what of the gas flow within the tube? Would the missile be pulled towards the wall of the tube?

Figure 45. 1/60th perspex model.

The problem of vulnerability to pre-emptive attack had been recognised from the outset, and reference was made in the original requirement to 'underground launch sites' without being specific about detail. Initial ideas were very vague, until a proper research programme was started at RPE Westcott. Initial ideas centred around some sort of 'U' tube arrangement, with the missile in one arm of the U. To study the dynamics of the gas flow, a 1/60th scale model of the rocket was used, together with jets of high pressure nitrogen gas to represent the rocket exhaust (it is not surprising that the men who worked on this part of the project lost most of their hearing). Perspex models of various configurations were made, with small woollen tufts to show the air flow.[1]

The next step was considerably more expensive. A 1/6th scale model U tube was to be built, complete with an acoustic lining, real rocket motors placed inside, and fired. Microphones would measure sound levels. Rather than excavate

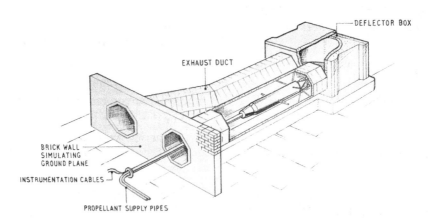

Figure 46. 1/6th scale model of the proposed underground launcher.

a deep hole, the U tube was built horizontally, on the ground, with a brick wall to simulate the ground plane, as shown in the drawing above. Octagonal concrete sections were built and fitted together, some of which can be seen at Westcott today, although, slightly surprisingly, there is no trace of the model silo itself.

The 1/6th scale missile could then be moved up and down and its Gamma motors fired. Various other effects could be investigated at the same time – for example, the effect of the rocket efflux on the concrete at the bend of the U tube. Low temperature concrete was preferred since it merely tended to melt; high temperature concrete fragmented, and the fragments were swept along the tube, gathering more as they moved along. Reading between the lines, some of the test runs could be somewhat alarming, as this excerpt from a report shows:

Figure 47. Remains of the model silo at Westcott as seen in 2010.

As was expected, the erosion caused directly by the impingement of the jets was small, but small particles removed from the surface were accelerated by the gas stream, and scoured the curved deflector plate downstream of the point of impingement. The effect was cumulative and persisted in the region of high gas speeds and where the deflector was concave towards the jets ... Pieces weighing two ounces or more were picked up about 50 yards from the launcher exit.[2]

But these were not the only problems. What was the effect of a one megaton explosion half a mile away? What effect would the shock wave have on the launcher and the missile, not to mention the personnel inside? The effect of 'radio flash' (now known as ElectroMagnetic Pulse or EMP) was unknown, as were the effects of neutron irradiation. Furthermore, the launcher had to accommodate the launch crew for a period of up to 24 hours after hostilities began. They would need living quarters inside the launcher itself.

Figure 48. The Gamma motors being fired within the U tube assembly.

There were turf wars from the outset. The Ministry of Works, part of the Ministry of Supply, would normally deal with constructional matters such as these, but the Air Staff wanted to handle it themselves. Aggrieved, the Permanent Secretary at the Ministry of Works appealed to the Treasury to adjudicate. Emolliently, the Treasury wrote back to say that yes, the Ministry of Works had a very good case, but they had decided to leave it to the Air Ministry since it was they who would have to use the end product. So who would the Air Staff appoint to build the launchers? De Havilland were not popular with the Ministry of

Supply, who found them difficult to deal with. In addition, de Havilland was in bad odour as a consequence of their over-spending (the Treasury at one time was considering sending in the accountants Cooper Brothers, predecessors to today's management consultants).

Figure 49. Another view of the model silo can be seen here, showing how the brick wall was used to represent the ground plane. (Chalked on the wall above the model missile are the words of a song from the then hit musical *My Fair Lady:* 'Wiv a Little Bit of Blooming Luck'.)

A minute in October 1957 said of de Havilland: 'It should be clear that design of underground sites is in no way the responsibility of this firm, which is already showing itself incapable of carrying the load that has been put on it.'[3] On the other hand, there were not many alternatives, and, in the end, the firm was given the job in lieu of anyone else. The evolution of the design can be seen in the various documents still in the National Archives, but the final version was given an airing in a presentation at de Havilland's offices in Charterhouse Square, London[4].

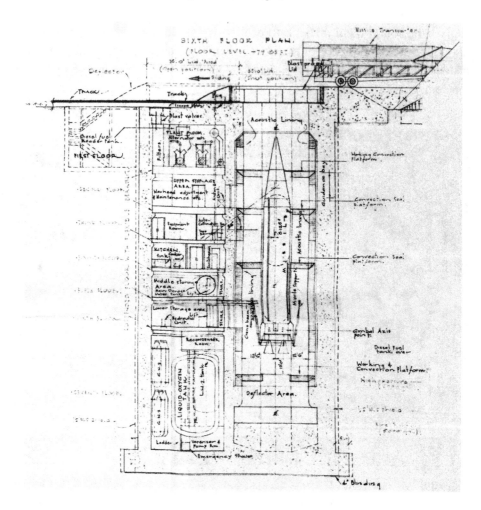

Figure 50. Architect's drawing of the 'underground launcher'.

The full specification can be seen in Appendix A, but Figure 50 makes the layout fairly clear. The final design is indeed impressive, as the architect's drawing shows (Figure 51). The figures of the men inside the launcher give some feeling for the scale of the launcher.

Starting at ground level, the emplacement is surrounded by a built-up bank, and the access road can be seen top right. The missile would have been brought on a transporter, then lowered into the missile shaft. The silo is drawn with the lid open, and the launch and efflux ducts can be seen. On the left is the entrance, with its airtight and blast proof doors.

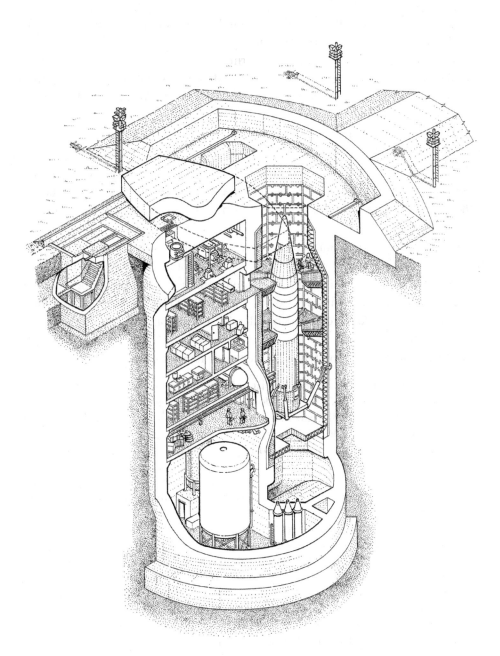

Figure 51. The 'underground launcher' reconstructed from the architectural drawings.

© English Heritage. From National Archives file AIR 2/17377, drawn by Allan Adams.

The cylinder is divided into two: one half houses the missile and the shafts, the other houses equipment and the living quarters.

The missile itself is held on a platform attached to the walls by springs or hydraulic cylinders, so that the relatively fragile missile would be insulated from the shock of nearby explosions. There are access platforms around the missile; the warhead would almost certainly be stored separately and would have to be fitted before launch. Around the walls of the shaft is an acoustic lining, product of the research at Westcott.

At the bottom of the launcher is the liquid oxygen tank and the kerosene tanks. Fuelling the missile in the confines of the launcher might well have been a rather hazardous procedure! This was one of the advantages of storable propellants.

Note that the whole launcher is surrounded with a ¾ inch mild steel liner – this, in conjunction with the lid, should act as a gigantic Faraday cage. This was to protect the launcher from EMP which might otherwise destroy all the electrical and electronic equipment inside. Looking at the thickness of the walls, building these launchers would have had a considerable impact on Britain's production of concrete!

A description of the prototype launcher, code named K11, says:

> Basically, the emplacement consists of a hollow re-inforced concrete cylinder, 66 feet internal diameter, extending downwards from ground level to a depth of 134 feet and divided internally into two main sections by a vertical concrete wall. One section houses a U-shaped tube, whose arms are separated by a concrete wall and are, respectively, the missile shaft and its efflux duct. The surface apertures of this U-tube are covered by a lid that can move horizontally on guide tracks. The other main section within the cylinder is divided into seven compartments, each with concrete floor and ceiling, for the various storage, operating, technical and domestic functions.
>
> The internal diameter (66 feet) of the concrete cylinder is determined solely by what is to be accommodated. Protection against [a 1 megaton explosion at ½ mile distance] is given by the lid and by the re-inforced concrete roof walls and foundations. The wall thickness will depend on the geological characteristics of the surrounding rock and may well be of the order of 6 feet. The depth of 134 feet is arrived at primarily to give sufficient clearance below the missile (itself 79 feet long) to allow for de-fuelling and re-fuelling the missile into and from the liquid oxygen and kerosene storage tanks located on the 7th floor.

The 400 ton steel lid could be opened in 17 seconds, and high pressure hoses would sweep it clear of debris first. Two firms, John Brown (SEND) [Special Engineering & Nuclear Developments], and Whessoe Ltd of Darlington, were commissioned to produce designs. Whessoe's report is dated March 1960, so that it is obvious that the design of the lid had not been finalised by the time of the cancellation. This, combined with the problems of finding a site for K11, meant that the schedule must have been slipping seriously. To have produced a fully working prototype by 1963 or 1964 on this basis looks difficult to achieve.

Brown's lid was 56 ft by 36 ft, weighing 600 tons. Both were hollow, although with strong internal bracing, and in Brown's case, 5 ft thick. The Whessoe lid used 3 inch thick steel plate top and bottom. The lids were to be capped with 6 inches of concrete. The rails on which the lid was to run were also of steel, in Brown's case 7 inches square in cross section, and for Whessoe, '12 inch overall width by 5½ inch rail width'.

John Brown's submission noted that

> the front edge will have an angled profile to act as a plough in the event of it being necessary to clear any rubble which may accumulate on the track. The two 36' sides will house pairs of four-wheeled bogies, and a continuous rack will run down the outside of these faces, engaging on a pair of pinions ...[5]

When considering the electrical drive mechanism, it was noted that the requirement stated:

> A cover weighing 600 tons has to be moved from the open to the closed position in 17 seconds and fine positioned to within half an inch within a few seconds of the end of this period. Alternatively, it has to be opened without fine positioning in 17 seconds. The time allowed for lifting the jacks is 5 seconds so that 12 seconds remain for the operation ...[6]

Provision was also made for keeping the lid free from debris, although the ground shock the silo was predicted to receive would be considerable. There are references in a de Havilland paper to 'Ground shock ... in a vertical direction, an instantaneous step velocity of 2¼ ft/sec is induced, which decays at a uniform rate to zero in one second ... in a horizontal direction, an instantaneous velocity of ¾ ft/sec is induced which decays at a uniform rate to zero in one second', whereas Whessoe notes that there are 'Acceleration forces on Ancillary Equipment equivalent to accelerations of $2g$'[7]. In addition, the rails would be exposed to heat from the fireball. The question this raises is whether the rails buckle under these loads and hence jam the wheels. A silo whose lid cannot be opened would not have been much use!

Other aspects of the design were to cause concern: the various effects created by a nuclear explosion nearby. Whilst the silo might survive the blast, there were concerns as to the effects of EMP and of high energy neutrons. The RAE Lethality Committee set to work to investigate these effects.

One effect, of course, is the thermal radiation or heating. Being underground, the launcher was relatively immune from this, although the lid and the rails would be exposed. This was not thought to be a significant problem. The high temperatures and energetic radiation produced by nuclear explosions also produce large amounts of ionised matter that is present immediately after the explosion. Under the right conditions, intense currents and electromagnetic fields can be produced, generically called EMP (Electromagnetic Pulse), that are felt at long distances. Living organisms are impervious to these effects, but they can temporarily or permanently disable electrical and electronic equipment. Ionised gases can also block short wavelength radio and radar signals (fireball blackout) for extended periods.

The occurrence of EMP is strongly dependent on the altitude of burst. It can be significant for surface or low altitude bursts (below 4,000 m); it is very significant for high altitude bursts (above 30,000 m); but it is not significant for altitudes between these extremes.

This was the reason for the steel liner to the silo: it would act as a Faraday cage, whereby the strong magnetic and electrical fields pass through the liner without affecting anything inside. For this to be effective, the cage must have no openings through which energy could leak. Studies were carried out on the consequences of such details as pipework into and out of the silo, and what effect they might have.

The main radiation hazard came from high energy neutrons, which would not only affect the warhead but also the crew inside the silo. One of the purposes of the lid was to act as a neutron absorber (the water within the concrete would help).

In addition, the equipment in the silo might be protected against ground shock, but the crew themselves could be seriously injured. There was even a proposal to keep a spare crew suspended in hammocks ready to take over. The missile itself would be suspended on hydraulic cylinders to act as dampers against the ground shock.

The work done at Westcott can be said to have validated the concept of the launcher (it was estimated at the time of cancellation that the Westcott work had cost £1.8 million). It is also of interest that the British design studies preceded any undertaken in the US, and Colonel Leonhardt, deputy commander for Installations, Ballistic Missile Installations, visited the UK to evaluate the design for the underground launcher. Dr Barry Ricketson of RPE travelled to America in 1959, as the minutes of a meeting noted: '… Dr Ricketson was at present in the United States giving the Americans the knowledge he had acquired in his studies of the problems of launching ballistic missiles from underground.'[8]

A similar research programme, which seems closely based on Westcott's work, was carried out in the US, where again a 1/6th scale model was tested. The Titan II silo design that resulted can be seen to have a family resemblance to the Blue Streak launcher (although similar problems lead to similar solutions). The last Titan silo was not taken out of commission until 1987, which tends to argue against any obsolescence of the UK design.

Within the silo, there would be a crew of three officers and five men per shift, and the crucial point of the OR for the launcher was that:

> ... the emplacement must be self-contained for an emergency period of four days (covering three days before an attack is expected and one day afterwards).

This is the pivotal issue for the silo concept: early warnings, launch times and the rest were irrelevant. Indeed, although much had been made of the fact it would have been impossible to launch Blue Streak between the time that an incoming attack was detected and the time when it arrived, this misses the point entirely. The UK would never have launched 'on warning'. There were no facilities for doing so. A man carrying the codes for a nuclear attack follows around the US President throughout his time in office, but in the UK there has never been such provision. There would have been very many times when the Prime Minister would have been inaccessible, such as when he was in his car on the way to Chequers, for example, and authority to launch would not have been delegated to the military.

The point of Blue Streak was to act as a deterrent, so that if the UK were attacked with atomic weapons, it would have the ability to retaliate for a period of up to at least 24 hours after the initial attack, although there is no reason why this period could not be longer. The UK might not then be a functioning society any more apart from this one vital aspect – its ability to strike back at its attacker. That was the point of deterrence. But would the launchers in fact be capable of this? Some thought so, when the effect and accuracy of Russian missiles was being discussed. Thus, in a Ministry of Defence memo in June 1958:

> This does not mean that a potential enemy attack with weapons achieving a c.e.p. of ½ mile would invalidate Blue Streak as a weapon. With a c.e.p. of ½ mile, one MT weapon delivered at each site would have a 50% probability of neutralizing a site. [This was an estimate based on the silo design and the known effects of nuclear explosions.] Thus to achieve an acceptable probability of destroying our retaliatory capability, a much higher ratio than one attack per site would be required; for example, more than 3 attacks per site would be required for a 90% probability and, if the reliability of the attacking weapons system is, say, 70% then more than 4 launchings against each target would be required. Alternatively, to give a 90% probability of putting a Blue Streak site out of action with 1MT delivered, a c.e.p. of

¼ mile would be required and again additional launches would be needed to offset the inevitable less-than-100% reliability. [c.e.p. is circular error probability; the chance of 50% of missiles arriving within this radius.]

A paper prepared in 1959 gives some idea of the projected costs:

The Ministry of Supply estimate that the total R&D cost of a below-ground Blue Streak would be £160–£200M, broken down as follows:

Underground Deployment	
100 sites at £2M per site	£200M
100 missiles at £0.5M per missile	£ 50M
10 years operating costs at £0.125M per missile per year	£125M
R&D	£160–£200M
Warheads	£100M
Total	£635-675M

Even with only 60 missiles deployed, which would have been the probable final total, these are impressive sums for 1959. But they were only estimates, based on no hard data, and the history of such projects made it very clear to everyone except the Ministry of Supply that all such estimates were usually wild underestimates. The Treasury, of course, was more cautious: in January 1958 it gave authority to start work on Australian facilities. However, it went on ominously, 'commitments on the part of the launchers related to "below ground" aspect are to be kept to a bare minimum and no work should begin on the "below ground" part of the launchers themselves.'[9]

Air Vice Marshall Kyle, Assistant Chief of Air Staff (OR), had to write to the Treasury:

At our meeting on Monday, 21st April [1958], in the Ministry of Defence, the requirement for underground launching for Blue Streak was queried and I confirmed then that this was the firm intention of the Air Council. From the attached note you will see that this has been made plain for well over a year and that the need for underground siting was envisaged in the original O.R. for the missile.[10]

Similarly, the Ministry of Supply argued that the idea of underground siting had been outlined in the 1958 Defence White Paper. The Treasury dismissed this by saying: '… the Treasury would never accept that a unilateral statement in a Defence White Paper committed the Government to any particular policy.' One does wonder then what the point of a Defence White Paper was. The idea that a Defence White Paper would be published without having been cleared by the Prime Minister is more than a little absurd.

In Australia, a U tube launcher was to be built in the side of a ravine to avoid unnecessary excavations. A similar idea was pursued at Spadeadam, where English Heritage has recently investigated the site:

A letter from the Ministry of Supply to the Treasury shows that plans were well-advanced for the construction of an underground ground launcher at Spadeadam by September 1958. Trial bore holes had been drilled during the summer and permission was sought to begin the construction of a full size silo at a cost of £690,000, plus a 15 per cent agency fee. Owing to the proximity of the bedrock to the surface and the great expense (and time) that would be incurred excavating a hole 150 feet deep, it was planned to dig a 30 feet hole through the overburden down to the bedrock. The base of the silo would be placed in the hole while the remainder of the structure would be above ground. It was also proposed to place the missile silo hole close to the Greymare Hill Missile Test Area so that advantage could be taken of its technical infrastructure.

A contemporary air photograph taken in August 1961 confirms that work had begun on the silo. It shows an excavation with disturbed ground to its north and traces of heavy vehicle tracks leading westwards back towards the southern end of the Greymare Hill complex. Following the cancellation in April 1960, all major civil engineering work was halted, proving that this work had taken place prior to that date. Subsequent to the abandonment of the project, the silo trials area was covered by a dense coniferous plantation, effectively hiding the site from view for over forty years.[11]

In the UK, a full engineering prototype launcher was to be built, but finding a site for K11 and indeed for the rest of the silos was not easy. In June 1957 a list of 92 possible sites had been prepared, but only by looking at a map for disused Ministry of Defence properties. In a stroke of lateral thinking, more than 40 roadstone quarries were looked at for suitability: after all, there was already a hole there, and roadstone was good and hard. Geologically, the silos had to be sited in hard rock or other fairly rigid material such as chalk. By October 1958, there was talk of Duxford for the site of K11, with alternatives at Odiham, Waterbeach and Stradishall. By January 1959, Castle Camp, Ridgewell, Sudbury, Raydon and Lasham had made their way onto the list[12].

Although there was talk of building clusters of six silos at a time, the sites had to be well separated by distance of several miles, so that one site would not be affected by attacks on other sites. But in February 1959, there was a change in policy: 'the first sites should be in the South of England, North of the Thames', and twelve disused airfields had been surveyed by March. These were Castle Camps in Cambridgeshire, Ridgewell in Essex, Tibenham and Hardwick in Norfolk, and Eye, Beccles, Sudbury, Metfield, Raydon, Bungay, Halesworth, and Horeham, all in Suffolk.

A problem emerged at Duxford, however: discovery of the water table 40 ft down and the resultant flow of water rendered it unsuitable[13]. Another site had to be chosen, and Upavon and Netheravon in Wiltshire were then earmarked for the job.

But now the Home Office intervened: they wanted the sites well away from any evacuation areas, and on the East Coast, so that fallout would be carried away by the prevailing winds. Upavon was dropped from the list of sites as a result. The 1959 election then intervened, holding up the progress, and after the election the emphasis changed: now the RAE was sent up to Yorkshire and Durham to look at the likes of Acklington and Eshott as locations for the K11 site. Similarly, the VCAS favoured Ouston or Morpeth.

Bircham Newton in Norfolk was also a strong contender, and the RAF went to investigate its geological suitability, as this memo from Wing Commander Wood indicates:

> I had a long talk with the Station Commander Bircham Newton (Group Captain Walford) during my visit yesterday. He was rather upset about some of the unavoidable mess (mostly vehicle tracks) which the Soil Survey Team have made on the airfield as he is trying to have the whole place look spotless for a visit by C.A.S. [Chief of the Air Staff] on 16th October [1959] – I undertook to see that everything was tidied up. The Group Captain also asked for guidance should he be asked about the holes on the airfield which will still be evident…[14]

Once the deliberations of the BND(SG) began to leak out, progress became even slower. A memo from the Ministry of Aviation to the Ministry of Works in January 1960 contains the paragraph:

> We are still not in a position to give you instructions to proceed on procurement for K.11 pending the completion of the current Defence review. There has been considerable correspondence at Ministerial level and it has been agreed that until the Defence review has been completed no new major commitments can be entered into on the Blue Streak project. Certainly K.11 is a 'major new commitment'.

One thing was certain: by the time of cancellation no site had been fixed upon, no excavation had been started, the design was not complete and the chances of K11 being operational by 1964 were looking increasingly remote.

---

[1] TNA: PRO AVIA 68/23. Work supporting development of an underground launching system for Blue Streak. RPE Technical Note No. 170. BWA Ricketson and ETB Smith.
[2] Ibid.

[3] TNA: PRO AVIA 92/18. Design of firing sites for ballistic missiles : policy.
[4] TNA: PRO DEFE 7/2247. Development of Blue Streak.
[5] TNA: PRO AIR 2/11115. Underground launcher: protective cover; design study.
[6] Ibid.
[7] TNA: PRO AIR 2/11131. Underground launcher: protective cover; design study.
[8] TNA: PRO AVIA 92/19. Design of firing sites for ballistic missiles: policy.
[9] TNA: PRO AVIA 92/18. Design of firing sites for ballistic missiles: policy.
[10] DEFE 7/2245. Development of Blue Streak.
[11] Cockcroft, W. (2006). 'The Spadeadam Blue Streak Underground Launcher Facility U1' *Prospero* **3** pp. 7–14.
[12] TNA: PRO AIR 2 14701. Blue Streak: selection of sites.
[13] Ibid.
[14] Ibid.

Chapter 7

Blue Streak – The Cancellation

All modern aircraft have four dimensions: span, length, height and politics.
— Sir Sydney Camm

Politics obviously play a very important role in a project such as Blue Streak. It was competing for resources, or to put it another way, for money, with not only other defence projects, but with Government spending as a whole. The Government was the customer, and it was the Government that would have to decide whether it was value for money. On the other hand, the Government is not one single entity; it is a mixture of departments all with their own agenda, and their own axes to grind. These agenda would often come into conflict, and that is what the story of the cancellation of Blue Streak illustrates.

The political decision by the UK to become a nuclear power in the late 1940s implied two interlinked technical policies: the development of the nuclear device itself, and the development of a credible means of delivery. Credible in this context can imply a variety of different concepts.

Primarily, the threat of the nuclear deterrent has to be credible to the opposition, and as far as the UK was concerned, this meant Russia. Secondly, the deterrent has to be credible to the armed services that have to deploy it, and also politically credible to the UK electorate. Thirdly, in the UK context, it had to be credible to the US, since the UK deterrent was perceived by the Government as essentially an adjunct to the US deterrent, and also since the UK wished to have some influence over US nuclear policy. This was not possible unless the UK had her own nuclear weapons. The 1957 Defence White Paper described the U.K deterrent thus:

> The free world is today mainly dependent for its protection upon the nuclear capacity of the United States. While Britain cannot by comparison make more than a modest contribution, there is a wide measure of agreement that she must possess an appreciable element of nuclear deterrent power of her own.

A bomb without a delivery system is of little use. In the early 1950s, the intended means of delivery was by a free fall bomb dropped from jet aircraft (the

V bombers, which were the Valiant, the Vulcan, and the Victor), which were to be further augmented with Blue Steel. However, technological advances meant that many other means of delivery became possible as the decade advanced.

Principal among these was the ballistic missile, which, above all, seemed to have one overriding advantage. This lay in its apparent invulnerability: once the vehicle was safely launched, it would be extremely difficult to intercept. But there were various ways in which ballistic missiles could be deployed, and as the various possibilities unfolded, each new system appeared to offer advantages over its predecessors. Thus it was possible to begin development of one system only to find half way through that another system was becoming feasible, and which threatened to supersede its predecessor. Technological advances during the 1950s were such that a new system could appear within a year or two of development having begun on an earlier system.

This was a particular problem for UK policy makers, those in the Cabinet and the Defence Ministry. For whatever reasons, development times in the UK were far greater than their US counterparts. UK policy makers were put into a position where they had to take a decision on a system which might take ten years to develop, with an expected service life of perhaps another ten years. Thus they had to look 20 years into the future, and with little hard intelligence as to the capabilities of the Soviet Union. Much of this intelligence was based on erroneous assessments of the industrial capacity and technological achievements of the USSR. The launch of Sputnik in particular led the West to think that the USSR was in many areas technically superior. In fact, the missile that launched Sputnik, the R7, was far too big and clumsy to be deployed operationally. However, Soviet nuclear forces from the late 1960s onwards would be a formidable challenge to a country such as the UK.

Political and Service tensions developed as a result of the development of potential rivals to Blue Streak. Proponents of one system often deliberately misused technical information to cast doubt upon another, and a good deal of the policy making was deliberately partisan. In other words, lobbying for a particular system was heavily influenced by particular Service departments who wished to control the means of delivery themselves and whose budget would benefit accordingly. There were also other Service factions who wanted as little money as possible to be spent on nuclear weapons, to free up the Defence budget for conventional weapons. Most of this is as true now as it was then!

The Operational Requirement for Blue Streak had called for a missile that could carry a megaton warhead over a distance of 2,000 nautical miles. The weight of the warhead then available meant that a relatively large missile had to be designed, and in retrospect, this was a mistake. Given the long development time for the missile, it might have been reasonable to assume that warhead design might have made significant advances in the interim. Indeed, at the time the design was 'frozen', Britain had not yet developed a fusion weapon. However, hindsight is a wonderful thing.

The use of a cryogenic fuel for Blue Streak was also a potential limitation on its deployment as the missile could not be kept in a 'ready to fire' state indefinitely. However, in this context it should be noted that exactly the same constraints were to apply to contemporary Russian and American designs such as the R7, Atlas and Thor. Furthermore the structure of this type of missile was relatively fragile, and extremely vulnerable if deployed on the surface. Hence the intention at the outset was to site the missile in 'underground launchers', as already described.

Compared with the US, development times for UK projects were very much longer. By the time of cancellation, Blue Streak had been under development for around 57 months with the first flight still some months away. The Thor missile, albeit smaller but of the same technological sophistication, was 13 months from inception to first launch. There are various reasons for this.

The first is that the Americans had much more prior experience than the UK: American missile development had proceeded almost uninterrupted since the war, whereas efforts in the UK had been much smaller scale and directed mainly at small defensive missiles.

A second reason was the means of procurement. In the US, specific teams were set up with considerable executive powers and, by comparison with the UK, almost unlimited finance. In the UK, the procurement ministry was the Ministry of Supply (later to become the Ministry of Aviation). A reading of the Ministry papers shows that the executive powers of the Ministry with regard to the industry carrying out the work were very much less.

Furthermore responsibility for the project was very much divided. The Ministry of Supply was the procurement Ministry. The Air Ministry would deploy the missile when it went into service. The Air Ministry, however, came under the control of the Ministry of Defence, who also then became involved. Finally, the RAE was to be the technical overseers of the project. Hence representatives of all these organisations, together with representatives from the firms, might all have been present at the various progress meetings. Such cumbersome bureaucracy cannot have helped the progress of the project. For example, an official in the Ministry of Defence wrote about the building of the facilities at Spadeadam:

I think the Minister of Supply ought to be shaken. It is up to him to warn us as soon as there is any administrative or financial difficulty to his not getting on as fast as he could with the project.[1]

The sending of memoranda back and forward (in the days before email!) from one Ministry to another must have been another time waster.

A third brake on the project was the Treasury, who took a much closer interest in Blue Streak than seems to be the case with many other defence projects. Thus in a minute to the Minister of Defence: 'During most of 1956 we were defending the very existence of Blue Streak against savage attacks by the Treasury.'[2] Such comments occur frequently in the Ministry of Defence files. By comparison, US resources were incomparably greater, and American engineers could afford to launch missile after missile until the design was a success. The UK did not have this luxury.

However, the feature of Blue Streak that was to prove the most controversial was the means of deployment in 'underground launchers'. These launchers evolved gradually from relatively simple ideas into what today would be termed missile silos, although the term was not then in contemporary British use. The design of these launchers would give almost as many technical problems as the missile itself, and their size and complexity would have created considerable construction problems for the UK

But Blue Streak was running into other financial difficulties, apart from the cost of the 'underground launchers'. In a sense, the Treasury's anxiety was justified, since the costs seemed to be open-ended. The Ministry of Supply seemed to be unable to make any realistic cost estimates, and the time was fast approaching when firm decisions as to silos and their location would have to be taken. In addition, even the Home Office was becoming concerned since their location affected civil defence decisions. The sheer size and scale of the silos was only just becoming evident: a site would occupy around three acres, and would have to be a considerable distance from any habitation. (One potential site at Bircham Newton was ruled out on the grounds that it was too close to the Queen's residence at Sandringham!) Stopping any such major project in its tracks is extremely difficult; very good reasons had to be found. There would be considerable political implications to cancelling such a major project.

Alternatives to Blue Streak did begin to emerge soon after development had begun. These were the Polaris submarine launched missile, and an air-launched missile code named WS138A, later to be known as Skybolt.

Polaris was a system developed by the US Navy. It used solid propellant motors, and the early versions had limited range and payload. However, a decision had been taken by the US Navy that since warheads would become much lighter as their design improved, the range/payload problem would be much less pressing by the time of deployment (warhead design in the US was considerably in advance of that in the UK).

Britain's first nuclear submarine was HMS Dreadnought, made possible once the US Navy had provided a design for a lightweight reactor. Since then there had been close co-operation between the two services, and Admiral Arleigh Burke, in charge of the programme in the US, was eager for the Royal Navy to acquire the Polaris system.

The problem was that Polaris was not particularly attractive to Whitehall. It would mean building quite a number of nuclear submarines (certainly more than the five, later reduced to four, planned after the Nassau agreement) and buying the missiles from America – providing America was prepared to sell them. Such an arrangement did not appear to be cheaper than Blue Streak, it would have taken just as long to get the submarines into service, it would mean spending valuable dollars on the missiles, and it would also appear to reduce Britain's independence as a nuclear power. Certain factions in the Admiralty had rather different views.

The Navy felt it had come out of the cuts in defence resulting from the 1957 Sandys' White Paper relatively lightly. But the deterrent and Blue Streak in particular was resented as it was felt that too much of the defence budget was being diverted towards it; money that could be used for new ships. However, attitudes began to change when the First Sea Lord, Mountbatten, was told of Polaris by Arleigh Burke, the originator of the system in the US. Mountbatten and Burke were old friends, and a rather clandestine correspondence began between them. Soon the correspondence became more official. Burke was a formidable proponent for the system: he invented rather poorly scanning clerihews for the system along the lines of 'move deterrents out to sea/where the real estate is free/and where they are/far away from me'[3] to hammer the point that a submarine on patrol does not need fixed bases, and where it can remain undetected and thus invulnerable. A faction within the Navy took up Polaris with enthusiasm during 1958, but realised that the big obstacle was Blue Streak. The UK could not afford yet another nuclear delivery system. Accordingly a campaign began with the Admiralty, as an internal memo shows.

> I was surprised and encouraged today that amongst those who are advising both the PM & the Chancellor there is a pronounced feeling that if we are to go on with the deterrent it should only be on the basis of Polaris.

I share your views that what we are most immediately concerned about is so to reduce the deterrent that we can maintain adequate conventional forces. I believe however that a decision to go for Polaris would give a large enough saving to guarantee the conventional forces we need. Further expenditure on Polaris would I think be less than future expense on Blue Streak plus fighter defence of the deterrent.

If I am right that we can get an immediate saving by taking the decision now, I feel that we should present the economic advantages of Polaris rather more strongly. Such a presentation would it appears fall on fertile ground.

Further papers indicate similar lobbying. In a note on a possible European IRBM from the First Sea Lord in February 1959, he comments:

Nevertheless, mainly thanks to the Treasury, it was possible to secure a chance to refer this point to Ministers on the grounds that, in the highly unlikely event of our NATO partners taking up the offer, HMG would be committed to the completion of Blue Streak with all that entails.

Later on in the note, commenting with reference to solid fuel motors, he says '… unless we can effectively answer them, the chances of upsetting BLUE STREAK may be considerably weakened.'

The Admiralty became even more excited when they discovered (and misinterpreted) a scheme for a UK ABM system: 'You would hardly believe it, but since sending you my note this afternoon we have unearthed further information which really does put BLUE STREAK out of court.'

The language used goes well beyond the simple evaluation of the merits of rival systems: it becomes distinctly partisan.

The Navy's efforts had not gone unnoticed by the Air Ministry either, as the following note in June 1958 from the Secretary of State for Air to Sandys shows:

The general conclusion that I come to is that the matter is of such fundamental importance and so complex that it might be more helpful to you if the Chiefs of Staff were asked to examine the requirement in all its aspects, strategical, tactical and technical, in the light of … the First Lord's paper, and then to put forward a considered military opinion to you.

Mountbatten, then First Sea Lord, was pushing hard for some form of report too, as a note to Sir Frederick Brundrett shows:

… we are all most anxious to see that the Powell enquiry is dealt with on the right lines, to be quite sure that it will lead to the right answer. This is a Defence question first and foremost, although it may have all sorts of secondary interests. We none of us can believe that Powell and two outside scientists can possibly arrive at the right answers if they have no Service views on the requirements represented at the Committee.

For this reason, we are all convinced that we must have adequate representation on the Committee from each of the Services; and that is why we decided that the three Vice Chiefs should sit on it.[4]

['Powell' refers to Sir Richard Powell, then Permanent Secretary at the Ministry of Defence, and a key figure in the story of the cancellation.]

This is a letter full of ambiguity: what is the 'right answer'? Presumably, to Mountbatten, this meant Polaris. And it is interesting that Mountbatten is pressing for Service representation on the committee.

Accordingly, Sandys minuted Powell in December 1958:

The Chiefs of Staff have no doubt been considering for some time the respective advantage and disadvantages from the British stand-point of basing our nuclear deterrent underground or under the sea.

I think we ought to have a discussion of this matter at an early meeting of the Defence Board. I should, therefore, be glad if you would let me have a summary of the views of the Chiefs of Staff as soon as you can after Christmas.

Powell submitted a reply to Sandys outlining the form he felt such an inquiry should take, and suggested its terms of reference as being: 'To consider how the British-controlled contribution to the nuclear deterrent can most effectively be maintained in the future, and to make recommendations.'

But then he had to push Sandys for further action:

In a minute of 23rd March I submitted proposals for setting up a study into the future of the British deterrent ... you agreed that this should be set in action but subsequently asked me to do nothing, in order to avoid casting doubt on the future of BLUE STREAK.

The Chiefs of Staff and Sir Norman Brook [the Cabinet Secretary] have recently asked me about this study. Both felt that it ought to go on, since the future of the deterrent is bound to come up again after an election, if not sooner. I think they are right, and should like to have your authority to proceed ...

In any event, a note from Sandys' office to Powell shows that his hand was being forced:

The Minister discussed with you this afternoon the proposed Study Group on the British deterrent. He felt that we had only recently reached our conclusions on the need and form of the British contribution to the nuclear deterrent. Little further information would be available, and in his view, the time was not yet ripe for a further study of this problem. He asked that if this matter was raised in your coming meeting at Chequers, you should say that he was considering setting up a Study Group, and you should leave this matter open. You agreed to discuss this further with the Minister after discussing it with Sir F. Brundrett and after your visit to Chequers.

But an internal Admiralty note written to the First Lord (the Earl of Selkirk) by the First Sea Lord (Admiral Sir Charles Lambe) in May 1959 gives another perspective:

> My predecessor also turned over to me the fact that the Minister of Defence had agreed to a team under the Chairmanship of Sir Richard Powell, to examine the pros and cons of three possible methods of providing the future British contribution to the Deterrent, namely Manned Aircraft, Ballistic Missiles (BLUE STREAK) and POLARIS. Though this had been agreed by the Minister, I understand that, just before he left for New Zealand, he ordered this investigation to be suspended, giving as his reason the fact that 'he did not wish the validity of BLUE STREAK to be questioned' ...
>
> As I see it, the present Minister of Defence [Sandys] will do all in his power to prevent any alternative to Blue Streak from even being considered. I am also certain that the new Chief of the Defence Staff [Mountbatten], when he takes office, will do everything in his power to see that the merits of Polaris are brought to the attention of HM Government. Domestically, I am certain that we in the Admiralty need a much clearer picture than we have at present of the probable repercussions of the Polaris programme on the rest of the Navy before we start any official pro Polaris propaganda. Indeed, I doubt it is right for the Navy to undertake any such propaganda at all. I believe we would be in a far stronger position if we were (at any rate, apparently) pushed into the POLARIS project rather than have to push it ourselves.[5]

BND(SG)

The events that finally led to the cancellation of Blue Streak began as a consequence of a meeting at the Prime Minister's country home, Chequers, in June 1959, referred to in Sandys' note above. The guest list for that weekend is quite impressive:

Harold Macmillan, Prime Minister;

Sir Norman Brook, Cabinet Secretary;

Sir Roger Makins, Joint Permanent Secretary to the Treasury;

Sir Frederick Hoyer Millar, Permanent Secretary at the Foreign Office;

Sir Patrick Dean, Chairman of the Joint Intelligence Committee (JIC);

Sir Richard Powell, Permanent Secretary at the Ministry of Defence;

Marshall of the RAF Sir William Dickson, Chief of the Defence Staff;

Admiral Sir John Caspar, Vice Chief of the Naval Staff;

Marshal of the RAF Sir Dermot Boyle, Chief of the Air Staff;

Lt. Gen. Sir William Stratton, Vice-Chief of the Imperial General Staff;

Lord Plowden of the Atomic Energy Authority.

This was indeed an august gathering: their brief from the Prime Minister was to try and look ten years into the future and plan for the changes that they saw coming. In the words of Norman Brook, the Cabinet Secretary, 'The purpose of the meeting is to put in hand a study of future policy ... This study will be undertaken by officials – the Prime Minister does not wish other Ministers to be troubled with it at this stage ...'[6]. There may well have been a subtext behind that last comment, perhaps along the lines that the Prime Minister wanted a relatively disinterested viewpoint for his future-gazing. The Civil Service and the military could also give a longer term view – they would still be there, implementing policy, long after the politicians had gone.

Their focus, as can be seen from the people present, was foreign policy and defence. A major issue at the time was the size of the Defence budget – indeed, the appointment of Sandys as Minister of Defence, and the resultant 1957 Defence White Paper, was intended as the first step in the rationalisation of defence spending. The intention was to keep defence expenditure at 7% of total Government spending, and indeed the deterrent was invoked as part of this. Nuclear deterrence could mean less money spent on conventional arms. In anticipation, the Air Staff had provided briefing papers on the various deterrent options for the assembly.

The deterrent at that time was being maintained by the V bombers, which would shortly be supplemented by Blue Steel. Soon bombers would be obsolete in the strategic role, and the only possible replacement available was Blue Streak. In the words (almost) of another Prime Minister, There Was No Alternative. Blue Streak might have been highly unpopular in Whitehall, but in the absence of a viable alternative it was either Blue Streak or no deterrent at all.

The RAF were not happy with the tone of the discussion when it came to deterrent policy:

> It was obvious in the Working Group's discussions that our sister services resent the overriding priority afforded the Deterrent (at present in RAF custody) under HMG's policy, and are covetous of the money and resources assigned to it. They have endeavoured to cloak these base motives by advancing arguments of expediency in the guise of military and political rectitude.[7]

It has been said that the Services often spend more time fighting each other than fighting the enemy, and this is quite a good example of that maxim. That atmosphere of inter-Service enmity (particularly between the Navy and the RAF) should be borne in mind when watching how events unfolded.

'Base motives' or not, at the end of their deliberations the Working Group 'invited the Permanent Secretary of the Ministry of Defence to consider further with the Secretary of the Cabinet the question of a separate inquiry into the means of delivery of the British contribution to the nuclear deterrent'[8].

Sir Richard Powell was the Permanent Secretary at the Ministry of Defence at that time, and having been given this brief, he wrote to Sandys, then Minister, about setting up an inquiry. As we have seen, Sandys was not at all keen for any inquiry.

Other matters intervened with the General Election of October 1959. During the run-up to an election much of Government is put on hold: Ministers have other pre-occupations, and there is little point in going ahead with projects if a change of government means they will be reviewed. There is always a hiatus in Whitehall as the new ministers come in and are brought up to speed on their department.

General Elections also give Prime Ministers the opportunity to reshuffle their Cabinet, and this one was no exception. Sandys was replaced as Minister of Defence by Harold Watkinson, a career politician who had been a businessman and had no great ideological position as far as defence was concerned – he set out to be a practical man, who would bring a businesslike approach to the department rather than an ideological one. His obituary in 'The Independent' newspaper says of him that:

> He was, already, a highly successful businessman and, like many before and after him (the late John Davies and today Sir James Goldsmith spring to mind), believed that businessmen could handle government far more efficiently and effectively than could politicians. He found out, however, that politics was an art of its own, and that the methods of man management that he had evolved for himself in business were ineffective when applied to the emotional, and often tortuous, handling of political affairs.[9]

Sandys himself was moved to a renamed Ministry of Supply – now the Ministry of Aviation – with the brief to 'rationalise' the aircraft industry. In many ways this can be seen as a demotion, or certainly a sideways move, given that he had been Minister of Supply in 1952 – so much so he asked Macmillan for assurances that his Cabinet seniority would not be affected.

It could be argued that Sandys had fulfilled his brief as Minister of Defence, and had taken his reforms as far as he could. His personal relations with some of the senior military figures had not always been good, and it was probably time for him to move on. Whether the ongoing Blue Streak saga was also a contributory factor is open to debate.

Sandys having gone, the way was clear to the setting up of the British Nuclear Deterrent (Study Group) or BND(SG) (known affectionately to the Admiralty in a later incarnation as the 'Benders', presumably from 'BNDS'). On the face of it, the composition of the Study Group was impeccable:

Sir Richard Powell, Permanent Secretary at the Ministry of Defence;

Sir William Strath, Chief Scientist at the Ministry of Aviation;

Sir Frederick Brundrett, Chief Scientist at the Ministry of Defence;

Sir Patrick Dean of the Foreign Office and Chairman of the Joint Intelligence Committee;

Mr. B.D. Fraser of the Treasury;

Vice-Admiral Durlacher, Deputy Chief of the Naval Staff;

Lt.-General Sir William Stratton, Vice-Chief of the Imperial General Staff;

Air Marshal Sir Edmund Hudlestone, Vice-Chief of the Air Staff;

Sir William Cook of the Atomic Energy Authority.

This was a high-powered group of men, and any conclusion they arrived at would be buttressed by the authority of their rank. Needless to say, their deliberations would have to be extremely confidential, since any leak could have considerable consequences. Having said that, it is clear that the Ministry of Aviation seemed to be 'out of the loop', despite their Chief Scientist being a member of the Study Group. After the report had been issued, the CGWL, Sir Steuart Mitchell, complained that

> Adequate opportunities did not occur during the drafting of their report by the Study Group for my Controllerate to brief you properly on the technical issues as they arose, nor to discuss with you the conclusions and recommendations of the report.
>
> I write this to say that now that I have seen the report I am seriously disturbed at the picture it presents in so far as the technical issues are involved, and that I disagree with some of the conclusions.
>
> I am having those technical aspects of the report which lie in my sphere examined (for the first time) in detail and will submit some comments on them to you in a few weeks' time.

Comments such as these from the CGWL show that the BND(SG) must have been distinctly selective in whom it chose to consult. It also shows how well they were able to keep their discussions under wraps.

Sandys himself seems to have had no prior warning either. The Treasury was delaying the authorisation of funds for further development, and as late as 25 January 1960, he was writing:

> Therefore, unless the Defence White Paper contains an announcement that Blue Streak is to be abandoned, which I regard as inconceivable, and which I would, of course, strongly resist, I must ask you to give the 'all clear' so that further serious delays can be avoided.[10]

The wording could, of course, be political disingenuousness, but it does not sound like the words of a man who has read the Study Group's report – or at least who has heard about their conclusions. The Chancellor, Derek Heathcote Amory, replied on 4 February:

> I do not think it would be reasonable, at a time when the future of the weapon is the subject of a searching review as a major question of defence policy, to accept that the programme should suffer no delay … I am afraid therefore that I still feel unable to authorise the further expenditure referred to …[11]

He also cited a previous hold up of funds (the Prime Minister's note of December 1958) as a precedent.

Another part of his letter caused one of the officials in the Ministry of Aviation to note:

> The Chancellor is stretching things very far when he says that the possibility of the weapon coming into service late has been one of the considerations necessitating the current review. The Chancellor, having always disliked the Blue Streak policy, might indeed almost be thought to have done his best, by imposing financial restrictions, to ensure that he would be able to say that the weapon would be late and therefore not worth having but in fact it is only the complete hold up of fresh capital expenditure in the last four or five months that has caused us to wonder whether the little elbow room that we had in the R&D programme would no longer prove sufficient.

So who on the Study Group could be seen as opposed to Blue Streak? The position of the Services is interesting. Firstly, the Army would have no strong views on Blue Streak one way or other, except in terms of cost. Blue Streak would take up a relatively large proportion of the defence budget, money that could be used for conventional weapons. The positions of the RAF and Navy are more interesting.

Certainly, sections of the Navy, led by Mountbatten, were campaigning hard against Blue Streak and in favour of Polaris. A memo from Lord Selkirk, First Lord of the Admiralty, illustrates this quite clearly:

> My aim last year was not only to make the Prime Minister, Minister of Defence, and other members of the Defence Committee aware of the potentialities of POLARIS, but also to check, so far as this was possible, the BLUE STREAK programme before it gathered momentum. We had some success. The decision at the turn of the year that BLUE STREAK should be allowed to proceed in 1959/60 was certainly accompanied by a growing realisation in the Defence Committee of its disadvantages and mounting costs.
>
> Since then, however, BLUE STREAK has become more firmly established and it looks, at the moment as if the 1960/61 Estimates discussions this Autumn may strengthen it further. If this should be so, its formidable cost, as shown in the draft paper you attached, will become a most serious threat to our hopes of increasing the size of our conventional naval forces, even it the total defence vote were to be fairly substantially enlarged.

We must carefully consider our tactics for dealing with this. Whatever help we may get from the new CDS [Chief of Defence Staff, Mountbatten], I believe that we must be prepared to make the running ourselves.

As I see it, the Government is unlikely to go so far as to stop BLUE STREAK unless there is something which can be put in its place as the future British controlled contribution to the deterrent. From what you say, we are unlikely to be in a position this Autumn, even if we were asked to do so, to present for consideration a substitute programme for POLARIS submarines. What then can we do?[12]

And, of course, there the Navy hit the nail on the head. There was, at the time the Study Group began its deliberations, no single well-developed system that could be put in Blue Streak's place. No British long-term possibility was even on the horizon, but there were possible American systems.

Polaris was showing great potential, but still had some way to go, and had other drawbacks, such as the need to build a fleet of atomic powered submarines from scratch. But during the few weeks that the Study Group deliberated, considerable progress was being made elsewhere on another missile – WS138A, or, as it would become better known, Skybolt.

Origins of Skybolt

In the words of the Air Ministry:

> Following the Interdependence meeting between the Heads of Government of the UK, the USA and Canada in September 1957, a Tripartite Technical Committee met in Washington in December 1957 and established a number of sub-committees to cover specified areas, one of which was strategic air to surface weapon systems. [This] subcommittee decided that an examination of the field should be undertaken by a joint RAF/USAF Task Group which met in Washington in April and November 1958.

The Task Group then issued a requirement for a missile which was circulated to industry, and a meeting was held to evaluate these proposals. To the astonishment of the British representatives present, proposals were produced from more than a hundred American companies. The meeting narrowed this down to 15 firms, who then gave brief presentations. From these, the Douglas design was then chosen.

The USAF did not want to be left out of the scramble among the US armed services for nuclear weapons systems. A system such as the new proposed missile would significantly enhance the capabilities of Strategic Air Command (SAC), although it was not envisaged so much as a strategic weapon, a 'city buster', but rather more as a tactical device. SAC's long range bombers would take several hours to reach the heart of the USSR and would have to overfly a

Figure 52. A Vulcan bomber carrying two Skybolt missiles.

good deal of hostile territory. The new missile was intended to be fired as the aircraft approached Russian borders, and the relatively low yield warhead was designed to suppress Russian air defences so that the aircraft would be able to deliver their multimegaton bomb load to Russian cities. Thus it was not a major weapon in the US arsenal, which already had Atlas, Titan and Minuteman ICBMs, as well as Polaris and the bombers of SAC. Instead, it was seen by SAC as a way of enhancing its credibility, and not being elbowed out by these other systems.

The initial requirement issued in January had been for a joint USAF/RAF missile with a range of 1,000 nautical miles, a weight of 10,000 lb and a c.e.p. of 3,000 ft. The warhead would have a weight of 600 lb and a yield of 0.4 Mt or 400 kT. A report on the progress of WS-138A, dated July 1959, was prepared by Group Captain Bonser of the Air Staff after a visit to the USA. WS-138A fitted British requirements perfectly. The range was such that strikes against almost any part of the Soviet Union could be launched without having to overfly hostile territory. Each Vulcan would be able to carry two missiles comfortably. Given that the US was to pay almost all the development costs, it would be an extremely cheap deal, and it offered the possibility of extending the useful life of the V bombers by several years.

Douglas estimated the overall cost of the programme to the RAF at between £42 million to £48.5 million, exclusive of development costs of some £2.5 million. The RAE considered that the predicted c.e.p. of 5,350 ft compared favourably with the Thor missile. WS-138A looked to be a bargain. In June 1959 representatives of the V bomber firms, Avro and Handley Page, as well as officials from the Ministry of Supply, visited Douglas to give them further details of the V bomber designs.

Skybolt was very different to Blue Steel or any of the other missiles carried by bombers since it was not intended to fly in the atmosphere, but was a fully ballistic missile. It had two stages, and was carried underneath the wings of the aircraft. Navigation was mainly inertial, but would include a star seeker sensor to improve accuracy (this is why it could not be carried in the bomb bay: the sensor needed a clear view of the sky). It was planned that the Vulcan would carry two missiles, one under each wing. There would be difficulties fitting the missile to the Victor (ground clearance was the issue), but the problem was not insuperable.

Thus by the time the BND(SG)[13] had begun its deliberations, Skybolt was well under way. Its development costs would be paid for by the United States. It would extend the useful life of the V bombers for years to come – perhaps to 1970 or beyond. It needed little or no new infrastructure. There was just one snag – how to cancel Blue Streak in favour of Skybolt and make it look convincing.

Early drafts of the Powell report are fairly neutral in tone. As the weeks go by the tone sharpens, and the final report devotes ten or more paragraphs discussing the vulnerability of Blue Streak, with the scenarios becoming increasing convoluted. Thus:

> If we assume that the Soviet attack would be made of ballistic missiles of an accuracy equal to that which we expect to achieve ourselves (0.55 NM) and that a warhead of at least 3 MT would be available, 95 per cent of the underground BLUE STREAK sites could be destroyed by between 300–400 Soviet missiles. Even allowing for the requirements of air-defence weapons for the protection of the Soviet homeland, and for the need simultaneously to pose a serious threat to the United States, we have no doubt that the Soviet stockpile by 1967 would be sufficient to provide these warheads for attack on the United Kingdom.

The crucial paragraphs read:

> As ballistic missiles cannot be recalled once they had been fired, the political decision to retaliate must also have been taken before this means of evading a pre-emptive attack can be adopted. For this tactic to succeed, authority would need be delegated to order nuclear retaliation on radar warning alone. We do not believe that any democratic government would be prepared to delegated authority in an issue of such appalling magnitude.
>
> This analysis shows that unless the political decision were delegated, the Soviet Union could carry out a successful pre-emptive attack on the BLUE STREAK sites, whether these were underground or on the surface; and that even if the political decision were delegated, the Soviet Union could still make a successful pre-emptive attack on the United Kingdom alone.

What of the vulnerability of the V bombers? The report has one paragraph:

> Four minutes are presently required to enable the V bombers, when operating from the planned dispersal airfields, to take-off and fly clear of a nuclear attack on their bases. Given 24 hours warning the V bomber force will be able to react in time to evade a pre-emptive attack made with missiles launched on a normal trajectory. If, however, the Soviets were to fire missiles on low trajectories from East Germany, the effective time for evasive action might be as short as three minutes. The Air Ministry believe that with improved techniques it should be possible to reduce the V bombers reaction time. But, in any event, the arrival of the Soviet missiles would inevitably be spread to some extent and some of the bombers would probably be able to escape. Furthermore, for short periods during a time of tension, it will be possible to reduce this risk by maintaining a proportion of the force on standing patrol. In either event, however, the capability of the V bomber force would be greatly reduced.

Whereas Blue Streak, in its hardened silo, would apparently not survive, 'some of the bombers would probably be able to escape'. To use a modern metaphor, this analysis does not seem to be using a level playing field. Furthermore, whilst some might be maintained on a quick reaction (four minute) alert, it will not be that many. Aircraft have to be serviced, crews have to rest. The three or four minutes mentioned is also the time from detection of the warning to the arrival of the missiles. The time between receipt of the warning from the radars and the order to scramble has not been factored in – and Bomber Command will not be keen to scramble unnecessarily. If the aircraft are scrambled, and the warning turns out to be false, the aircraft have to be turned back, landed and refuelled – taking them out of the picture until all this is done.

This analysis suited the Air Staff very well. It was, after all, the RAF who would operate Blue Streak once deployed, but the option of Skybolt now opened up a new range of possibilities. Rather than operating from a hole in the ground, the RAF would revert to its traditional role of flying aircraft. The V bombers would be given a whole new lease of life. As far as they were concerned, Skybolt was far preferable to Blue Streak.

In this context, the letter which Watkinson wrote to Sandys contains an interesting comment:

> The Chiefs of Staff have been considering their attitude to Blue Streak and have now given me their unanimous advice that they find Blue Streak, as a fire first weapon, unacceptable. I am afraid Dermot sold the pass here to begin with.[14]

Dermot refers to Dermot Boyle, Chief of the Air Staff, and it is clear from Watkinson's comment that Boyle was not at all reluctant to see Blue Streak go.

The Foreign Office representative (and more importantly, the Chairman of the Joint Intelligence Committee), Sir Patrick Dean, did send a note dissenting from the Study Group's conclusions, but the Foreign Office had no great stake in the outcome – their main interest lay in the preservation of a British deterrent, which the report provided.

And not surprisingly, it was the Treasury representative that was the most vociferous in his opposition. One of the key phrases in a memo he wrote to the rest of the Study Group is: '... surely we would not decide to equip our troops with spear proof shields if we know that by the time we have made their shields the enemy is going to have fire arms' – in other words, the Blue Streak silo may be adequate now, but it won't be by the time the missile is deployed. Elsewhere in the same memo he says, 'I think we ought to "show our working" somewhere', which is presumably a reference to the 300–400 missiles.

As an advisor to the air staff put it:

> Mr Fraser's letter is clearly directed to killing at Blue Streak in two stages – first comes the politically more acceptable proposition that we should abandon underground deployment for Blue Streak. When this is accepted, the need to have the weapon at all can be questioned.

And so we must then try and see why the report comes to the conclusion that it does.

The oddest feature is that prior to the report, there are many estimates made by the Air Staff and the Ministry of Supply as to the survival rate of Blue Streak in a silo. To give one example, one Air Ministry memo gives the survival rate of Blue Streak when attacked by a one megaton warhead missile thus:

c.e.p.	0.5 nm	1 nm	2 nm	3 nm	4 nm
Above ground	0%	0%	6%	29%	50%
Below ground	50%	84%	96%	98%	99%

Quite how these figures were arrived at is not clear. Above ground survival rates are easy enough to estimate, but the underground figures depend on the survivability of the silo, and it is never made clear what assumptions are being made in all these various estimates. The other major factor is the c.e.p. of the missile – obviously the more accurate, the higher the 'kill rate'. The Powell report assumes a c.e.p. of 0.55 miles, although with no apparent justification. Interestingly, the chairman of the JIC is a member of the committee: the estimate of the accuracy of Russian missiles obviously does not come from him, and, more interestingly, he writes a note subsequent to the publication of the report to say he does not agree with its conclusions.

It is clear from all these memos and estimates that the supporters of Blue Streak thought that they had an unassailable case; that underground-based Blue Streak would survive a Russian attack relatively unscathed. The first time this assumption was queried was in a rather obscure paper written for the equally obscure Air Ministry Strategic Scientific Policy Committee in October 1959. The paper is unsigned, but bears all the hallmarks of Solly Zuckerman, Chairman of the Committee, later to become Chief Scientific Adviser to the Ministry of Defence. The important section is hidden away in paragraph 17:

> The vulnerability of static bases to ballistic missiles depends upon the accuracy and power of the attacking weapons. The operational reliability of any missile system also affects the number of missiles required to knock out any target given a specified accuracy. Intelligence estimates of present Russian missile accuracy and known results of USA test firings indicate that there is no technical or scientific obstacle to the achievement of an accuracy of some 2500 ft [0.42 nautical miles] for comparatively reliable 500/1500 mile range missile systems. In the case of the USSR we should expect this level of accuracy to be achieved at the latest by 1970, and possibly sooner.

He then goes on:

> Given that the enemy knows through his intelligence services where our fixed installations are, and making the prudent assumptions that he can launch a surprise attack of sufficient speed and accuracy, it must be concluded that by about 1970 at latest, and possibly before, the USSR could neutralise all fixed UK static bases, whether they be airfields or above ground or below ground missile sites.[15]

This is true in a very limited sense. The underground sites could be destroyed if the scale of the attack is great enough. The question then comes whether the Russians could or would launch an attack on this scale. The Powell report maintained that 300–400 3MT missiles with a c.e.p. of 0.55 miles could destroy the launchers. The more important questions were whether the Russians were capable of launching such an attack, and whether they would actually do so.

Professor William Hawthorne was also a member of the same committee, but he was already a silo sceptic, as a note to CGWL, Sir Steuart Mitchell, in 1956 indicates: 'I can imagine a few "impregnable" subterranean fortresses being built at enormous expense, but not many, since politicians may find them hard to justify.'[16]

The picture is further complicated by the close relationship between Zuckerman and Mountbatten. In 1959 Mountbatten became Chief of the Defence Staff, and Zuckerman was a highly influential Government scientific adviser.

Zuckerman was adept at finding his way around the 'corridors of power' being offered the post of Minister of Disarmament in the first Wilson administration (he refused). He was certainly heavily involved in many of the negotiations about Polaris and Skybolt with the US Government, and had very decided views on military matters and the deterrent. As well as being an active proponent of Polaris as against Blue Streak, he is supposed to have wielded considerable influence, in conjunction with Mountbatten, in the decision to cancel the TSR2 aircraft. It is interesting therefore that the first suggestion of vulnerability comes from someone associated with Mountbatten.

All these arguments would soon become academic (although objections to the report would rumble on for two or three months), since the Chiefs of Staff used the conclusions of the report as a pretext for a letter to Watkinson, the new Defence Minister:

> You stated in a minute dated 5th January, 1960, that you wished to obtain a decision of the future of BLUE STREAK at the first Defence committee meeting after the return of the Prime Minister, and you asked the views of the Chiefs of Staff on the matter.
>
> We attach no military value to BLUE STREAK, and we recommend the cancellation of its further military development for this purpose, together with the planned deployment.[17]

The reasoning was that if Blue Streak was vulnerable to a Soviet attack, then it would have to be regarded as a 'fire-first' weapon. Politically, this was completely unacceptable. Earlier in the same paper, they had said,

> We need a new strategic nuclear weapon system ... in about 1966, but since we regard BLUE STREAK as a 'fire-first' only weapon we do not consider it meets this need. We therefore recommend the cancellation its further development as a military weapon. We also recommend the cancellation of the planned deployment.

The 'fire-first' argument comes directly from the BND(SG) report[18], which contained the fatal lines:

> ... authority would need be delegated to order nuclear retaliation on radar warning alone. We do not believe that any democratic government would be prepared to delegated authority in an issue of such appalling magnitude.

It can be argued that the conclusions of the Powell report were, at best, disingenuous, at worst, downright dishonest. Some immediate objections were ruled out by pointing to the terms of reference ('Our main object has been to examine the technical and operational factors') which said that the study should not take political factors into account. This neatly avoided one very obvious objection.

In the scenario postulated, the Soviet Union would launch a surprise attack on the UK using 300 to 400 missiles, each armed with 3 megaton warheads and having an accuracy of 0.55 miles c.e.p. This attack would be sufficient to destroy the Blue Streak silos with the missiles inside, and thus the UK would not be able to launch a counter-attack – in other words, we would not be able to deter the Soviet Union from launching such an attack. The Soviet Union could thus attack the UK unscathed.

But could anyone on the study group have justified a scenario whereby the Soviet Union launched an attack of 1,200 megatons, or 1.2 gigatons(!), on the UK whilst America stood by and watched? That America would allow its troops and servicemen stationed in the UK to be annihilated? That the launch of 400 Soviet missiles, even if it later became clear that they were not aimed at America, could occur without the American authorities launching at least some missiles in retaliation? It could also be argued (as some in the Ministry of Aviation later tried) that such an attack would also be a self-inflicted wound for the Soviet Union – the fallout generated from such an attack (made all the worse by the explosions being ground bursts), coupled with the prevailing westerly winds, would render Europe and most of European Russia completely uninhabitable.

There is also a further assumption, which is that all 400 missiles are serviceable, ready to fire, and are launched (and arrive!) successfully. If 400 warheads are needed to take out Blue Streak, how many missiles would Russia have to fire in order to achieve this? What sort of 'safety factor' would be needed to ensure that 400 arrived on target?

It was all very well to say that such considerations were not within the terms of reference if the scenario postulated is clearly absurd, and it is certainly arguable that a scenario whereby the Soviet Union launches more than 400 missiles at the UK without American retaliation (and thus a global war) is absurd. There is the further point that 400 missiles would represent a very considerable portion of the Soviet arsenal. Would they be prepared to devote such a proportion merely to take out the deterrent of what was, in Cold War terms, such a minor opponent? Indeed, taking the serviceability point, nearer 600 missiles would be needed. Sir Patrick Dean, as Chairman of JIC should have been able to give an estimate of the number of Soviet missiles based in Europe, and the answer would probably have been nowhere near the number assumed in the report.

It would also be extremely difficult to invent any plausible political situation whereby the Soviet Union would launch such an attack on the UK in absence of an attack on itself. Britain would never launch first, since they knew all too well how vulnerable the UK would be to any nuclear attack, let alone one on this scale, and the Russians would know that too. A British attack would be, in the jargon, a 'second strike', which was the point of the silos being able to survive the initial strike. But what political situation could see the Soviet Union launching 400 missiles at the UK and the UK alone? Such a scenario would also have to assume that the attack would be launched in the sure and certain knowledge that America would not become involved.

Moreover, the report stated that:

> The current Joint Intelligence Committee assessment is that we should get strategic warning of at least 24 hours before any heavy Soviet attack on this country. There is therefore no need to maintain our deterrent forces constantly at maximum readiness in order to guard against a "bolt from the blue" attack. But even if there is a period of rising international tension before the outbreak of hostilities, there is still a possibility that the enemy will attempt to achieve tactical surprise in the timing of his attack. The longer the period of such tension the greater the scope for tactical surprise, because deterrent forces cannot be maintained indefinitely at maximum readiness.[19]

So there is no surprise attack, no 'bolt from the blue', but, on the other hand, the Russians were supposed to be capable of co-ordinating the launch of several hundred missiles all within half an hour of each other without anyone noticing.

But what of the 'technical' factors? These too can be challenged, and indeed they were. 1.2 gigatons of nuclear explosion almost certainly would have taken out the Blue Streak silos, but there were a considerable number of caveats to this scenario too.

Firstly, we will assume the attack happens during what was called, rather euphemistically, a 'period of tension'. During such a time, it is probable that some of the missiles would be fuelled and ready to launch. Because of the restrictions involving liquid oxygen, this would not be very many. The probability would be that the missiles would be rotated: after so many hours, the missiles that were fuelled up would be stood down and other missiles made ready in their place. Even those missiles armed, fuelled with kerosene, but not liquid oxygen, could be launched within about 15 minutes at the very most.

In order to ensure that all the Blue Streak silos were destroyed before any missiles could be launched, the attack would have to be very carefully co-ordinated, such that all the attacking missiles arrived within a very short space of time, of the order of 15 minutes. There are problems to this.

The first is that the attacking missiles will be launched from a variety of launch sites, and will have a different time of flight. Thus their launch times would have to be very carefully co-ordinated. There is also the problem that around six or so missiles would be allocated to each silo. This brings the problem of 'fratricide', when another warhead in the vicinity of a 3 megaton blast might well be destroyed. Further, once the first explosion has taken place, there will be a very considerable disturbance of the atmosphere, to put it mildly! An incoming warhead that meets the shockwave from such a blast would be thrown off course, so to speak. The chance of it still arriving at its aiming point is remote. Yet if the attack is staggered to allow for this, then it does give windows of opportunity, small though they may be, for a Blue Streak launch to occur. The whole point of making such an attack is to ensure that no retaliation is possible.

But let us assume that the scenario is technically possible, that such an attack is possible, and that it would take out all the silos. What of the alternative possibility: the V bombers armed with Skybolt?

There was no intention, at this stage, that the RAF would mount standing patrols in the same way that SAC did. The best Bomber Command would have to offer would be aircraft on Quick Response Alert (QRA). These aircraft could be scrambled very quickly when the order was given. Unless the crew was strapped in, with the engines running (which was a possibility, although such a state could not be maintained for long), it would take some minutes before the crew could get into the aircraft, start the engines, taxi to the runway, take off and get clear of the airfield.

There were half a dozen V bomber bases, all in eastern England. During 'periods of tension', the aircraft could be dispersed to alternative airfields – up to 24 were planned. The facilities at some of these dispersal airfields were not entirely adequate, and such a dispersal could not be maintained for any period of time due to the lack of maintenance facilities. Even so, with 300–400 missiles at their disposal, the Soviet Union could launch a dozen or so at each airfield, and, QRA or not, the chances of an aircraft surviving such an attack, even in the air, is remote.

Sixty Blue Streak silos or sixty bombers on airfields? Which would stand the greater chance of survival? In the face of an attack on such a scale, the probable answer is that neither would survive, and in the event of an attack on a lesser scale, I would put my money on the silos.

There is also a rather more subtle psychological point in favour of aircraft rather than missiles, which usually gets oblique mentions in such discussions, but which did appear to loom large in minds of officials and politicians. The launch of a missile is irrevocable. Once the button has been pressed, there is no way of recalling the missile or even of destroying it in flight. Aircraft, on the other hand, give something of what might called a 'time cushion' – they can be sent on their way, but it will still be a matter of some hours before the pilot finally presses his button to launch his missile. This reduces the chance of 'launch by mistake' – in other words, it might appear that the UK has been attacked, and so it is imperative to react. Suppose the button is pressed – and the report of the attack turns out to be false. If you have pressed the button in a missile silo, you have started a nuclear war. If you have scrambled aircraft, they can be recalled. This would explain a good deal of the Whitehall dislike of Blue Streak. When Polaris took over from the bombers, there was still that time cushion. There was not the same sense of urgency to get in your response before your deterrent was destroyed.

But to return to the Powell report: once circulated, it drew one of two responses, depending which side a person was on, and by this time, there were few who were neutral. If someone was against Blue Streak, then the report was read and used as conclusive evidence that Blue Streak should be cancelled. So many eminent people can hardly be wrong. Someone in favour of Blue Streak faced an uphill task in trying to defend the missile. Even though the conclusions might be absurd, the reputation of the project had been tarnished – perhaps irrevocably so. (Another way of telling which side someone was on was whether they referred to 'fixed sites' (against) as opposed to 'underground launchers' (in favour)!)

Sir Steuart Mitchell's response is typical of many of the incredulous memos which came from the Ministry of Aviation:

> The vulnerability of underground Blue Streak is, in my view, grossly overstated in the report. As a result, Blue Streak is presented to Ministers as exclusively a fire first weapon, with all the serious disadvantages thereof. This, in my view, is based on misunderstanding of the technical facts.
>
> The report, in my view, accepts much too light heartedly the technical concepts of the WS.138A weapon system. History is strewn with weapon system concepts in which USAF expressed total confidence and which they pursued with great ardour for a year or two and then totally dropped. WS.138A today is where Hustler was four years ago except that the confidence and enthusiasm in WS.138A today is less than in Hustler four years ago. Hustler today is dead. To any technician familiar with the problems of a ballistic missile fired off a fixed site, the thought of putting our deterrent shirt on an American plan to fire such things off an aeroplane and having them in Service operations in the UK by 1966 or soon after as stated in the report is alarming. In my view the plan proposed in the report is dangerously unsound.[20]

'Seriously unsound', 'grossly overstated', 'alarming', 'dangerously unsound' – these are strong words indeed in this context, not phrases in common use in Whitehall. These are from the Government's chief expert on guided weapons! And he goes on in a further memo to say:

To provide ... 375 sites and missiles, together with roads, services, centralised communications; to give training and firing drill to crews; and to bring the whole edifice to such a pitch that they can fire the whole lot within a few seconds of the correct moment is, by any yardstick, a gigantic undertaking. It amounts in fact to an effort about six times bigger than the Blue Streak effort which it is trying to pre-empt. This aspect is dismissed in the report as simply being 'within the Russian capability' ...

All these factors, in our view, added to such a gigantic total that while by itself it may be technically feasible yet it is not a practical probability. We think that the very size of this pre-emptive effort is so big as to constitute a deterrent in its own right – providing Blue Streak is deployed underground.

All this is virtually ignored in the report. ...

If our politicians are at all competent, it is difficult to conceive of a war with Russia involving the UK but not also involving the USA.

If however such a situation came about, e.g. Middle East oil, it is most unlikely that Russia would pre-empt our Blue Streaks. She would surely fight such a war with conventional weapons (and win it easily this way), knowing that we know that if we "fired first" with Blue Streaks resultant Russian retaliation would destroy the UK.

Pre-empting Blue Streak in such a war is probably the last thing the Russians would do because if he did it would leave himself open to a 'fire first' blow of shattering size from the US.[21]

And then with regard to the arguments about the relative vulnerability of the V bombers and Blue Streak, Sir Steuart wrote yet another memo:

Blue Streak is condemned because of alleged inability to withstand 300 3MT rockets directed against 60 sites. It is therefore fair to consider what 300 3MT bursts could do to our V bomber force.

The Air Ministry claim that they can get "4 bombers airborne from an airfield in 4 minutes", – presumably four minutes from the local order to scramble.

The distance travelled by a typical V bomber from the point of take-off (i.e. from point where airborne) is

After 1 minute airborne flight ... 6 n. miles.
After 2 minute airborne flight ... 13 n. miles.
After 3 minute airborne flight ... 20 n. miles.

The disabling radius of a 3 MT airburst against a V bomber is approximately as follows:

V bomber on ground ... 14–15 miles (not tethered since it is about to take off)

V bomber airborne between zero height and 12,000 ft ... 12 miles

One 3 MT burst over the airfield ... would therefore disable not only all aircraft still on the airfield but also all airborne bombers within 12 miles of the burst.

Further, it would be reasonable to suppose that the Russians, in addition to the airburst directly over the airfield would at the same time lay on two additional airbursts along the direction of the bomber flight path.

It will be seen that the combination of one airburst over the airfield and two airbursts about 14 miles away and about 30° to the axis of the runway would disable not only all aircraft still on the runway but also all aircraft that had become airborne during the preceding 3 minutes. If these rocket bursts occurred at any time within the period up to 6 (not just 4) minutes before the order to scramble, it is difficult to see how any significant number of the bombers could escape.

I do not know the number of airfields which the V bomber force would use under dispersal conditions, but if one supposes the number to be, say, 50 at the most, then to carry out the attack outlined above would require a total of $3 \times 50 = 150$ missiles.

If the Russians were in any doubt as to the direction of take-off of the 7 bombers, they could remove the ambiguity by placing two other airbursts 14 miles at 300 from the other end of the runway to cover the reverse direction of take-off. Thus, for 50 airfields, would require a further 100 missiles, thus bringing the total missiles up to 250, which is still below the 300 postulated for the knocking out of Blue Streak.

Further, a number of the dispersal airfields are understood to be sufficiently near each other for the damage radius of a burst on one airfield to overlap on to the neighbouring airfield, thus reducing the weight of attack required.

Finally, it should be realised that the bomber bases in East Anglia would only get 2½–3 (not 4 minutes as commonly stated) warning by BMEWS of the Russian 1000 mile rocket if fired on a low trajectory from satellite territory.

[BMEWS: Ballistic Missile Early Warning System. There was a station at Fylingdales on the North Yorks Moors.]

This was just one of many objections – and this from the Government's principal expert on missiles. Sandys was writing similar memos as late as March.

So, if the report was so obviously flawed, why was it put forward, and why was it accepted so uncritically? One reason was that the answer it gave was the answer that a lot of people wanted. Never mind the argument, look at the conclusion. Another reason was that the Study Group had been composed of some very eminent men, and there was the obvious appeal to authority – a group as eminent as this could hardly be wrong.

But in one sense the argument had already been lost. Blue Streak had had sufficient doubt cast upon it that revival became impossible. (Similarly, later, when the US Secretary of Defense Macnamara cancelled WS.138A/Skybolt, but still offered it to Britain, Macmillan rejected the offer on the grounds that 'the lady's virtue has been impugned'. The same was happening to Blue Streak.)

Indeed, the battle went outside Whitehall with the publication of an article in the Daily Mail on 4 February 1960. The headline ran 'Blue Streak is a damp squib' and went on:

> This is the key problem – one of the most critical defence questions this country has faced – that Mr. Harold Watkinson, a senior cabinet minister, was put into the Ministry of Defence last October to solve ...

His brief from the P.M. was short and to the point. First, Britain must continue trying by all means within financial reason to remain an effective nuclear power.

Second, an attempt to do this by striking a new balance between the runaway costs of the deterrent and our obvious weakness in conventional forces.

With this in mind, it follows inevitably that the first file called for by Mr. Watkinson in the Ministry of Defence was labelled 'Blue Streak'.

In the days of his predecessor, Mr. Duncan Sandys, this file was the sacred cow. Mr. Sandys had made up his mind unalterably that Blue Streak, secure in its underground cells, was the answer for Britain.

It is said that throughout last summer Mr. Sandys faced increasing pressure from financial and military experts to think again. Apparently, he was obdurate. Only in his successor has a ready listener been found.

The most disturbing aspect of the cost of Blue Streak was strengthening military opinion that this fabulously expensive weapon would be secure underground only for a few years at most.

Once the Russians could guarantee the accuracy of their rockets within half a mile, Blue Streak would be as vulnerable as today's Thors which stand plain for all to see fixed on their surface-launching pads in East Anglia ...

The most interesting feature of this article is: where did the journalist get his information? To be as well informed as this suggests a leak from someone close to the top – although there is nothing new in that. And, further, although Blue Streak is portrayed as obsolete, no alternative is mentioned.

Other newspapers also picked up on the political implications of the struggle: another article mentioned the constant redrafting of the Defence White paper, and noted of Sandys that '... if he wins this Cabinet battle his personal standing among Ministers will be immensely enhanced. If he loses, his resignation cannot be ruled out.'

But despite all the arguments between Ministries and their civil servants, as we have seen, the Chiefs of Staff short-circuited the whole debate in their letter to Watkinson saying that they attached no military value to Blue Streak as a weapon, and recommending its cancellation. This was to put Watkinson in a very difficult position. He had only been Minister for Defence since the October 1959 election, and in his first Cabinet post (he would disappear from the Cabinet in the reshuffle known as 'the night of the long knives' in July 1962). Irrespective of all the papers from the Ministry of Aviation, and from Sandys, a Cabinet colleague who had been in successive Conservative cabinets since Churchill's election victory of 1951, and who had been the originator of the project, Watkinson had now received a memo from the Chiefs of Staff to say they had no confidence in the weapon. There was little he could then do. If his Chiefs of Staff told him that they attached no military value to a project then Watkinson was in no position to argue. He wrote to Sandys on 9 February and, unusually, the word 'personal' is handwritten at the top. The letter reads:

Dear Duncan,

Rippon and Strath will have told you how things have been developing in your absence.

The first development is that WS.138A seems to be doing well. You will have seen the messages sent to your Ministry by the Mission, which show that it has now been approved by the Department of Defense ...

This leads on to Blue Streak. The Chiefs of Staff have been considering their attitude to Blue Streak and have now given me their unanimous advice that they find Blue Streak, as a fire first weapon, unacceptable. I am afraid Dermot sold the pass here to begin with.

If then it is open to us to obtain an American weapon on acceptable terms, we are faced with a disagreeable choice. Either we must go on with Blue Streak in the knowledge that the Chiefs of Staff advise against it as a weapon. Or we must cancel it in favour of an American weapon, with all that may be involved in the way of losing the ability to develop missiles on our own. No intermediate course seems to be feasible as I understand from your department that if Blue Streak is to go on at all there is no sensible way in which any significant sum could be saved. I am not sure that they have really thought this out enough, but you will know better than I about this.

This then is the choice so far as spending defence money is concerned. It may be that the Minister of Science will conclude that he can justify financing the development of Blue Streak and converting it into a project for space research, primarily from civil funds. I have put this proposition to him but I should doubt whether he can find the money.

All this presents us with a difficult choice and I am not yet clear what it is best to do in the national interest. In order to help me to form an opinion I have been asking your department for information about the consequences of stopping Blue Streak. Your department is directly in touch with the Minister of Science's office about the cost of the space research programme.

I should very much like to know what you think, as soon as your people have finished setting out the consequences of stopping Blue Streak, or any possibility of saving something from the wreck.[22]

'Dermot' referred to in paragraph five was the Chief of the Air Staff, Sir Dermot Boyle. Blue Streak was intended for service in the RAF, yet even the head of that Service had rejected it. Watkinson's implication is that Boyle's withdrawal of support for Blue Streak was the precipitating factor. The phrasing of the letter is also interesting: Watkinson realises the implications for Sandys, yet cannot argue against the advice given to him by the professionals.

What were the motives of the Chiefs of Staff? Crudely, they could be summed up as follows.

The Army had no real interest one way or the other. Their only real interest was to keep the cost of the deterrent as low as possible to allow more room in the military budget for new equipment (tanks and the like) for conventional forces. If a cheaper alternative to Blue Streak were available, they would vote for it.

Mountbatten, as mentioned, had other motives. He was, in addition, Chief of the Defence Staff and a man with a considerable Whitehall network. Cancelling Blue Streak, to which he was opposed anyway, opened the way for the Navy to acquire Polaris submarines, which they did in the mid-1960s. His successor as First Sea Lord, Sir Charles Lambe, was also pushing hard for Polaris.

Boyle, of the RAF, also saw new opportunities for his service. Not only would the V bombers be given a fresh lease of life, there was a window of opportunity for the RAF to acquire further aircraft to supplement and replace the V bombers. Proposals were well advanced at the time of the cancellation of Skybolt to modify the VC-10 airliner to enable them to carry the missile.

Apart then from the Ministry of Aviation, Blue Streak had no Whitehall or Service support. Hence it was to go.

But there were other political considerations to the cancellation. Two of the most important were the political dimension of the cancellation, particularly given the cost to date, and the implications for relations with Australia.

Since the late 1940s, there had been considerable co-operation between the UK and Australia in matters of weapons development. The UK had devices to test but no room in which to test them; Australia had the room but did not have the devices. Thus Australia provided testing sites for atomic weapons (the Monte Bello Islands, Emu Field, and Maralinga), as well as the Long Range Weapons test site at Woomera. All Blue Streak test firings were to have been carried out at Woomera, and many facilities, funded jointly by the UK and Australia, were nearing completion. Thus the Foreign Office in particular was concerned about the impact of the cancellation on the Menzies Government and on Australian public opinion.

As to the cost of the project so far, there was a salvage option: to continue Blue Streak not as a military weapon but as a satellite launcher. This would help deflect much of the political criticism, but it was an option that had not been thought through very clearly – in particular, the cost implications.

Now the decision went to the Cabinet Committee on Defence, and the minutes of its third discussion on 6 April concerning Blue Streak read as follows:

> THE PRIME MINISTER said that the first question for consideration was whether the provisional decision … to abandon the development of BLUE STREAK as a weapon should now be confirmed. There were two main issues to decide:-

(a) Would it be militarily acceptable to rely on the V-Bombers, with SKYBOLT, rather than on BLUE STREAK, as our strategic nuclear force from about 1965 onwards?

(b) Was it reasonable to assume that SKYBOLT and eventually POLARIS (if we needed it) would be made available to us by the Americans on satisfactory terms?

THE MINISTER OF DEFENCE said that ... the general consensus of opinion was that, in circumstances other than a surprise saturation attack, the V Bombers equipped with SKYBOLT would have certain advantages over BLUE STREAK. The main considerations leading to this conclusion were political rather than scientific or technical. The Bomber force had qualities of mobility and flexibility which were useful for conventional operations as well as for the nuclear deterrent. It had the advantage that it could be launched on a radar warning without an irrevocable decision being taken to launch the nuclear attack itself.

THE MINISTER OF AVIATION agreed that there would be certain financial and political advantages in depending on the V-Bombers and SKYBOLT rather than on BLUE STREAK for our strategic deterrent force in the later 1960's [*sic*]. From the military point of view, there was no marked advantage one way or the other. In these circumstances he would concur in the decision that the development of BLUE STREAK as a weapon should be abandoned.

... The Americans had indicated their willingness to make SKYBOLT available unconditionally, except for the suggestion, which we might be able to persuade them to modify or abandon, that specific reference should be made to its use for North Atlantic Treaty (N.A.T.O.) purposes. It should be possible to reach a similar understanding as regards POLARIS (on which, however, no immediate decision was required) ...

THE PRIME MINISTER said that the Committee's discussion showed that their provisional decision to abandon BLUE STREAK as a weapon could now be confirmed. The next question to be considered was whether its development should be continued for scientific and technological purposes. The officials' Report showed that there were only two alternatives:

(a) to cancel BLUE STREAK completely; even if this were done immediately, there would be unavoidable nugatory expenditure of about £72.5 millions, of which £22 millions would fall in 1960/61.

(b) to adapt it as a space satellite launcher at a cost, including the development of a stabilised satellite, of about £90–100 millions.

The advantages and disadvantages of these two courses could not be wholly assessed in material terms. To cancel BLUE STREAK would involve dislocation of industry, difficulties with the Australians, heavy charges, the loss of the potential value of a large British rocket for space research or other purposes and the abandonment of that part of the work already done which was relevant to the development of a satellite launcher; but it would curtail expenditure in the longer term, and make resources available for other purposes. To develop BLUE STREAK as a space satellite launcher would be much more costly... and the Ministry of

Aviation had not been able to consult the firms concerned about whether the £90–100 millions launcher and satellite programme would in fact be practicable ...

THE CHANCELLOR OF THE EXCHEQUER said that an immediate decision should be taken to bring all further work on BLUE STREAK to an end. The nation's resources over the next few years would be inadequate to meet all our existing commitments. Since there was no suggestion that any other project should give way to the development of BLUE STREAK as a space satellite launcher, he did not see how the heavy expenditure involved could be met. The programme was estimated to cost over the next four or five years some £75 millions more than the cost of immediate cancellation; past experience suggested that this figure might be considerably increased and that other defence projects, for which no provision had yet been made, would eventually come forward to take the place of expenditure saved on BLUE STREAK. The national economy would benefit from the industrial and man-power resources made available by the complete cancellation of BLUE STREAK ...

Summing up, THE PRIME MINISTER said that the Cabinet should be informed of the decision to cancel the development of BLUE STREAK as a weapon and invited to consider whether this decision should be announced in terms that all work on BLUE STREAK should cease completely or that further consideration was being given to its development as a space satellite launcher. If the latter alternative were adopted it would be desirable for a final decision to be taken if possible within the next few weeks.

The Committee took note that the Prime Minister would arrange for the Cabinet to be informed of the decision to cancel the development of BLUE STREAK as a weapon and of the terms in which this decision might be communicated to Parliament and to the Government of Australia on the alternative assumptions that -

(a) all further work on BLUE STREAK should cease;

(b) consideration should be given, in consultation with industry and the other interests concerned, to the adaptation of BLUE STREAK as a space satellite launcher.

The decision having thus been taken, it fell to Watkinson to make the announcement in the House of Commons on 13 April. He rose to read a statement which ran thus:

Blue Streak.

1. The Government have been considering the future of the project of developing the long-range ballistic missile Blue Streak, and have been in touch with the Australian Government about it, in view of their interest in the joint project, and the operation of the Woomera range.

2. The technique of controlling ballistic missiles has rapidly advanced. The vulnerability of missiles launched from static sites, and the practicability of launching missiles of considerable range from mobile platforms, has now been established. In light of our military advice to this effect, and of the importance of reinforcing the effectiveness of the deterrent, we have concluded and the Australian

Government have fully accepted that we ought not to continue to develop, as a military weapon, a missile that can be leached only from a fixed site.

3. To-day our strategic nuclear force is an effective and significant contribution to the deterrent power of the free world. The Government do not intend to give up this independent contribution, and therefore some other vehicle in due course will be needed in place of Blue Streak to carry British-manufactured nuclear warheads. The need for this is not immediately urgent, since the effectiveness of the V-bomber force as the vehicle for these warheads will remain unimpaired for several years to come, nor is it possible at the moment to say with certainty which of several possibilities or combinations of them would technically be the most suitable. On present information, there appears much to be said for prolonging the effectiveness of the V-bombers by buying supplies of the airborne ballistic missile Skybolt which is being developed in the United States. H.M. Government understands that the United States Government will be favourably disposed to the purchase by the United Kingdom at the appropriate time of supplies of this vehicle.

4. The Government will now consider with the firms and other interests concerned, as a matter of urgency, whether the Blue Streak programme could be adapted for the development of a launcher for space satellites. A further statement will be made to the House as soon as possible.

5. This decision, of course, does not mean that the work at Woomera will be ended. On the contrary, there are many other projects for which the range is needed. We therefore expect that for some years to come, at least, there will be a substantial programme of work for that range.[23]

The Opposition based their first attack on the grounds of waste of large sums of public money, which Watkinson was able to counter with the argument that Blue Streak would be developed as a satellite launcher. It was a useful point for the Opposition to seize on, as it was an issue which could cover its own internal divisions about the deterrent.

But the satellite launcher option was a useful defence, indeed the only defence open to him, which raises the question: how genuine a statement was this? Did Watkinson and the Ministry of Defence really want a satellite launcher, or was this statement merely a political fig leaf? Given the enthusiasm (or lack thereof) with which the subsequent Cabinet committee greeted the topic, the suspicion is that the fig leaf is the correct answer. Certainly the initial reaction among those in the House was quite vigorous: George Brown was the then Shadow Defence Secretary, and demanded an immediate emergency debate, which, however, was not forthcoming. (Sandys' absence from the House was noted by Jim Callaghan, who seized upon it to say: 'I was commenting that it was a little unfair that the Minister of Defence should have to face all this music, and I was wondering where the Minister of Aviation is and when he is going to resign.')

Given the costs of the project at the time of cancellation, the Opposition managed to force a later debate. Whilst Brown might have been a good speaker, what he said at the debate does not read well today. This is partly because, like all Opposition speakers in any debate, he had not had the Civil Service back up and briefings that Ministers have. It is also interesting to note that he seems to have had some inside information on 'fixed sites', but although he makes a great show of saying that he himself had advocated dropping the system, he sidesteps making any justification.

Although Watkinson, as Minister of Defence, opened the debate, Sandys was obviously the target for the Opposition. He gave a very straightforward speech in reply, even if he might not have been entirely convinced by his own side's case. He was able to undercut Brown by resurrecting a quote from the time of Thor, when Brown had said that what the UK needed was its own missile with its own warhead, which is what Blue Streak had been. The other notable part about his speech is that, by Commons standards, it was not particularly partisan: he laid down the facts as he saw them, and did not attempt to make political capital from the decision.

But with the cancellation now official, interest within the Ministries of Defence and Aviation turned swiftly to Skybolt. The Navy was still not happy that Polaris had not triumphed, as can be seen from another internal Admiralty memo. It comments of Watkinson:

> Skybolt lay ready to his hand (he thinks) as a blood transfusion to keep the V bombers effective from 1965–1970. ... Our trouble is that the Minister has been advised by interested parties, in very optimistic terms, about Skybolt's state and prospects. I would almost say that he has been led up the garden path. I would warn you that some of the advisers he will bring to you with him are bitterly anti-Navy.

It is ironic that 20 months later, in December 1962, Skybolt itself was cancelled by the US (as predicted by Brundrett and CGWL), and the UK had to negotiate hard to obtain Polaris. This meant that the 'deterrent gap' was now stretched to the late 1960s, and while the Polaris submarines were being built, the deterrent was being carried by V bombers with free fall bombs and the short range Blue Steel stand-off missile.

From the outset, the British Government had been warned that Skybolt was very much in the development stage, and there was no guarantee that it would actually be deployed by the USAF. The veteran Labour MP, Sydney Silverman, referred to Skybolt in the House of Commons in June 1960 thus:

Would it be a fair summary of what the right hon. Gentleman has told the House to say that the result of his negotiations in the United States is that what he has really done is to buy a pig in a poke with a blank cheque?

Whilst Silverman was making his attack mainly on party political grounds, there was more than a grain of truth in his comment.

Polaris did serve the UK well for nearly 30 years (although its mid-life upgrade, Chevaline, was also a source of controversy), and carrying the deterrent offshore leads to the argument that the mainland itself is no longer a target. Given however the number of NATO and US nuclear bases in the UK, that argument rather falls down. The cancellation of Blue Streak, and the reason given, meant that land-based missiles were never again an option for the UK deterrent. As to what purpose the UK deterrent was to serve, however, is another question.

The Powell report may have been ingenuous in its conclusions, but in the end, Blue Streak was indeed cancelled as a military weapon. Was this the right decision?

The answer to this can only be 'yes'. Skybolt, had it been deployed, would have been almost as effective a deterrent for a good deal less money. Deterrents are there for political reasons: the whole point of them is that they should never be used! Britain's deterrent was a perfect example – there was no way that it would be used without America becoming involved, and indeed that was one of the points of it.

As to the vulnerability issues, the RAF was considering a follow on to the Vulcan as a Skybolt carrier – in the form of the Vickers VC10 airliner.[24] This is not as absurd as it sounds: airliners are designed to stay in the air for long periods and to have very short turn round times. The point of Skybolt was that the carrier aircraft would not need to fly anywhere near the enemy defences – and so an airliner would have been an ideal vehicle for the purpose. Proposals were put forward for a system of standing patrols (not possible with the V bombers) using 42 modified VC10s.

As it turned out, Skybolt was cancelled and Britain was offered Polaris. The submarines were designed by the Admiralty, and the whole project was carried out exactly on schedule and within budget – an achievement that had eluded the Ministry of Supply for a decade (and is still eluding the Ministry of Defence today). Polaris served the country well in its function as a deterrent, staying in service until the mid-1990s.

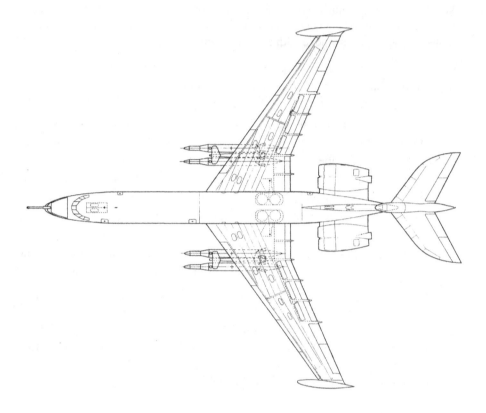

Figure 53. Plan view of VC 10 airliner modified to carry four Skybolt missiles.

Blue Streak, even in its silos, would have seemed outdated by around 1970, if not earlier. It would not have been difficult to have built and designed a substitute which was smaller and cheaper (warheads had become very much lighter in the interim), and which could also have been based in the same silos, but given the pace and cost of the project up to 1960, how much the silos would have cost, and when they would have been finished, is a very open question.

This may seem to be an extended exposition of the cancellation in what is, in the main, a book on the British rocketry programme, but it had a very considerable impact on the future of Blue Streak as a satellite launcher, and thus by extension, on any potential British space programme.

At the time of cancellation there was still a good deal of development work to be done on Blue Streak. If the missile had not been cancelled, then the cost of this would have fallen on the defence budget. Furthermore, there would have been a number of test and development launches needed at Woomera, and the facilities there would also have been charged to the defence budget. Building a launcher from a fully developed Blue Streak would have been relatively cheap. The upper stages would have been Black Knight derivatives, and the development of Black Knight itself had not cost a great deal. The major cost would have been in building the interface between Blue Streak and the upper stages (a further expense which would have been worthwhile would be the uprating of the RZ 2 motor from the 137,000 lb thrust for the missile to 150,000 lb – this would allow for heavier upper stages). Thus a quoted cost of £65 million for a Blue Streak satellite launcher in 1960 might have been reduced to a tenth of that by 1964, making such a launcher far more probable. What such a launcher might have been used for is the subject for later discussion!

Unpopularity of Blue Streak

However, one further major question is left unanswered. The motives of the Services, the Ministry of Aviation, the Air Ministry and the Treasury appear obvious enough. What is not obvious is why the Ministry of Defence and Powell himself took the position they did.

There are several possible scenarios.

The first is that Powell himself, possibly in concert with other senior civil servants in other departments, felt that the project was insupportable. Although he did not have the authority himself to cancel it, he could set up circumstances that gave others the opportunity. Thus if the Treasury and the Chiefs of Staff were to object sufficiently, then the new and relatively inexperienced Minister, fresh to the Cabinet, had little choice. It is interesting that the major attack on the project was only mounted after the October 1959 election, when Sandys was moved from Defence to Aviation. It would also mean keeping the details of the report from Sandys at the Aviation ministry for as long as possible, which seems to have been the case.

Another possibility lies not with Powell but with Watkinson and Macmillan, as indicated in the Daily Mail article. Under this scenario, Watkinson is appointed by Macmillan with a specific brief to ensure the cancellation. But why would Macmillan want to do this?

A possible answer lies not in the cost, but the timing of the cost. Expenditure on Blue Streak would reach its peak from 1960 to 1965. Although expensive in terms of capital cost, its running costs were (or appeared to be) extremely low. Both Skybolt and Polaris were considerably less expensive to buy (the development costs would have been covered by the Americans) but their running costs were very much higher. Flying aircraft, or running submarines, does not come cheap. However, these running costs would not have been incurred for several years to come, which would be well beyond the lifetime of the Macmillan Government.

A further answer might lie in the silos themselves. It is curious that the estimates of the vulnerability of the silos were never questioned one way or the other. Although the Air Ministry and the Ministry of Supply had done the calculations as best they could, the design was still a paper one (indeed, the design of the lid, one of the most crucial points in any silo design, had not been finalised by the time of the cancellation). In a sense, this makes a nonsense of the whole argument: since no one actually knew the exact strength of the silos, the debate was in many ways so much hot air. But they would raise political difficulties. They would certainly consume a great deal of civil engineering resources, and the political impact of such large and controversial structures in Conservative constituencies should not be overlooked. Indeed, the Home Office under Butler had come into the argument at one stage, requesting that the silos be situated on the east side of the country so that, given prevailing westerly winds, fallout from an attack would to taken away from the UK. In addition, for Civil Defence purposes, Butler wanted the sites well away from centres of population.

But however well-disguised the real reasons were, and no matter how much the papers conceal the true motives, a very revealing letter was written a year after the cancellation. A Technical Sub-Committee was to be set up for the BND(SG), and Zuckerman wrote to various eminent scientists, inviting them to join. One was Sir Robert Cockburn, who had been working for the Government in various capacities since the war, and at one time had been CGWL at the Ministry of Aviation. He wrote back to Zuckerman, and one paragraph of his letter reads:

> Blue Streak was cancelled because it was not politically viable rather than because it could be pre-empted. The scale of pre-emption was admitted to be of the order of 3,000 megatons. Supporters of the system argued that this was so excessive that pre-emption could be ignored in practice. The argument was not accepted and vulnerability was advanced as the main reason for cancellation. The real reasons were more fundamental although still not clearly appreciated. I suggest no British

statesman could visualise exploiting a deterrent threat which if mishandled could only lead to the annihilation of the whole country; nor could he believe that a threat involving such consequences would be taken seriously by an opponent.[25]

In other words, once missiles are fired, they cannot be recalled. And with missiles, which are seen as potentially vulnerable whilst on the ground, the incentive is to fire early. Bombers can be recalled, and they do not need to fire off their missiles until it is certain the UK has been attacked. The same is true of submarines lying undetected in the Atlantic. This, probably more than anything else, reflected the true reason why Blue Streak was cancelled.

The cancellation is a graphic example of how Whitehall can work. What is of more interest is the study of how much policy was made by officials and how much by Ministers. Ministers rely on officials for advice: how impartial was that advice? Civil servants themselves have opinions. Furthermore, the documentary evidence that survives tends to suggest that a good deal of policy was not made on paper, but in briefings, and that papers were presented with a particular pre-determined slant or viewpoint (although there is nothing new in that!). Ultimately, it might be said that the correct decision was made, but that the evidence presented was misleading, and the motivations of the various participants were, to say the least, often concealed.

---

[1] TNA: PRO DEFE 7/2245. Development of Blue Streak. CW Wright 21 February 56.
[2] TNA: PRO DEFE 7/2245. Development of Blue Streak.
[3] TNA: PRO ADM 1/27389. Polaris/Skybolt.
[4] TNA: PRO DEFE 19/11. British nuclear deterrent.
[5] TNA: PRO ADM 1/27389. Polaris/Skybolt.
[6] TNA: PRO AIR 8/1961. Study of future policy: the Brook Committee.
[7] Ibid.
[8] Ibid.
[9] The Independent, 21 December 1995.
[10] TNA: PRO AVIA 66/1. Guided weapons and electronics: Blue Streak.
[11] Ibid.
[12] TNA: PRO ADM 205/202. The British nuclear deterrent.
[13] There are various National Archive files for the BND(SG), but in the main, this account draws on AIR19/998 and AIR 19/999.
[14] TNA: PRO AVIA 66/1. Watkinson to Sandys, 9 February 1960.

[15] TNA: PRO DEFE 19/11. British nuclear deterrent. THE NUCLEAR DETERRENT - 1970 AND AFTER. Note by the Air Ministry Strategic Scientific Policy Committee, 5 October 1959.

[16] TNA: PRO AVIA 54/2146. Blue Streak development: Joint US/UK Advisory Committee; technical correspondence.

[17] TNA: PRO AIR 19/998. British Nuclear Deterrent Study Group. Memorandum for the Minister of Defence by the Chiefs of Staff 5 February 1960.

[18] Ibid.

[19] Ibid.

[20] TNA: PRO AVIA 65/910. Deterrent policy: CGWLs actions.

[21] Ibid.

[22] TNA: PRO AVIA 66/1. Watkinson to Sandys, 9 February 1960.

[23] Ibid.

[24] See for example AIR 20/10925. VC10: deterrent role.

[25] TNA: PRO DEFE 19/87 Report by the British Nuclear Deterrent Technical Sub-Committee. Sir Robert Cockcroft to Zuckerman on being asked to serve on BNSG technical sub-committee. 7 March 1961

Chapter 8

BSSLV

One of the ironies of the Blue Streak saga is that the UK spent a very considerable sum on its development and on the Europa satellite launcher as part of ELDO, sums of the order of £200 million at 1950s/1960s prices, with absolutely nothing to show for it at the end of it all.

However, over a period of around 15 years from 1957 onwards, designs and proposals for an indigenous satellite launcher based on Blue Streak came and went. The design which most nearly came to fruition was known inelegantly in official files as the Blue Streak Satellite Launch Vehicle, or BSSLV for short. This chapter looks at the various designs proposed over the years. Not surprisingly, since Blue Streak, Black Knight and Black Arrow were the only liquid fuel ballistic rockets being developed by the UK, most of the designs tended to revolve around these vehicles, but in addition liquid hydrogen stages were often proposed. Alternative HTP designs could have been produced from de Havilland's Spectre or Armstrong Siddeley's Stentor motors, but were never considered by the RAE or Saunders Roe, where most of the designs originated.

The design of Blue Streak was finalised by around the start of 1957. At the same time, Black Knight had been set in motion, and it might be useful to give some approximate data for each vehicle.

	Blue Streak	Black Knight
Diameter	10 ft	3 ft
Thrust	300,000 lb	16,400 lb
Weight	196,000 lb	12,000 lb

Blue Streak was around 15 times more massive than Black Knight, and this highlights one of the difficulties which designers of the time faced. The most efficient way to design a satellite launcher is to match the size and performance of the individual stages, but instead designers of the period were taking rockets designed for entirely different functions and putting them together into a launcher, however mismatched they might be. This might be less efficient, but it saved time and money.

But to step back for a moment, the origins of the BSSLV are both obscure and interesting.

It is well known that the Pentagon was interested in reconnaissance from satellites in the mid-1950s, but was held back by a legal consideration. Aircraft from one country could not overfly another country without permission – to do so would be violating the second country's airspace. What had not yet been legally established is the height of a country's airspace – did it end at the top of the atmosphere or stretch out to infinite space? And how would 'top of the atmosphere' be defined?

Thus one of the reasons why the satellite proposed as part of the 1957 International Geophysical Year would be launched by a civilian rocket was so that it could not be described as military in nature by the Soviet Union. Its purpose would be entirely scientific, but at the same time it would set a legal precedent.

All this, of course, became moot with the launch of Sputnik 1. The Soviet Union could no longer complain if American satellites passed over Russian territory, and had no way of knowing what instruments or cameras the satellites were carrying. Over the next few decades, America was to spend literally billions of dollars on satellite reconnaissance. In March 1967, President Johnson was quoted as saying,

> … we've spent thirty-five or forty billion dollars on the space program. And if nothing else had come out of it except the knowledge we've gained from space photography, it would be worth ten times what the whole program has cost. Because tonight we know how many missiles the enemy has and, it turned out, our guesses were way off. We were doing things we didn't need to do. We were building things we didn't need to build. We were harboring fears we didn't need to harbor.'

For comparison, the GDP of the UK in 1967 was $111 billion.

But it was not only America that was interested in reconnaissance. So was Britain.

> Towards the end of 1954 the D.R.P.C. appointed a working party to consider the problems of long-range reconnaissance. This working party reported in November 1955 and one of the recommendations in the report was that the most promising line of research and development would be an orbiting satellite to carry optical reconnaissance equipment.[1]

Desmond King-Hele of the Guided Weapons Department, RAE, was given the task of writing a report outlining the advantages and disadvantages of reconnaissance by satellite. RAE GW reports are usually very prosaic documents, designed to convey technical information only. It comes as something of a surprise, therefore, to read King-Hele's introduction to his paper[2].

> Escaping from the earth, this 'dim vast vale of tears', has long been one of man's recurrent dreams, a dream enshrined in primitive myth and exploited in many later European writings, among them Dante's famous *Divina Commedia*, which includes visits to all the known planets, the sphere of the fixed stars, and beyond, even to the empyrean. Neither Dante nor most of his successors could specify a realistic means of propulsion, and it is only since the advent of the modern rocket motor the dream has shown any possibility of being fulfilled. Now that plans are afoot in Britain for a long-range ballistic rocket missile, the first step towards fulfilment – an unmanned satellite circling the Earth – has advanced beyond a possibility to a logical development.
>
> The equally recurrent military need for continual reconnaissance also appears to be satisfied by a satellite vehicle, provided its launching is not denounced as a warlike act. If launched in high or middle latitudes it will inevitably pass over enemy as well as friendly territory, and much better photographs should be obtained from a satellite than from high altitude aircraft 'peeping over the frontier' through hundreds of miles of haze.
>
> Bringing the satellite safely back to earth would be almost as difficult as placing it in an orbit; so the satellite is here assumed to be expendable, i.e., it is doomed at best to die in a brief blaze of glory, lighting up the night sky to the wonderment of some remote tribe, and at worst to drop with dull thuds in recognisable pieces on an enemy city.

Some of his rhetorical flourishes are not entirely successful: 'to drop with dull thuds' strikes a note of surely unintentional bathos, and political correctness today might cause 'the wonderment of some remote tribe' to be reconsidered. After this rather ornate introduction, the prose becomes more mundane and more suited to the matter under consideration.

The problems as he saw them were firstly, providing adequate propulsion and structure without exceeding the stringent limits on weight; secondly, methods of guidance, control and power generation; and thirdly, recording and transmitting back to Earth a picture of interesting territory which the satellite passes over.

King-Hele was aware that the satellite would not be able to provide 'live' data that is, when it was above the territory to be photographed, it would not be able to relay the information back directly, as the receiving station would be over the horizon and thus out of radio contact. What is slightly more surprising is that he rejected the idea of recovering film directly from the satellite after re-entry on the grounds that atmospheric decay of the orbit would be too unpredictable. The possibility of commanded re-entry did not seem to have been considered at all.

As to the resolution obtainable:

> A continuous strip record would be taken while over interesting territory, by infra-
> red photography. This record would be processed or otherwise stored until
> transmission back to the ground, which would take place over friendly territory at
> low data rate. For an orbital altitude of 200 nautical miles quite good resolution,
> 100–200 ft, should be obtainable with a camera of about 50" focal length, and a strip
> 40–50 miles could be covered.

Power supply for the satellite was another problem. Among King-Hele's
suggestions was:

> A third and more direct method is the Bell solar battery, in which electricity is
> generated directly by light falling on thin strips of pure silicon impregnated with
> traces of boron. The first models of the solar battery had an output of about six watts
> of exposed surface but by improved techniques output has now been raised to about
> 11 watts (11% efficiency of conversion). The standard 2000 lb satellite assumed here
> has a surface area of about 150 sq ft, so the area requirements are within the realms
> of possibility. The present high cost of the Bell device (quoted elsewhere as around
> half a million dollars per kilowatt) would however have to be reduced.

Interestingly, given the provisional status of Blue Streak at this date, King-
Hele suggested that:

> It will obviously be advisable to profit as far as possible from work on the Blue
> Streak surface-to-surface ballistic rocket missile, and accordingly the type of
> propulsion and structure at present proposed for Blue Streak are used here.

He obviously took this further in a later Guided Weapon Department report
of May 1957, entitled 'The Use of Blue Streak with Black Knight in a Satellite
Missile'. This was the first time that a full analysis of Blue Streak as the first
stage of a satellite launcher had been carried out ... five months before the launch
of Sputnik! His calculations showed that the Blue Streak/Black Knight
combination could put a payload of 2,280 lb (very close to 1,000 kg) into low
earth orbit. The conclusions of the report were that:

> It should be possible to place quite a massive satellite vehicle in an orbit by using a
> three-stage missile which consists of (1) the first stage of Blue Streak, (2) the first
> stage of Black Knight and (3) the final satellite. Some structural strengthening would
> be required, in Blue Streak especially: increases in structure weight of 1000 lb on
> Blue Streak and 100 lb on Black Knight have been assumed here. Under these
> assumptions, and with an 850 ft/sec advantage from the earth's rotation, the satellite
> payload would be about 2400 lb for 200 n.m. orbital altitude and 2200 lb for 400
> n.m. orbital altitude.[3]

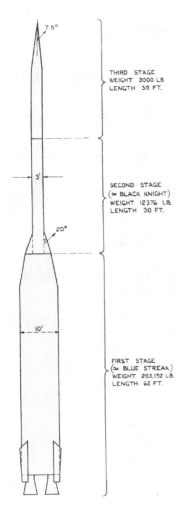

Figure 54. King-Hele's proposal.

That is a remarkably accurate prediction, made all the more impressive since the two vehicles in question were still in existence only on the drawing board! King-Hele's proposal can be seen in Figure 54. Looking at the design more closely reveals that the 3 ft diameter Black Knight sits uneasily on the 10 ft diameter Blue Streak. Making a one ton satellite that would fit into a 3 ft payload fairing might also be a challenge (for comparison purposes, the average saloon car is about a ton in weight).

In 1957, there was also considerable security surrounding both Blue Streak and Black Knight, and it was not until 1959 that Geoffrey Pardoe of de Havilland was able to present various satellite launcher designs, using the same ideas, to a meeting of the British Interplanetary Society (BIS). The design that emerged as the favourite perhaps took the answer to the geometry problem a little too far: instead of being a slim 3 ft cylinder, the upper stage had become a 10 ft sphere![4] There is considerable logic behind this design, but there was also a practical difficulty relating to cost and facilities, which was much less obvious.

Saunders Roe had designed their test facility at High Down for a 3 ft diameter vehicle; the stands could be relatively easily adapted for designs up to 54 inches Beyond that, there would, apparently, have to be considerable rebuilding. The cost of this was held up as an objection to design after design, until Black Arrow with its two metre diameter lower stage was produced – without any objection from anyone!

Although there was no specific requirement for a satellite launcher, RAE had considerable autonomy and pressed on with studies. The minutes of the 'First Meeting in Ballistic Missile Division to Discuss the Development of Satellite Vehicles' are dated February 1958[5]. These meetings were obviously a follow on from King-Hele's paper, and discussion centres around Blue Streak. Curiously, the minutes of a later meeting in October 1958 mention that:

> Mr. Longden reported that de Havilland had been doing satellite performance estimates. He understood that using only Blue Streak and Black Knight they could get a ton into orbit.
>
> Mr. Cornford said that this work should not be encouraged.

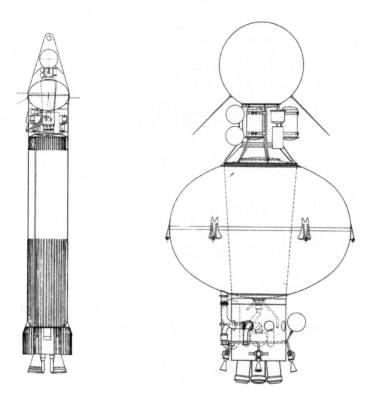

Figure 55. Pardoe's design for a Blue Streak launcher. The complete vehicle is shown on the left; the somewhat unusual second stage is shown in enlargement on the right.

Unfortunately, the minutes give no explanation as to why!

In May 1959, Harold Macmillan was asked in the House of Commons by Mr Richard Fort (Conservative Member for Clitheroe; sadly, he was to die in a car accident four days later): 'if he is yet in a position to make a statement about space research'.

This was probably a 'planted' question so as to give the Prime Minister an opportunity to make his statement. As part of his answer, Macmillan said:

> Meanwhile, however, design studies are also being put in hand for the adaptation of the British military rockets which are now under development. This will put us in a position, should we decide to do so, to make an all-British effort.
>
> I have asked my noble Friend, the Lord President of the Council, in consultation with my Right Hon. Friend the Minister of Supply and other Ministers concerned, to exercise general supervision of these new developments.

He then went on to say:

> I cannot give any figure of the cost of using the British rocket should it be decided to do so. What we are doing now is to spend a substantial but modest sum, more in hundreds of thousands than in millions, first for the design of the instruments, which is the first element, and secondly, for making the necessary designs for modifications of the British military rocket should it be decided when the time comes to use it for this purpose.

There were some sceptics in the House:

> Mr. Chetwynd [Labour member for Stockton-on-Tees]: Is the Prime Minister satisfied that there is intrinsic value in this scientific work or is it just an attempt to keep up with the Joneses?

To which the Prime Minister replied in true Macmillan style:

> I am not by nature or, I am afraid, by education very favourably inclined to swallow all that the scientists tell me, because, alas, I do not understand it, but I am impressed by the universal opinion of these very distinguished people whom we have consulted, and I feel certain that in this scientific instrument work it is clear that Britain ought to play her part in this advancing scientific effort. As to the method of launching these instruments, what I felt right to do, and what, I think, the House will think sensible, is to make preliminary work at not very large expense which will put us in a position to use our own rocket should, when the time comes, we decide to do so.

As a result of the announcement, Dr Robert Cockburn (who at that time was CGWL) at the Ministry of Supply was asked for design studies for a rocket and for the satellite itself by the Office of Lord President of the Council, Lord Hailsham, who was also acting as Minister for Science[6]. The RAE set up a series of panels to produce a detailed design late in 1959, resulting in the brochure produced by Saunders Roe, entitled 'Black Prince'.

It was obvious that Saunders Roe had been involved before the announcement had been made. In a meeting back in April, 'Saunders Roe tabled and presented several schemes for RAE consideration and discussion'.

As a result of all this work, a brochure for a launcher was ready in time for the military cancellation[7]. The brochure was prepared by Saunders Roe and RAE, and the vehicle is called, rather optimistically, Black Prince. Its design is very conventional: Blue Streak as the first stage, a 54-inch Black Knight second stage, with the Gamma engine uprated to 25,000 lb thrust, and various configurations suggested for a third stage, depending on the mission. This would also use HTP and kerosene, with a small four chamber rocket motor, designated the PR.38, from Bristol Siddeley.

The weight breakdown for the launcher in the initial design brochure was:

	All burnt weight	Propellants	Thrust
First stage	12,500 lb	173,000 lb	270,000 lb (sea level)
Second stage	1,520 lb	16,000 lb	25,000 lb (vacuum)
Third stage	500 lb	2,400–4,300 lb	2,000 lb (vacuum)

This would mean that the upper stages would be about 1/10th of the vehicle weight – a distinctly inefficient design. The vehicle would have a first stage diameter of 10 ft; the upper stages 54 inch or 4½ ft. It would be nearly 98 ft tall.

But the brochure points up another flaw in the whole project. Different versions of the third stage were to be tailored to three specific missions: a low earth orbit of 300 nm 'for experiments in stellar U.V. spectroscopy', a 300/7000 nm orbit 'enabling investigations of the Earth's radiation belts to be undertaken', and a 300/100,000 nm orbit for a 'Space Probe'. Such imprecise missions do not engender confidence, nor answer the question: what was the point of Black Prince?

However, as part of the announcement of the cancellation of Blue Streak as a weapon in April 1960, stress was laid on its conversion to a satellite launcher. To be cynical, it could be said that this was a useful ploy to deflect criticism; it was argued by Watkinson, the Minister responsible for the cancellation, that the considerable expenditure to date had not been wasted, since Blue Streak was now being used for another purpose. The hollowness of this argument can be demonstrated by the fact that the expenditure to date had been in the order of £60–80 million; another £60 million would be needed to complete the project, and yet no Ministry was prepared to put up more than a few million pounds out of their own budgets. Watkinson, against the advice of his civil servants such as Sir Edward Playfair, was prepared to offer £5 million, but this would have to come out of existing budgets – not an easy option. Similarly, Hailsham was not

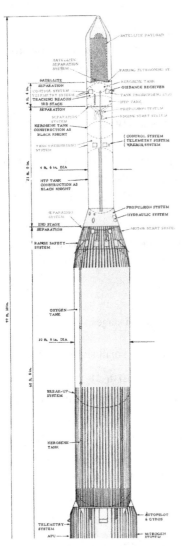

Figure 56. From the Saunders Roe brochure for 'Black Prince'.

prepared to raid the science budget, and there seemed no obvious other source of finance, unless the Treasury were to fund it over and above existing expenditure – which was highly unlikely!

This is shown very clearly in a letter from Hailsham to Sandys (now Minister of Aviation) in April 1960, just after the announcement of the cancellation of Blue Streak as a military weapon.

> The Advisory Council on Scientific Policy said in its last report that to leave vital scientific needs unsatisfied 'in order to shoulder the crippling cost of a large programme of space exploration on a purely national basis would be, in the Council's view, the grossest folly.' At that time, of course, we were going ahead with Blue Streak for military purposes and the Council said 'we do not consider that the technological advantages likely to accrue from the development of a British rocket and satellite programme designed for civil purposes over and above the effects of the existing military missile programme, would be sufficient to justify the very considerable expenditure involved'.

Using vehicles already developed for military purposes was one thing, but for the science budget to take on all the development costs of a launcher would be a very different matter. If Blue Streak had gone ahead as a military missile, then most of the development costs would have been covered by the defence budget, but that was no longer an option. Certainly the cost of the programme could not be justified from a very limited science budget.

There was little military interest too: the Chiefs of Staff, in conjunction with the DRPC stated in December 1959 that 'The Committee agreed that, while satellite research was important nationally, there were not at present any strong reasons to justify the spending of Defence funds on it.'

On the day the announcement of the cancellation was made, CGWL at the Ministry of Aviation (Sir Steuart Mitchell had returned to the post) wrote to all the firms involved – Rolls Royce, de Havilland, Sperry and others – and mentioned: 'The Government is now considering whether to continue Blue Streak in the role of a Satellite Boost Vehicle, but it is unlikely that a decision will be reached on this until about June.' Optimistically, he noted that 'The intention would be to complete development by early 1964 with the firing of not more than 8 rounds, at a rate of approximately 3 per year'. (There was a curious British usage of the time which referred to the launches as 'rounds', rather as if they were shells or bullets.) At a meeting between RAE and Saunders Roe in May, the firing programme was outlined thus:

> The initial 1st stage firing would be planned for October 1961. Thereafter there would be three firings in 1962, three in 1963 and one in 1964. The following more detailed firing breakdown was given: -
> (a) 1st firing – As Blue Streak F1 but simplified ...
> (b) 2nd firing – As above.
> (c) 3rd firing – Plus separation bay and dummy 2nd and 3rd stages.
> (d) 4th and 5th firings – As above.
> (e) In 1963 – 6th firing using live 2nd and 3rd stages plus a small simple satellite.

In the event, not even the first flight had occurred by early 1964!

Later in April Mr Syme of the Ministry of Supply in Melbourne, Australia, noted:

> The programme would involve firing three Blue Streak missiles of current design, three modified ones carrying dummy upper stages, and two complete three stage versions, making eight firings in all, extending from the first firing in mid 61 to the last in early 64 at about four monthly intervals.

And the Ministry of Aviation had similar visions:

> ... [the] first missile on the launcher by April 61, static firing in July and first flight October 1961. Thereafter flights in March, July and October 1962 and 1963, terminating with the eighth flight in March 64.
> Flights one and two would be unmodified Blue Streaks; firings 3, 4, and 5 would carry dummy upper stages; firings 6, 7 and 8 real upper stages. In addition, the Black Knight plus third stage require firing from Area 5 [the Black Knight launch site at Woomera]. Three flights suggested between September 1962 and May 1963.

The latter reference is to separate flight testing for the second stage. Thus the first complete vehicle would be F6 in March 1965.

To further these considerations, a Cabinet committee was set up, chaired by the Cabinet Secretary, Sir Norman Brook. Not surprisingly, he took advice from the Government's Chief Scientist, Sir Solly Zuckerman, who did not have a high opinion of Blue Streak and its technology. A flavour of the advice he gave can be seen in a quote from Brook: 'The main advantage would be to keep a leading position in liquid fuel technology, but this was an obsolescent skill.'

The value of his advice can be gauged from the consideration that derivatives of both the Thor and Atlas rockets, somewhat older than Blue Streak, are still in regular use 40 years after this advice was given, and there are still no large satellite launchers that do not use liquid fuel in at least part of their design. To be fair, however, from the military point of view he was quite correct. Then there was the question of cost: Zuckerman pointed out that 'the whole UK R&D expenditure is £500M [per annum],' of which £238 million was for defence, whereas the launcher would cost £15 million per annum, with a total cost of £64 million. (Much later, in June 1966, Zuckerman was to say of the launch of the first French satellite: 'We could have anticipated and greatly exceeded this achievement had we decided in 1960 to adapt Blue Streak for this purpose and added Black Knight to it as a second stage'. It is a pity he had not been more forceful at the time.)

Senior civil servants are not noted for their extravagant use of language, yet in all the files that survive from the Cabinet Office and the Ministry of Aviation, no enthusiasm for the project ever shines through. Instead, Blue Streak, even in a civilian guise, is seen as an unfortunate inheritance from a military programme, and the very strong impression is that the project is maintained for political reasons rather than through any intrinsic merit.

But to return to money: cost estimates for aerospace projects are a notorious minefield, and the launcher was to prove no exception. Hasty initial estimates gave £35 million.[8] This was solely for vehicle development costs. But there would be other costs as well – the vehicles themselves (around £2 million each) and the satellites that would be carried. This pushed the total estimate up to £64 million by July – although this included in addition the cost of three satellites, which the early estimate did not. An analysis was also done of the unit cost of a launcher, once development was complete, and these estimates were in the region of £1.8–£2.1 million, depending on the number of launches per year.

After the cancellation, a Missile Conversion Committee was set up, and the Ministry of Aviation gave some costings for the programme:

1960/61	–	£31.0 m
1961/62	2 firings	£15.8 m
1963/64	3 firings	£13.4 m
1964/65	3 firings	£9.7 m

If the same programme of work were spread over five years, then:

1960/61	–	£31.4 m
1961/62	2 firings	£12.4 m
1963/64	2 firings	£10.2 m
1964/65	2 firings	£10.2 m
1965/66	2 firings	£9.1 m

In total, the four year programme would cost £69.9m; the four year programme £73.3m.

But, noted the Treasury gloomily,

> ... the Ministry of Aviation ... have very grave doubts about the five year programme on technical grounds. They are not at all sure that a firing every six months is in fact a viable arrangement. [ELDO managed only one launch a year between 1968 and 1971, with one gap lasting 16 months.] It would inevitably mean that many of the expert staff would be twiddling their thumbs for some time between each firing.

Later, the memo summarised the position by saying:

> ... the most that can be said is that the five year programme might put off by a year the (? evil) day when the scientists would be able to come along with their next big leap into space.

The difference in US and UK resources, or the proportion of their resources that they were prepared to devote to space research, is shown up quite starkly in a conversation recorded by Zuckerman in December 1960 with the US Assistant Secretary for War, who suggested that the UK joined the US in space work. 'When I told him our total [annual defence] budget for R and D was £220M, he immediately replied that, even if we put all our defence R and D money in, we were obviously not starters in a US/UK collaboration programme in rocketry and satellites'[9]. If £220 million was not enough, how far would £15million take the UK?

Funding for Blue Streak itself after the cancellation was kept at a 'tick over' rate, and although the Chancellor had initially agreed to provide money for the project, the beady eye of the Treasury was always on it. But this points up another factor: the lack of any firm decision. This was summed up in an internal Ministry of Aviation paper soon after the cancellation:

> The current BLUE STREAK programme will have run down to the level needed for the development of a satellite launcher by about the middle of July. A decision on the future of BLUE STREAK must therefore be taken within the next few weeks.
>
> The possible courses are:
>
> (a) To cancel BLUE STREAK completely.
>
> (b) To undertake a programme of space research based on the joint Anglo/Australian development of BLUE STREAK as a satellite launcher.
>
> (c) To undertake a programme of space research as at (b) and to explore the possibilities of co-operative programmes with European and Commonwealth countries.
>
> (d) To take no final decision on space research, but to continue the development of BLUE STREAK at the minimum level while the possibilities of co-operation with Europe and the Commonwealth are explored.
>
> If course (d) were adopted, then interim measures would be needed to ensure that this country retains the ability to develop a satellite launcher and undertake a space research programme, until the necessary decisions are made. The extension of the current BLUE STREAK contracts on a month-by-month basis could not last for more than 2 or 3 months, since it would be unsatisfactory and uneconomic, and would prevent contractors from entering into the longer term commitments involved in the supply of materials. Contracts would probably have to be extended for at least 12 months. In practice course (d) would be difficult to present to Parliament and the public, particularly if it resulted in the abandonment of space research after a year or so.

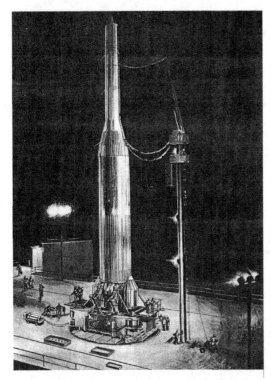

Figure 57. Artist's impression of Black Prince from the Saunders Roe brochure.

But, as we can see, course (d) is what the Government ended up with. Often a lack of any firm decision is worse than any of the decisions that could have been taken. Immediate cancellation would have saved money, pressing on with the project would have produced an end product.

There was no room in the military budget despite Watkinson's offer. There was little money for anything: the Ministry of Aviation had been given £50,000 for the initial design study for a satellite launcher, and they were told by the Treasury in April 1960 that if they want £250,000 for wider studies then they must apply for it. More realistically, Sir William Downey asked for another £35,000 in May. This is hardly an extravagant budget. In addition, money had to be provided for the 'tick over' contracts at the likes of de Havilland and Saunders Roe.

So in May meetings began between RAE, Saunders Roe and Bristol Siddeley to continue the development of the design. Mention was made of an intention to ground launch the upper stages as part of the development programme. Saunders Roe set about the business of designing the second stage in detail and producing a test tank structure. De Havilland began the design of the interstage section, and Bristol Siddeley was keen to press on with the new engine for the third stage.

The RAE took the design very seriously: just as there had been a series of panels for the missile on guidance, propulsion and so on, a series of similar panels were set up for Black Prince. However, in July CGWL wrote a memo decreeing that:

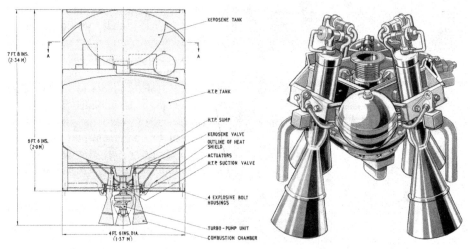

Figure 58. The third stage of the proposed Blue Streak launcher (left), which would have been powered by the PR.38 motor (right).

It has been decided that the proposed 3-stage satellite launching vehicle based on Blue Streak, Black Knight, and a third stage, and the subject of current project design studies, shall be known as the BLUE STREAK SATELLITE LAUNCHER.

The names Black Prince and Blue Star, which have been used semi-officially, are not to be used.

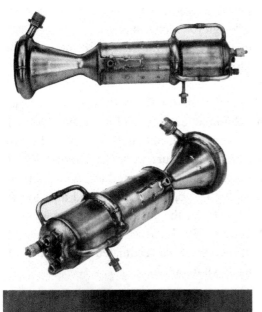

Figure 59. The third stage would have been powered by the PR.38 motor. Here a chamber for that motor is being test fired.

But even so, there was continued Government vacillation on the programme. The then Minister of Aviation, Peter Thorneycroft, had visited both Canada and Australia, and, as part of his mission, had attempted to interest both countries in a Commonwealth satellite launcher.

Canada's reply was that she was already investing enough in space; privately the Canadians told the British that they were spending considerable sums on Alouette, a Canadian satellite to be launched by NASA, and on various sounding rocket programmes. Indeed, they admitted that the total expenditure was far more than publicly acknowledged, and no other commitments were possible. Similarly Australia was not interested other than supporting the work being done at Woomera. With any New Zealand contribution likely to be 'modest', and with South Africa no longer part of the Commonwealth, such a joint programme seemed unrealistic.

However, whilst the Commonwealth expressed no interest, France did. Very soon after the initial cancellation, French officials made various discreet contacts with British diplomats to ask about Blue Streak. However, they were as much aimed at finding out design information for their own ballistic missile project as enquiring about a joint satellite launcher programme.

While the indecision dragged on, RAE and Saunders Roe continued with the development of the BSSLV. On a meeting in December 1960, RAE mentioned that the second stage diameter of 54 inches was producing control problems, and that they would prefer a larger diameter stage. Given existing facilities at

High Down, the greatest diameter possible would have been 58 inches and Saunders Roe were asked to produce a design for such a stage.

The third stage design is shown in Figure 58, with the PR.38 motor on the right. Some test chambers for the motor were constructed and fired, as can be seen in the photographs[10] in Figure 59.

However, the initial performance calculations for Black Prince gave a figure of around 1,750 lb in low Earth orbit. Communications satellites needed to be in a much higher orbit – preferably geosynchronous, although RAE and the Post Office had been interested in 8-hour and 12-hour orbits as alternatives. In addition, Woomera is poorly placed for launching such satellites, with its range restrictions necessary to prevent overflying of populated areas, and also too far south for geosynchronous launches. Estimates of the cost of an equatorial launch site were in the region of £20–30 million. Black Prince's payload decreased rapidly with altitude and with launch direction: it was estimated that for a 2,500 nm orbit at 40°W of N, the payload was down to 600 lb.

A Liquid Hydrogen Stage

One way round this limitation was to use a high energy upper stage, and so at the same meeting in December 1960 Saunders Roe were asked to consider a liquid hydrogen third stage with 4,500 lb of propellants. The brochure they produced was up to their usual high standard, laying down the problems clearly, and describing the solutions equally clearly. This design and that of RPE has been described in the chapter on rocket motors, so no more about the designs themselves need be said here. The cost of developing a liquid hydrogen stage for the BSSLV was put at between £5.5 million and £7 million.

Saunders Roe did hedge their bets somewhat at the instigation of RAE: the design might have been tailored to the BSSLV, but it was also drawn up with the possibility of mounting it on the French second stage which was now being negotiated. The prolonged European negotiations were slowly getting somewhere, although it would take many more months before any clear shape would emerge. At a meeting in March 1961 at Cowes, the RAE had to tell Saunders Roe that it had now been definitely decided that the second stage would be of French design. Saunders Roe and Bristol Siddeley were to continue with the design for the third stage, with particular emphasis on liquid hydrogen. At the same meeting, Saunders Roe reported that manufacture of the second stage tank structural test specimen was about 25% complete. Although it might seem that the work on the HTP second stage had been wasted, it carried through first to the 54-inch Black Knight and then to the second stage of Black Arrow.

In a further Design Study Progress meeting in July, the chairman noted that the situation was very vague (which, given the ELDO negotiations, was probably an understatement). He hoped that Saunders Roe would play a part in any future design studies and that they would continue to maintain a design study team. At the same meeting, Saunders Roe presented a brochure for the HTP/kerosene third stage, which would involve both high thrust and low thrust stages; a 2½ hour period of low thrust was mentioned, and in addition, there was a further report on their liquid hydrogen stage. But this is effectively where work on Black Prince, or BBSLV, comes to a halt. Instead, attention turned to ELDO and Europa.

It might be thought that the creation of ELDO would have finally put the lid on any further thoughts of a BSSLV, but RAE was still doing its best to resurrect the idea. Throughout 1963 and 1964, meetings were being held between RAE and Saunders Roe on 'medium energy upper stages'. This is a euphemism for HTP stages. In a meeting in June 1963, the chairman of the Working Party, H.G.R. Robinson of RAE, stated that 'a study of upper stages for Blue Streak were needed as a back-up of the system studies for the E.L.D.O. launching vehicle, in case the latter was not available, or unsuitable'[10].

The RAE was considering designs for circular equatorial orbits of either 13,740 km or 36,060 km height – i.e. 12-hour or 24-hour orbits. This would need an apogee motor for a final stage. In a follow up meeting, Saunders Roe were told that they would be given design contracts for 'An Investigation on British End Stages in combination with the Blue Streak Launcher Vehicle', and 'in view of possible ELDO

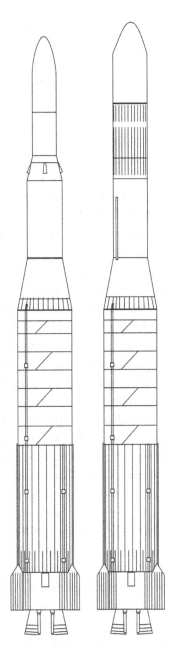

Figure 60. Blue Streak with Black Arrow upper stages (left), compared with Europa (right).

involvement, RAE recommended that the proposed body diameter be 2 metres (i.e. 6.5 ft.)'[11]. Although Saunders Roe duly undertook the design studies as requested, they never went past the paper stage. On the other hand, the link to what would become Black Arrow becomes more obvious.

During 1963 and 1964, the design for what would become Black Arrow was evolving. Amongst all the imperial measurements of feet, inches and pounds is a first stage diameter of 2.0 m! This mixing of units not only seems odd but also would seem to have no connection with Blue Streak. However, the ELDO vehicle was being designed with a French second stage, which was also to be of 2.0 m diameter. The idea was then that if the Blue Streak interstage were to be suitable for the French vehicle, it would also be suitable for Black Arrow. In the event, there was a problem, since the Blue Streak side of the interface was much wider; the French stage was to have a skirt that would mate with the lower stage. So this odd metric feature was included in case, as seemed possible in 1963, ELDO did not go ahead, or, for whatever reason, was not a success.

Figure 60 shows a comparison between Blue Streak with Black Arrow mounted on top and Europa. The 54–inch diameter of the payload shrouds may have been a problem for a conventional satellite, although not for a communications satellite, where the payload might be quite small. The other problem was that Blue Streak might have been struggling to lift the 40,000 lb weight of Black Arrow.

But the design for Europa did materialise, and it might then be thought that all further suggestions for a solely British launcher might have died forever. However, there was one enthusiast for space exploration in the House of Commons, the Conservative MP Neil Marten, who asked a Question in the House about the possibilities of the Blue Streak/Black Knight combination. This might well have been a put up job, but it gave RAE the excuse to work the figures for a Ministerial reply, a task they set to work on with eagerness.[12]

This was in March 1968, when Europa, despite problems with its second stage, still looked viable. Two versions were considered. The first employed the first stage only of Black Arrow, with four of the eight chambers and one of the turbopumps deleted. The payload estimate was for 1,800 lb in a 200 NM polar orbit (750 kg at 500 km); however, extended nozzles on the motors (as in the Black Arrow second stage) resulted in an improvement on this of 'some 20% to 25% giving a performance appreciably greater than ELDO A'. Adding the third stage of Black Arrow – the Waxwing motor – would increase payloads 'by about

200 kg'. This would put the maximum payload up to 1,100 kg. Development costs were put at £1 million, and the unit cost per vehicle at £1.5 million.

A follow-up note gives some interesting comparisons between ELDO A and the Blue Streak/Black Arrow combination.

	Black Arrow Stage	ELDO A 2nd Stage	ELDO A 3rd Stage
Vacuo S.I.	250.5	280.8	292
Burnable Props. (kg)	13256	9800	2498
Mass of stage (kg)	1134	2389	737
Structural efficiency	0.079	0.196	0.200
Ideal velocity*	6244	4489	4610
Ideal velocity**	4636	2366	2714

* from stage itself (m/s)   ** from stage when used as part of multi-stage vehicle

This shows the potential of a Blue Streak/Black Arrow combination. It also shows up very clearly the structural efficiency of Black Arrow as opposed to the ELDO second stage.

Equally interesting are the costings, for what they are worth:

HSD for work on Blue Streak:	£100k
Rolls Royce for work on Black Arrow motor:	£400k
Westland – Black Arrow modifications	£100k
Motor bays	£30k
Ground equipment modifications:	£80k
Modifications at Woomera say	£150k
Total	£960k – say £1000k

[*sic* – total is actually £860k!]

Cost of first flight vehicle: Blue Streak £1000k
Black Arrow £400k – say £1500k total.

To carry out the modifications and launch one test vehicle for £2½ million seems a touch on the optimistic side.

There were, of course certain snags. To do this in parallel with Europa would have been politically unacceptable (but it would have been interesting to see the reactions if the Blue Streak/Black Arrow combination had reached orbit by 1970). There was also the question of payload – but the Perigee Apogee System (PAS) design could have been adapted from ELDO to give a geostationary capability.

At around the same time, Saunders Roe produced their own brochure for a Blue Streak/Black Arrow combination. The skirt to the engine bay was flared out to match the Europa interface, as originally intended. They went through all the permutations with their usual thoroughness, considering eight chamber and four chamber variants, two- and three-stage versions, and also reviving the liquid hydrogen third stage possibility. Their payload calculations were on the optimistic side, however, since they estimated that Blue Streak and Black Arrow together could put as much as 3,000 lb in low earth orbit, and even a few hundred pounds in a geosynchronous orbit.

It is interesting to note that Saunders Roe were obviously thinking of communications satellites as an application for the launcher. Under the payload shroud the satellite is sketched with two solid fuel motors: one which would convert a low Earth circular orbit into a highly elliptical geotransfer orbit, and the second of which would then act as an apogee motor to convert the elliptical orbit into a circular geosynchronous orbit. In this context, it should be noted also that although the US was prepared to sell launchers to other countries, this offer was subject to considerable restriction and would almost certainly not have included commercial communications satellites. Britain's military communications satellites, Skynet, were launched by the US only as a result of the close military ties between the two countries. Thus Britain might have been able to produce a low cost launcher for communication satellites – but not a very powerful one.

However, there was a considerable divergence in views between the establishments and the Ministries. The RAE and its associated establishments were constantly producing ideas based on Blue Streak, yet it was obvious that the Ministry of Technology, as the Ministry of Aviation had then become, was firmly set against any idea of a British-based launcher, and certainly, as already mentioned, it would have been politically unacceptable whilst ELDO was still in existence. It would also have been financially unacceptable as far as the Treasury was concerned.

Blue Streak with a Centaur Upper Stage

Late in the Blue Streak saga, HSD published a brochure which was interesting technically, even if the chances of the British Government being interested in it were remote. The brochure has the look and feel of one put together in a hasty or cursory fashion – all the text is in block capitals – and does not really do justice to the proposal, except in the artwork.[13]

The proposal was for a Blue Streak launcher with an American Centaur D1 upper stage, built in Europe under licence (rather cannily, the brochure says 'Europe' rather than 'Britain'!). Optional French L17 strap-on boosters were proposed as an optional extra. As to the payload, the brochure states:

Performance in geostationary orbit:

- without strap-on boosters	650–700 kg
- with two L17 boosters	870–920 kg
- with four L17 boosters	1000–1050 kg
(grouped in two pairs)	

More tellingly, it goes on to say 'Typical payload ranges are quoted – actual capability for specific payload requires detailed study of optimised trajectory and earliest permissible fairing jettison time' – in other words, the figures quoted are estimates rather than being the result of any precise calculations. They do seem reasonable, and the brochure says 'performance capability is higher than the proposed Europa III' – which was true up to a point. There is a drawback in the sense that the vehicle has been stretched as far as possible, and had really reached the limit of its performance.

The proposed launch site was Kourou in French Guiana, which, like the rest of the proposal was technically feasible but politically completely impractical.

One technical side note: Centaur was the only other rocket stage, apart from Atlas and Blue Streak, built using pressurised stainless steel 'balloon' tanks. Centaur was originally designed to go on top of Atlas, hence the similarity in construction. In that sense, Blue Streak and Centaur were well suited. The Centaur stage had some teething problems, but by 1970 was a well-tested and proven design.

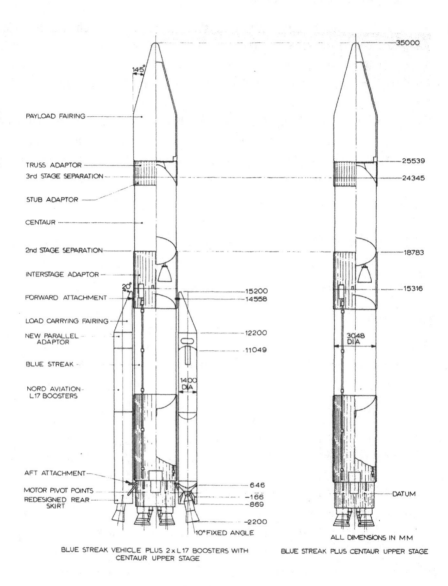

Figure 61. HSD's proposal for mounting the American Centaur stage on Blue Streak.

The other major advantage of the proposal was that by comparison with the ELDO design, and any possible Europa III designs, all the components were flight-proven, and the bugs ironed out. But interesting though the idea might have been, it was a product without a customer. The UK was determined to have nothing more to do with launchers; any new European launcher would be French led, and use of an American stage, even built under licence, would have been a non-starter.

---

[1] TNA: PRO AVIA 65/352. Space research vehicle design and performance: potential of reconnaissance satellites.

[2] TNA: PRO DSIR 23/25364. Preliminary assessment of an earth satellite reconnaissance vehicle (RAE TN GW 393).

[3] TNA: PRO AVIA 6/19852. Technical report G.W.455. The Use of Blue Streak with Black Knight in a Satellite Missile. D.G. King-Hele and Miss D.M. Gilmore. May, 1957.

[4] TNA: PRO AIR 20/10573. Blue Streak: possible use as satellite launcher.

[5] Science Museum Library and Archive

[6] TNA: PRO AVIA 65/351. Space Research Vehicle Design and Performance: British Studies.

[7] Saunders Roe Technical Publication 435.

[8] TNA: PRO AVIA 66/4. Blue Streak as a satellite launcher.

[9] TNA: PRO DEFE 7/1397. UK space research programme. 19 December 1960

[10] RAE Communication Satellite Study. Working Party No. 2 meeting of 10 June 1963. (Science Museum Library and Archive)

[11] RAE Communication Satellite Study. Working Party No. 2 meeting of 17 June 1963. (Science Museum Library and Archive)

[12] TNA: PRO AVIA 13/1351. Space policy: Black Arrow review 1967.

[13] TNA: PRO DSIR 23/39853. Blue Streak/Centaur launcher.

Chapter 9

The European Launcher Development Organisation – ELDO

The political, technical and financial fiasco that would become ELDO grew from an act of political cowardice by the British Government. In an attempt to deflect some of the criticism that he knew would come its way after the Blue Streak cancellation, Watkinson had announced that development would continue as a satellite launcher. This was a disastrous move from several points of view.

Firstly, it deflected very little criticism. Very few people were interested in Blue Streak as a satellite launcher, but they were interested in the effect the cancellation might have on the Government's defence policy. Secondly, the popular enthusiasm for a satellite launcher was small to non-existent. Thirdly, even civilian development was going to cost a great deal of money, and fourthly, there was no demand for a satellite launcher. It looked very much as though it might become a white elephant before it was even built.

One way round the problem was to try and find partners: Thorneycroft had tried the Commonwealth but had received little concrete help. The French, on the other hand, were very interested, as this note from Selwyn Lloyd, then Foreign Secretary, shows:

> The French Ambassador raised with me on the 8th of July the question of Anglo-French co-operation in the development of Blue Streak as a space project. He said that the reply to the French *aide-mémoire* had been communicated to the French Government and that he himself had spoken about the matter to the French Minister of Defence when he had been in Paris.[1]

Thorneycroft was anxious to find partners for the launcher, and so a draft document was quickly put together as the basis of an offer to the French:

1. All firings to be done at Woomera.
2. Offer to divide satellites for initial programme equally between partners, each paying their share of the cost.
3. The French to make the third stage boost.
4. French technicians to be associated with the completion of BLUE STREAK and

BLACK KNIGHT, and to be given all design information and know-how in a return for a payment of £X for five years.[2]

Not everyone was happy with a bilateral deal with the French; Edward Heath, then Lord Privy Seal, wrote to the Minister of Aviation thus:

> Although I realise that the French are more likely than anyone else in Europe to make a useful technical and financial contribution to the development of a European launcher, I do not feel it is politically possible now, having made an approach to so many European countries to turn round and tell the Europeans that we propose to enter into an exclusive bilateral scheme with the French ...
>
> The French may of course try to steer us towards a bilateral scheme: if so, we should have to think again, and very hard: for I see substantial political objections to some exclusive Anglo-French scheme ...[3]

Despite Heath's objection, a party of French engineers visited London on 29 September, moving on to Farnborough on 30, September followed by a much longer four day visit in November, when the itinerary included Hatfield, Ansty, RPE, Spadeadam, Cowes and London.

Thorneycroft summarised the results of the visits in a note to the Prime Minister:

> The French have replied on Blue Streak. Essentially they have said 4 things -
> (a) That they will join us in an approach to other countries in Europe.
> (b) That they remain uncommitted at this stage.
> (c) That they want to make a good slice of the composite rocket themselves.
> (d) That it will all take a long time to arrange.
> (a) is excellent, (b) and (c) are understandable and (d) must be avoided at all costs.[4]

Then it was the turn of the British to go to Paris in January 1961. The record of the meeting begins with the sentence: 'M. Pierratt opened the discussion by stating that the French had received instructions from top level on Wednesday last to work at a joint solution with the British for a European space launcher' ('top level' being interpreted by the British in this context as meaning General de Gaulle himself).

The meeting then went on to discuss the details of the design. The French, at this stage, were leaning towards the idea of a solid fuel second stage deriving from their military programme. The British technical representative, Dr Lyons from RAE, was

> ... disappointed ... not so much that the payload was less, but because it was less flexible in terms of a change in diameter. A 1.5 meter second stage would not produce a good three stage build-up if in later years a Hydrogen third stage was considered. The solid fuel motor would have been a cheaper option, since it was already being developed for the military. The liquid fuelled proposal would use

UDMH and N2O4, although the French admitted when asked what experience they had with these propellants: '... very little indeed. They had fired some engines of a research size for very short durations only'.

The most surprising point about the technical exchanges is how seemingly ill-prepared the French were. The first contacts had taken place more than six months ago, the first technical visits more than three months before, yet obviously no detailed consideration had been given to the design at all. If even such a basic point as whether to go for a solid or liquid fuelled design had not been considered in any depth, then there was a great deal of work still to be done.

There was an interesting coda to the meeting. To quote:

> French said they would try to work out these costs this evening and on Saturday morning. They would have to go carefully into the savings to be made on the military investment and the question of British help in this field was of major importance.
>
> Twinn said we fully recognised this, and in order to be as helpful as possible we would like the French to define in more detail than previously just how we could help with the military programme.[5]

The items were listed in a separate table:

MILITARY INFORMATION EXCHANGE ITEMS

Structure	Methods of manufacture
	Damping
	Stage separation
Control	Fuel sloshing
	Coupling between structure and control: frequencies
Guidance	Comparison between inertia and radio guidance
Equipment	Refrigeration and cooling
Re-Entry Head	Stabilisation, orientation
	Rate of spin, methods of spin control
	Heat flux
	Materials
	Equipments concerned with operating the bomb

These are exactly the questions one might expect, although if the French were intending to produce a solid fuelled missile, some of the items such as sloshing would become redundant. It is unlikely that Britain could have given much help on large solid motors. Some of the other items were ones which had given particular difficulty to the British only two or three years before – inertial guidance and re-entry in particular.

Sir Steuart Mitchell's comment on the re-entry head read as follows:

The design of re-entry head which we finally ended up with for Blue Streak is:-
(a) Of British origin.
(b) It is now joint UK/US information.
(c) It is agreed by the US to be much better than their designs as regards
    invulnerability and US has now copied it.
(d) As regards invulnerability it is so advanced that neither the US nor ourselves can
    conceive a counter to it.[6]

Writing to Solly Zuckermann in a memo entitled 'Possible Transfer to French Government of Military Technical Information on Blue Streak', he also notes that:

Re-entry head.
Radar Echo. Information on this is mostly Top Secret and would be of great value to the French. The most advanced work in this field is British and is acknowledged by the US to be ahead of their work. It is thought that future US warheads may be based on this British work.
Release of this information would be contrary to I and II of para 3 in that it could provide an enemy with a ballistic weapon against which we see no defence and it would prejudice American weapons. It is desired to draw particular attention to this point and it is recommended that this information should not be released.
To provide a line of defence on which any technical conversations might be conducted it is suggested that we take the line that -
Details of shape, weight, dimensions, etc. of the Blue Streak re-entry head cannot be discussed as they contain "atomic" information.
Decoys. Information on these would contravene I and II of para 3 above. This is a sensitive Top Secret field in which we are well ahead of the USA who accordingly would be apprehensive if we released information to the French.[7]

He also made the point that '... the re-entry head design is highly specific to the weapon parameters ...'

It is clear, however, that the French regarded the military information as something of a *quid pro quo*:

... they wished to make a political point of associating the exchange of military information with the cost for the space launcher, and they wished to make the inter-dependence clear at Ministerial level by presenting both cost and military exchange papers to Mr Thorneycroft at Strasbourg.[8]

The British delegation was less than happy with this idea: 'The representation at Strasbourg was not the channel at which we would prefer to deal with this.'

Mitchell wrote a minute ('French Proposals for 2nd and 3rd Stage') for the Minister, summing up the position to date. An interesting comment was that 'since August [1960] no approach to UK firms to start design of the second and third stages has been permitted, partly to avoid compromising our negotiating position in Europe'. Using a French second stage would increase development time, and CGWL felt that this 'now gives enough time to develop a liquid hydrogen 3rd stage'.

He also had these comments to make on the French proposals:

> ... if the French chose a liquid motor for the 2nd stage, and if they followed their present lines of development, the performance of their 2nd stage would again be appreciably lower than that of Black Knight.
>
> This is due to the fact that the French have not developed high performance turbo pump fuel systems on UK or USA lines and, for their weapon development, are not prepared to face up to their technical complexities. They intend to use the cruder method of gas pressurisation of the fuel tanks as a means of pumping the fuel. The resultant penalty in tankage weight is considerable.
>
> As a result of the above, the conclusions as to French 2nd stage performance are as follows:
>
> The French agree that there would be a loss of performance, but argue that it would not be great ... We, with much more experience, consider that the penalty would be considerable ...[9]

There was a way round this performance loss: replacing the planned HTP/kerosene third stage with one using liquid hydrogen. His suggestion was:

> Participation by other European countries in the Space Club is essential. Hence I suggest that development of the liquid hydrogen 3rd Stage should be offered to a consortium of European countries with some UK technical participation in the development teams.

His final conclusion was that

> We are in favour of proceeding with a French 2nd stage and a European 3rd stage, recognising that by so doing the completion date for the European launcher will be delayed by perhaps 1½ years and that the total costs may rise by perhaps 10–15%.

There followed some very rapid writing of proposals which would be put to other European countries. The end result was a long brochure, describing the vehicle and its missions in considerable detail[10]. The introduction provides a useful summary of the history of the project to that date:

> After the British Prime Minister's Statement in May 1959 that an investigation would be made of the adaptation of British rockets for satellite launching, extensive studies of the capabilities of Blue Streak, in combination with other rocket stages, have been made by the United Kingdom Ministry of Aviation at the Royal Aircraft Establishment, Farnborough.

The later proposal that a satellite launching vehicle system based on a Blue Streak as a first stage should be developed as a joint European and Commonwealth effort, has recently caused these studies to be extended by joint Anglo-French investigations into a design incorporating a French second stage.

The original British proposals were put to representatives of European nations at Church House, Westminster, London on the 9th and 10th January, 1961. At Strasbourg, during the week of 30th January to 3rd February, a preliminary description of a joint Anglo-French proposal was presented for the consideration of representatives of a number of European nations.

One of the guiding principles of the United Kingdom studies was the minimisation of cost, particularly capital cost, and thus the greatest possible use should be made of existing equipment and facilities, including the rocket ground testing and development facilities at Hatfield and Spadeadam, in England, and also the launching and other range facilities at the Weapons Research Establishment, Woomera, Australia.

France, on her side, has undertaken an extensive national programme of basic studies and development of ballistic missiles. The French proposal for a second stage, later to be described, is closely related to this programme in order, again, to minimise cost, delay and technical uncertainty.

This brochure contains outlines of the jointly proposed satellite launching vehicle and its systems as they stand at February, 1961. The opportunity has been taken since the Strasbourg meeting to bring the proposals into accord with the latest technical information. The assessment work which will lead to a full design study is by no means complete, depending as it does considerably on the parts of the work to be undertaken by the European nations involved. All aspects of the combination of the French second stage with Blue Streak have not yet been completely examined. The brochure, therefore, contains the joint Anglo-French proposals as far as they have gone, and where the necessary work has not been completed, the parallel work done on the original British configuration has been referred to. In the absence so far of an alternative proposal for the third stage, the third stage described is the original UK proposal.

The brochure went on to describe the capabilities of the launcher:

(i) A large satellite weighing between one and two thousand pounds in a near circular, near earth, orbit. This satellite would be space-stabilised with a primary purpose of making astronomical observations above the earth's atmosphere.

(ii) A smaller satellite of several hundred pounds weight, moving in an eccentric orbit out to two or three earth radii, for the investigation of the earth's gravitational, magnetic, and radiation fields, and the constitution of the earth's outer atmosphere.

(iii) A satellite of the order of one hundred pounds weight, in a highly eccentric orbit reaching out to about 100,000 miles, to carry instruments for the study of the sun's atmosphere.

These aims have been subsequently extended to cover the possible launching needs for Satellite Communication Systems and this has led to the consideration, in addition, of circular orbits at several thousand miles altitude.

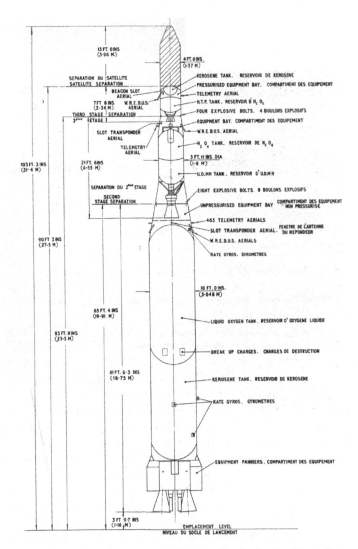

Figure 62. The Anglo-French proposal. This is effectively the Black Prince design with a French second stage.

The first three objectives are taken from the Saunders Roe brochure for Black Prince, published a year previously. It does highlight an absurdity of the programme: £60 million for three satellites does seem excessive. The communications requirement is new, and the Saunders Roe liquid hydrogen stage was optimised for just such a role. The main problem was that even 5,000–6,000 miles was still too low an orbit for communication satellites. RAE and others tried looking at 8-hour or 12-hour orbits, but it is only the geostationary orbit which is of any practical use.

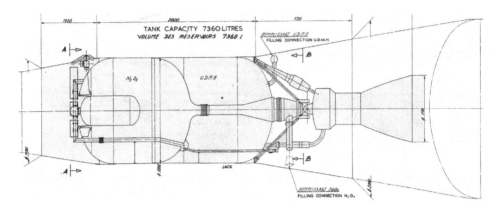

Figure 63. The French proposal for the second stage (the final version would be very similar, except that the one large chamber would be replaced by four smaller ones).

The brochure then went on to describe the vehicle in more detail:

> The original British proposal for the second stage was to use a modified form of the ballistic research vehicle Black Knight. This has now been replaced by a proposed French second stage making use of techniques currently under development in that country. This stage will be propelled by a liquid propellant engine using Nitrogen Tetroxide and UDMH with, a sea level thrust of 25 tons (32 tons vacuum) and vacuum Specific Impulse of 276 seconds. The vehicle tanks contain approximately 7 tons of propellants and are pressurised by means of a solid propellant gas generator. The single thrust chamber is gimbal mounted for control in pitch and yaw. Roll control is achieved by means of auxiliary jets mounted at the top of the vehicle ...
>
> Studies indicate that it is possible to inject a satellite into orbit using the proposed two stage combination but this would necessitate a long coasting period after perhaps 90% of the second stage propellants had been burnt, followed by a relight of the second stage engine to inject both satellite and empty second stage into orbit. This approach introduces problems of relighting the engines under zero acceleration as well as the necessity for ensuring correct orientation of the second stage at engine relight.
>
> Though such problems have been solved in other satellite launchings, the two stage vehicle would give considerably reduced payloads and would be unable to put any payload into higher orbits. The preferred approach is therefore to introduce a small third stage rocket. This is sometimes referred to as a vernier stage. The engine of this stage, working at a relatively low thrust level of between 1000 lb and 2000 lb would be started during separation from the second stage and would continue to burn through what would otherwise be the coasting period, cutting off when orbital altitude and velocity had been achieved. The low weight of the third stage structure and engines, compared with that of the relit second stage, affords considerable improvement in payload weight into low orbits and makes possible the injection of payloads into very high orbits.

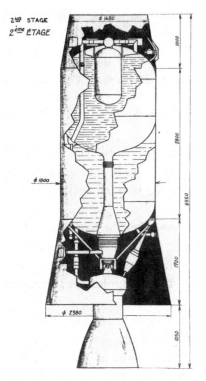

Figure 64. Another view of the proposed French second stage.

The British proposal for a third stage engine is a four chamber design, each chamber pivoted about one axis for steering. It would use hydrogen peroxide and kerosene. With low thrust and four chambers, very high nozzle expansion ratios, of 1000 : 1, are possible without undue chamber size and length ...

It is possible to meet the several orbital requirements by exchanging satellite payload weight for propellent weight in the third stage whilst maintaining constant the overall weight of the third stage plus satellite payload at some 5000 lb; that is, the third stage incremental velocity can be increased at the expense of payload. The tank volume is altered to suit the orbital mission allowing the remainder of the third stage, including the engine, and all equipment, to remain sensibly unchanged ...

For the configuration just described, with a take-off thrust of 300,000 lb weight a satellite of 2,160 lb may be put into a 300 mile circular polar orbit.

Corresponding payloads for elliptical polar orbits, both with perigee heights of 300 miles, and apogee height of 7000 and 100,000 miles, are respectively 910 lb and 320 lb.

For a typical high altitude equatorial orbit (launched near the equator) at, say, 5000 miles altitude; a payload weight of 700 lb is calculated.

These are 'nominal' payloads making some allowances for weight growth of the launching vehicle. It would be prudent, however, to assume that actual payloads would be perhaps 200 lb less than these nominal values to allow for unforeseen contingencies.

The negotiations were not easy. Enthusiasm for the project in Europe was very limited. Indeed, in May 1961, Thorneycroft asked Mitchell what needed to be done to go ahead with an all-British launcher, and received a reply saying that it would be quite straightforward with the original HTP design, with or without the liquid hydrogen stage. Even Australia seemed to be making difficulties, and Thorneycroft took the unusual step of writing to Mitchell to ask whether it was feasible to launch Blue Streak from Spadeadam!

The conclusion to his hastily written paper (the full version can be found in Appendix A):

Spadeadam is technically both feasible and attractive. From the cost point of view, it is approximately the same as Woomera, and is much cheaper than any alternative.

It must be accepted, however, that some cut-downs on to UK territory would inevitably occur if we fire from Spadeadam. The chance of serious damage to life and property from such cut-downs are numerically small.

The risk of damage to foreign countries, or to shipping, is negligible.

The crucial point is the political acceptability of the risk in the UK Hitherto this has been regarded as unacceptable, and it would be no less now than when previously considered. My advice is that the risk is appreciable and should not be accepted.[11]

As Mitchell says, the crucial point is political acceptability. The thought of launching a rocket as large as Europa from an inland site in Britain is one which should fill any politician with horror. The repercussions from an accident would be horrendous.

There is also another technical point. Mitchell describes the launch direction as 'North 15 East', or 015° in modern parlance. To be restricted in launch direction in this fashion very much reduces the value of the site (and this also applied to Woomera). Different satellites fulfilling different roles need different orbits. It certainly would be useless for communication satellites.

Fortunately, agreement was reached with the Australians, and Woomera would indeed become the launch site for the first ten launches.

Although several European countries sent delegates, most of Thorneycroft's efforts were devoted to persuading the German and Italian Governments to join the project. Both countries were reluctant; the Italians wanting to reserve their money for their own national programme. Belgium and the Netherlands were willing to participate, but their contributions would be small. Denmark had taken part in the discussions but decided in the end not to join, but again any Danish contribution would not have been very significant.

After further protracted negotiations, Germany agreed to join, and would build the third stage; the Italians would provide the satellite fairings and the Satellite Test Vehicle (STV). Thus the final membership of ELDO consisted of the UK, France, Germany, Italy, Belgium and the Netherlands, with Australia making the seventh member.

The cost of the programme was split thus:

Britain:	38.8%
France:	23.9%
West Germany:	22.0%
Italy:	9.8%
Belgium:	2.9%
Netherlands:	2.6%

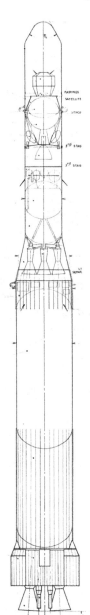

Figure 65. The inital design for the ELDO launcher.

Australia would make no direct contribution, but would instead develop the Woomera launch site.

ELDO came into formal being in March 1962 by a Convention which was signed by the seven Governments and which came into force on 29 February 1964 after ratification by the signatory states. The headquarters were in Paris, and it was governed by a Council that had two representatives for each member state. The Council was assisted by an International Secretariat under the direction of a Secretary General, with two Deputy Secretaries General, one in charge of technical affairs and the other of administrative affairs. The staff of the Secretariat amounted to around 180 people in 1965.

But while design work for the new launcher had started, ELDO itself was already running into serious political trouble. Indeed, it would spend most of its existence staggering from crisis to crisis, either technical, financial or political.

By 1964, the design of the vehicle had finally been decided[12], and work was beginning on the design and construction of the upper stages. The French then dropped something of a bombshell by stating ELDO A was inadequate, that it should be dropped, and that the organisation's efforts should be directed towards a new launcher, ELDO B.

There were immediate objections from the other member states, mainly on the grounds that an entirely new upper stage would be technically demanding and take several more years to develop, whilst in the meantime, nothing else would be happening. Blue Streak had already been successfully tested, and work was proceeding on F4, which was Blue Streak with dummy upper stages. Under the French proposal, there would be no further launches for some years until the new upper stages had been designed and developed. To be fair to the French, if ever there was a time to go for a design that was far more capable, this was it, but given that it had already taken four years to get to the point of beginning work on ELDO A, the reluctance of the other countries was understandable.

But the French took their objection to ELDO A one step further: they refused to provide any further funding. This did produce quite a serious crisis: without an agreed budget, all work would grind to a halt by the middle of 1965. Negotiations with the French proved difficult: the British representative referred to what he called 'decisions handed down from Mount Olympus' – in other words, a decision taken on high, and presumably a reference to General de Gaulle, which the French ELDO representatives could do little about. One junior minister, Austen Albu, described the situation thus: 'Whatever the merits of the case we are in fact being blackmailed by the French.'[13]

The British Government had by now become actively hostile to ELDO, and there were hopes that French intransigence might bring about the collapse of the organisation.

> From the economic point of view, the safest course would still appear to be to decline any further financial obligations beyond our share of the original £70 million on ELDO A, to which we are already committed. It has certainly not been demonstrated that a firm stand on these lines will involve serious dangers to those policies on which it is really important that we should have our neighbours' support [referring to other members of ELDO]. Such action might indeed gain us enhanced respect in the more responsible sections of our neighbours' administrations...
>
> If however the feeling of Minister's colleagues is against such risk of friction to neighbourly relations as a firm stand might involve, the next best course would be to take the line that on present evidence Britain
>
> is not prepared to depart from ELDO A as originally conceived,
>
> is unwilling to proceed to completion of the ELDO A programme until more adequately costed,
>
> as regards ELDO B no commitment could be considered until much more information was available.
>
> There are many who consider that if Britain takes a position along these lines ELDO will die a natural death, without Britain having to plunge the dagger. First Secretary, however, will appreciate that such a happy outcome cannot be guaranteed and that the more moderate course must carry the risk of a lingering British involvement in these unrewarding activities.[14]

There were attempts at a compromise. One was to proceed with what was called ELDO A(1+3), to keep the programme going whilst work began on ELDO B. This was a proposal to put the German third stage on top of Blue Streak – hence the (1+3) designation. This, it was thought, could put 300 kg into a 500 km orbit. Use of an apogee motor would enable payloads to be put into highly elliptical orbits, which might suit some of the proposed European Space Research Organisation (ESRO) requirements. Given that the German stage was the least well developed part of Europa, this too was somewhat optimistic.

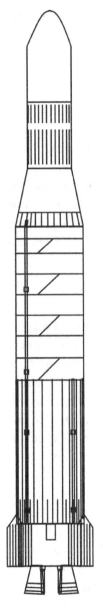

An ELDO document[15] described the proposal thus:

> The 1+3 programme would provide for development of basic techniques, establishment of facilities, and experience by personnel as a foundation for the ELDO B programme including proof of the first stage and engines; development of throttled engines, live-stage separation; instrumentation, safety, nose-fairing and STV separation, and inertial guidance. This is work which can only be carried out in a vehicle based on Blue Streak.
>
> The studies so far undertaken, necessarily limited by time, give the Secretariat good grounds of assurance that the programme is technically feasible.
>
> The payload performance of the 1+3 vehicle is strongly dependent on the 3rd stage performance and empty mass. Its round to round variation will be somewhat smaller than that of the original three-stage vehicle. For a 300 km orbit, the upper limit of payload performance is about 500 kg ...
>
> This payload performance would have application to:
>
> a) missions requiring light satellites, e.g. for navigation, meteorology or geodesy,
>
> b) ESRO requirements for the launch of small satellites, i.e. those within the launching capacity of Thor Delta.

It was a proposal that died together with ELDO B. A 500 kg payload is quite respectable, but whether it would be worth using a launcher as expensive as the (1+3) scheme is debatable. A sketch of the proposed vehicle is shown on the left.

The (1+3) programme was intended to run in parallel with the ELDO B development, but ELDO B was abandoned as a result of the Intergovernmental Conference in July 1966. Instead, a new five year programme was drawn up, starting in January 1967 at an estimated cost of 331 MMU. (1 MMU was effectively the same as 1 US dollar, so at the then rate of exchange this was a little less than £120 million.) 10 MMU were set aside for 'studies and experimental' work – ELDO B was not entirely dead yet.

Figure 66. The '1+3' proposal.

On the other hand, the rejection of ELDO B left the organisation with a vehicle that had very little purpose. In order to salvage something from the wreck, the Perigee Apogee System (PAS) was put forward. This consisted of two solid fuel motors and a communications satellite. The system would be put into orbit by Europa, then the first solid fuel motor would be fired to put the satellite into a highly elliptical geosynchronous transfer orbit. The apogee motor would convert the elliptical orbit into a circular orbit.

A geosynchronous orbit required a launch site close to the equator – and Woomera was too far south. The launch corridors from Woomera were very restricted by the centres of population below the flight path. ELDO set about finding an alternative site, and the two main contenders were Kourou or Darwin, and, as we shall see, Kourou was chosen.

So a new launch site for Blue Streak was built in the depths of the South American jungle. The last launch of Europa from Woomera was F9, after which Australia left the organisation. A non-flight model Blue Streak, known as DG, was taken out to Kourou to test the facilities. F11 (there was no F10) would be the first launch from South America, the first with the PAS operational, carrying a communications satellite for France, and the last ever launch of Blue Streak and Europa.

The Political Failure of ELDO

But in parallel with this crisis, another was developing; this time within the British Government. ELDO had been set up by the Conservative Government under Macmillan, and Douglas Home, Macmillan's successor, was not in office long enough to bring about any major policy changes. The new Labour Government under Harold Wilson took a very different view of the organisation, aided by the Civil Service, who had always been opposed to ELDO, and saw their chance to cancel it. The Treasury memo on space to the new Chancellor, Jim Callaghan, in 1964 is an interesting read (it is reproduced in its entirety in Appendix A)[16].

Part of Wilson's rhetoric at the 1964 General Election had revolved around the idea that the Conservative Government had been out of date and out of touch, as opposed to a more dynamic Labour Party. The phrase 'the white heat of the technological revolution' is attributed to him after his speech at the Labour Party Conference at Scarborough in October 1963. Like many such catch phrases, he did not say it quite in this form; it has been slightly paraphrased. (His actual words were 'In all our plans for the future, we are re-defining and we are re-stating our Socialism in terms of the scientific revolution. But that revolution cannot become a reality unless we are prepared to make far-reaching changes in economic and social attitudes which permeate our whole system of society. The Britain that is going to be forged in the white heat of this revolution will be no place for restrictive practices or for outdated methods on either side of industry.')

Wilson felt that the Conservatives had committed themselves to some extremely expensive technological programmes such as Concorde, TSR 2, ELDO, and so on, and that the scientists, engineers and technicians involved should instead be working in private industry, helping to produce up-to-date goods for both the domestic and export markets. A number of aviation projects were cancelled in 1965, but Concorde and ELDO proved to be more difficult – they were international projects, and the Government had signed treaties which were hard to break. If the British Government were to cancel the projects, then they would be repudiating the treaties and be liable for damages – and the cost of the damages might well soak up any saving (although it is probable that the Government ended up spending far more on both Concorde and ELDO than it would have paid in damages as a result of cancellation in 1964).

Having discovered that pulling out of ELDO might be more trouble than it was worth, the British began employing other tactics. One was a demand for a reduction in its share of the budget. The argument being used here was that other countries were benefiting from developing new technologies, whereas developing Blue Streak was fairly routine work and nothing new was being learned.

Another tactic was to become excessively legalistic as to the nature of the work being carried out – whether it was part of the 'original programme' or not. The British Government had signed up to the programme as agreed in the original convention, and nothing else. Any change to the programme – for example, the perigee/apogee system – could then be opposed on the basis that it was not in the original agreement. These problems became more acute as costs rose, and new budgets had to be negotiated. Finally, the British Government effectively withdrew on the basis that it was interested in developing the technology that went into the satellites rather than the launchers. This withdrawal was *de facto* rather than *de jure*, as we shall see.

An example of the British attitude can be seen in a memo concerning the French and ELDO B:

> If ministers accept the Chief Secretary's view that the UK should not participate in the ELDO programme as proposed by the Minister of Aviation, it will be important to handle this in such a way as to minimise political repercussions. I do not doubt that if the UK delegate were to stand up at the beginning of the conference and announce crudely that the UK is to withdraw from the organisation, there would be an unfortunate reaction among other members. But as stated earlier, the French have themselves called the whole future of the Organisation into question by insisting that its programme should be radically recast and that until this is done the French financial contribution will be restricted. It ought to be possible to take advantage of

this to throw most of the onus for the collapse of the organisation on to the French. One need not say in terms that the UK regards ELDO as undesirable. All that would be necessary to say, as I see it would be that the UK are not prepared to depart from the concept of ELDO A as originally conceived and that they are not even willing to proceed to completion of this programme until it has been more adequately costed. As for ELDO B they could not begin to consider a commitment in principle on the basis of the inadequate information about the cost, technical validity and economic prospects of the project so far available. This, one hopes, should suffice to bring about the demise of ELDO.[17]

Such behaviour was also guaranteed to irritate Britain's partners. It was Britain, after all, that had worked so hard persuading these countries to join ELDO, and now, halfway through the programme, it was Britain that was working to destroy the organisation. There is an interesting letter on the subject in the ELDO archives:

It is unrealistic and wasteful to attack either the British decision of April 1968 not to contribute to the ELDO overrun and not to participate in post 1971 rocket development, and the decision of the Four [that is, the four nations remaining in ELDO: France, Germany, the Netherlands and Belgium] to make sure that Europe possesses rockets for putting into orbit both near and geostationary satellites. What is at issue is purely the British contributions for 1969, 1970 and 1971 of a total value of about £M17.

In April, Mr. Wedgwood Benn announced the decision mentioned above, but stated that of course the UK would carry out its commitments as agreed in 1966 and earlier, evidently referring to these £M17. It seems, however, that the UK expected the immediate collapse of ELDO following the British announcement with dissolution liabilities, the British share of which could at most come to £M10 after 1968. This expectation may also have been partly responsible for the hesitations expressed about the supply of Blue Streak.

By the time of the Bonn Conference it had become clear that the Four wished ELDO to continue. This led on the one hand to British assurances on the supply of Blue Streak which were accepted by the others, and to the proposal made by Wedgwood Benn at Bonn that if the British liability to ELDO was reduced (the figure of £M17 was not mentioned then, but subsequently, especially at the ELDO Council on 29 November) the UK would put that much and more into application satellites, reversing the UK decision of April on this issue. Although the UK was helpful at Bonn on several issues (use of launchers, unifications), acceptance of this protocol was made a necessary condition for lifting the reservation the UK had put on this and several other points.

In fact the British proposal was never seriously considered because the others
(i) saw no sense in an applications programme without a launcher;
(ii) if the UK switched resources from the technically uninteresting production of Blue Streak to application satellites, it would benefit her and no-one else;
(iii) the others were not in a hurry on applications;

(iv) they thought the UK was keen on applications anyway.

In these circumstances, the UK, on December 18, 1968, in a letter to the Ministers, released herself unilaterally from the commitment by regarding ELDO's austerity plan T9 as different from what agreed in 1966. This pretext, although possibly justified on the narrowest legal basis, shocked the others by its patent conflict with earlier British statements. They fear the effect of this unilateralism as a precedent and certainly are asking whether such legal devices could be used equally in any other technologically risky long-term programme. The very basis of European technological co-operation has been undermined by this step through its fundamental shaking of confidence. The UK's fitness as partner for any future enterprise is now being questioned even by her closest friends. Note that in this painful development there has never been any advice on the £M10 first presumably evaluated in April and now offered as a present, as no commitment is now said to exist.

All European technological co-operation in space, and possibly elsewhere, will be ruined by this destruction of trust. The severity of the step does not seen to have been understood in the UK, and is of course totally divorced from the merits or otherwise of ELDO.[18]

And even as ELDO was falling apart, the French Government was still pressing Britain on its participation, to which a memo from within the Department of Trade and Industry commented that:

The fact remains that there is little to be gained by the ELDO Secretariat or by the other ELDO members from making a fuss to keep us in the organisation. Legally we can argue the toss. Politically we can point up the logic of our position. And financially the organisation will be no worse off by our departure.[19]

In other words, the Government had washed its hands of the organisation, and there was very little in practical terms that the other countries could do. ELDO was finally wound up in 1972, and the British Government has never participated in any part of the Ariane project that followed.

Those in the ELDO Secretariat were well aware of its weaknesses, but had their hands tied. An exposition of the situation was given to the Royal Aeronautical Society in February 1968 by Dr Iserland of ELDO:

The difference between the technological task of ELDO and a political, economic or scientific task of other organisations, showed up from the start: time is a prime factor in technological achievements and, in particular, in space missiles.

When the Convention was signed in 1962, it was decided, therefore, not to wait until its ratification by the Governments, which took place only in 1964. To enable work to be started immediately, a Preparatory group was instituted as a part-time body with the responsibility of specifying and co-ordinating the work and preparing for the functioning of the Organisation on entry into force of the Convention. During this period, each member country started the work under its own authority and at its own financial risk by placing the contracts. To avoid discontinuity, the Convention also stipulated that after ratification, the authority for the contracts would remain with the Governments for the Initial Programme and that direct contracts between ELDO and the firms would necessitate the consent of the member state.

Paradoxically, this laudable desire for speed to start the work, characteristic of the technical nature of the enterprise, resulted, after 1964, in a factor slowing down unnecessarily the speed of progress. Since ELDO did not place the contracts itself, it was not vested in the authority of the 'overlord' which is essential in carrying out efficiently an industrial programme.

Strictly speaking, with the kind of organisation imposed for the Initial Programme, the executive lines for co-ordination and management of any part of the development programme of the Europa I launcher were as follows: if the central technical group in the Paris headquarters, known as the Secretariat, found it necessary to define or to specify some technical requirement, it would have to approach the appropriate ministry of the country in which the equipment was built: if this ministry accepted the need for it, it would pass the recommendation to the establishment which was entrusted with the supervision of the national work on behalf of the Government – in the UK this would have been Farnborough. This establishment would then specify the technical requirements to one or two different firms.

This already long process of imposing in 'open-loop' an already chosen solution is still relatively straightforward, compared with the process of agreeing on a technical solution where information had to go up and down this long ladder several times simultaneously in one or two different countries, first to find a technical solution and then to implement it. Needless to say, this strict formal line could not always be followed and technical features often had to be by ELDO representatives with some representatives from specialised establishments or industrial firms. However, a pragmatic practice which does not follow the agreed formal lines of control and financial authority can obviously lead to confusion at the risk of offsetting thereby, the advantage of the direct approach. The alternative, to avoid too long information lines, was to agree on a solution in meetings or Working Groups. Since ELDO had no authority to arbitrate a solution (not being the 'overlord'), a kind of unanimity rule had to be followed, which consisted of convincing the representatives of each country and firm of the correctness of the suggested decision or to find a compromise, which was not necessarily technically optimum. There then still remained the long channel for implementation through the various steps mentioned before.

It will easily be imagined what difficulties are encountered in co-ordinating by such means, for example, the definition of a common electrical circuitry throughout the three-stage launcher.

In some instances, the frightening complexity of this type of co-ordination had a direct influence on the choice of some technical solutions. Here is one example:
When it had to be decided whether a central sequencer for all flight events should be adopted for the complete vehicle or else individual sequencers for each stage, relaying each other after exchange of signals at the cut-off of one stage to initiate the start of the next one, it was judged that only the solution with individual stage

sequencers had any chance of practical achievement, the central sequencer being outside the possibilities of such a remote co-ordination set-up in view of the numerous events, intimately related to details of each stage, which it would have to control. Without judging which solution is superior on strictly technical grounds, it can certainly be stated that the partial failures in our last two launches have some relation to that choice insofar as the light-up of the second stage did not occur in both cases because of incorrect functioning of the second stage sequencer, while the first stage operated with its own sequencer. Perhaps with a central system, we might either not have launched at all or else had correct sequences throughout the two-stage flight ...

The first critical period developed when ELDO presented the budget for 1965. The new estimate of cost to completion of the Initial Programme was for approximately 400 million MU [MU = Monetary Unit, effectively equivalent to 1 US dollar] i.e., twice the original amount estimated in 1961 before the creation of ELDO. Consultations between the member countries became necessary according to the Convention. They took place early in 1965. France suggested to stop the development of the Europa I launcher and to proceed directly to the development of a more advanced and more powerful launcher with upper stages based on liquid hydrogen/oxygen.

The critical period lasted for about three months, during which an extensive study of the French proposal was made, before it was decided to continue the development of Europa I. The effect of this period can still be felt, as it slowed down the work and resulted in delays; delays which amounted to considerably more than the period of uncertainty itself. For the consultations of 1965, ELDO had prepared proposals for follow-up programmes after the Initial Programme. Decisions on these later programmes were, however, postponed by the member countries until 1966. This fact did not help to speed up the work on the Initial Programme after the crisis had been resolved.

Consultations between member countries resumed in 1966, but this time at ministerial level. It was now the turn of British Government to express doubts about the technical and economic validity of the Europa I launcher and to be concerned about the increasing costs. A second critical period began, and it took three sessions of the Ministerial Conference from April to July 1966 to resolve the crisis.

The comments about the flight sequencers (electronic systems that produce the signals to initiate events during flight) are interesting: 'only the solution with individual stage sequencers had any chance of practical achievement'. Certainly the failures of F5 through to F8 can be put down to exactly that cause: the flight sequencers sending the wrong command at the wrong time, which might not have happened had there been one sequencer for the complete vehicle. It is a considerable indictment of the organisation that it was forced to adopt an engineering compromise which could well have led to the loss of five consecutive launches.

The attitude persisted as far down as the individual launch teams. Alan Bond (later to be designer of the UK spaceplane HOTOL) was the Rolls Royce performance engineer sent from England to monitor the engine performance on the F8 round, and has this to say about his experiences:

> The Rolls Royce team at Woomera was under the very capable management of John Bowles. The insular nature of the various teams was striking from the start, not only internationally but also between the vehicle and propulsion teams from the UK.
>
> I am not implying any animosity, there certainly was not any. In fact there was a palpable sense of doing something very important which everyone was very proud of. But there were cultural barriers to communication, except through the regular management channels. In the whole four months of the campaign, the conversations I had with members from the French, German and Italian teams could be counted on the fingers of one hand.
>
> This was in complete contrast to the experience I had seven years later as part of the JET fusion research project where the integration of the international teams was very close. JET went on to be a world beating success and a demonstration of what collaboration can achieve.[20]

But there were other failures in ELDO, within the organisation itself. ELDO was very much a political construct, designed to cope with all the wheeling and dealing that went on in a multinational organisation such as this. The failure lay on the technical side.

Each country provided its own part of the vehicle, and acted independently. Thus the British set up Blue Streak as the first stage, and then the French would come along and add their stage on top, then the Germans would come with their stage, and finally the Italians would fit the payload and payload shrouds. There was no one in overall technical command. The Secretariat could only make recommendations to member states, with exhortations such as these:

> Following the F7 trial [F7 being the seventh flight], the Secretariat tried to inculcate a greater awareness of the need for better technical discipline and control of operations during a trial. Meetings and discussions took place with Member States on the subject of inspection and defect reporting in particular. During the F8 trial, some improvement was obvious, but it is still apparent that these disciplines are not accepted as having the importance attached to them which the Secretariat would wish. The supply and control of spares was also still far from satisfactory in the upper stage areas.[21]

Despite the exhortations, matters did not improve, as the report on the failure of the eleventh launch, F11, shows:

> Two main points provide the basis for the failure of the project.
> – the poor organisation of the management system as a whole;

– the technical difficulties of the third stage and its equipment.

The management system established since the beginnings of the Organisation has proved its ineffectiveness.

There exists a certain confusion about the respective roles of the national agencies, the Secretariat, and industry. With regard to the internal structure of the Secretariat, levels of authority are not sufficiently clear. Some firms are badly organised and have not shown a sense of responsibility. Finally, political problems have too often taken precedence over the technical problems and cost-effectiveness of the project.

In these conditions, the Secretariat was unable to play its proper piloting role, which resulted in an unflightworthy launcher and the abnormally high cost of the programme.

Without going into detail, the main technical problems lie in the third stage. Its design is complicated and its wiring needs to be thoroughly revised. Its integration has been particularly deficient. Three major systems in this stage have net been qualified: the sequencer, the middle skirt separation system, and the guidance computer. The latter, moreover, which is a prototype product, is not flightworthy.

To guarantee an adequate level of reliability, it is necessary:

– to achieve by appropriate tests the integration of all the on-board electrical systems of the third stage and to demonstrate their electromagnetic compatibility;

–   to reorganise the Secretariat in order to transform it into an efficient management tool and provide it with unquestionable technical competence, so that it may play its proper role in discussions with industry.

– to rationalise the Secretariat industrial arrangements to enable a satisfactory solution of the interface and integration problems.[22]

F11 had been launched in November 1971. Ten years after the initial Anglo-French proposals, after eleven launches and literally hundreds of millions of pounds, the vehicle was still not, in the words of the report, flightworthy. Even so, ELDO still hoped to continue with Europa:

Following the failure of F11, the ELDO Council set up a EUROPA II Project Review Commission on 18th November 1971.

This Commission's terms of reference were to propose corrections to the programme from both the technical and organisational points of view and to indicate the consequences of these corrections for future launchings.

The aims that the Commission sought to achieve were the following:

– to determine the technical, administrative and financial conditions for ensuring a substantial probability of success for the next EUROPA II launch within reasonable time limits, or to conclude that this is not possible;

– to propose a fresh target plan for launchers, launches and payloads from F12 to F16 inclusive.[23]

But the Germans failed to make much progress on the redesign of the third stage. The launch of F12 was put back until October 1973 (the F12 Blue Streak arrived at Kourou in April), but it soon became apparent that ELDO was going nowhere, and in May 1972, the F12 launch was cancelled, Europa II abandoned, and ELDO was wound up at the end of the month.

---

[1] TNA: PRO AVIA 66/4. Blue Streak as a satellite launcher.

[2] Ibid.

[3] Ibid.

[4] Ibid.

[5] TNA: PRO AVIA 65/1708. Blue Streak satellite launcher development: European co-operation.

[6] TNA: PRO AVIA 66/7. Blue Streak satellite launcher project: Pt B.

[7] Ibid.

[8] TNA: PRO AVIA 65/1708. Blue Streak satellite launcher development: European co-operation.

[9] TNA: PRO AVIA 66/4. Blue Streak as a satellite launcher.

[10] Anglo-French technical proposals for the development of a satellite launching vehicle system / Propositions techniques Franco-Britanniques pour la mise au point d'un porteur de satellite; document elabore par le Ministere de l'aviation du Royaume-uni et le Ministere de l'air francais.

[11] TNA: PRO AVIA 66/8. Blue Streak satellite launcher project: Pt C.

[12] Historical Archives of European Union, Florence, ELDO 118. Technical Definition of the Initial launching System. October 1963.

[13] TNA: PRO EW 25/52. Review of UK space policy including European Launcher Development Organisation (ELDO).

[14] Ibid.

[15] HAEU, Florence, ELDO 828. Alternative Development Programme.

[16] TNA: PRO T 225/2765. Ministry of Aviation space programme: future policy.

[17] TNA: PRO EW 25/52. Review of UK space policy including European Launcher Development Organisation (ELDO). FR Barratt, 26 March 1965.

[18] HAEU, Florence, ELDO archives.

[19] TNA: PRO AVIA 92/259. UK withdrawal from European Launcher Development Organisation (ELDO).

[20] Private communication.

[21] HAEU, Florence, ELDO archives.

[22] Ibid.

[23] Ibid.

Chapter 10

Europa

Europa was a three-stage vehicle (the Perigee Apogee System of Europa II was sometimes claimed as a fourth stage), the first stage being a modified Blue Streak, which had an overall length of 60 ft 4 inches (18.4 m) and a diameter of 10 ft (3.05 m). The take-off mass was 197,500 lb (89.4 tonnes), which included 126,000 lb (57 tonnes) of liquid oxygen and 56,300 lb (25.5 tonnes) of kerosene fuel. At the first stage engine cut-off, the mass remaining, including residual propellants and tank pressurisation gas, was 15,200 lb (6.8 tonnes).

The liquid oxygen was held in the upper tank, which was a pressurised shell without stringers or frames, constructed of welded stainless steel sections with a minimum thickness of 0.019 inch (0.48 mm). The kerosene tank was also of stainless steel but strengthened with internal frames and stiffeners and external longitudinal stringers. The two tanks were separated by a deep domed diaphragm built up of welded segments, and similar domes closed off the top and bottom ends of the tank assembly. The liquid oxygen tank was fitted with anti-slosh baffles.

The propulsion bay was attached to the tanks by a short cylindrical skirt structure which was designed to transfer the thrust loads to the tanks above. In the propulsion bay were two thrust beams from which the rocket engines were gimballed, and the ends of these beams, as well as transmitting the thrust loads, acted as support points for the vehicle in the launcher stand. The bay was of light alloy construction, and housed the combustion chambers, the propellant pump assemblies, the engine control systems, as well as the hydraulic pumps and servo-jacks for swivelling the combustion chambers. The bay was closed at the rear end by a heat shield to keep out exhaust gases and radiant heat from the rocket exhausts. Two panniers were placed on either side of the propulsion bay to house more equipment and instrumentation.

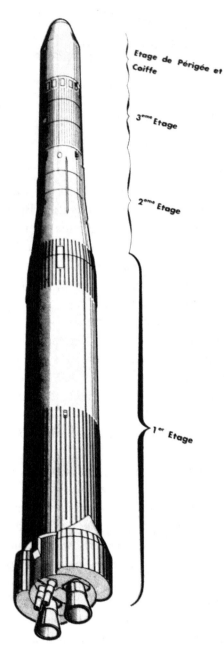

Etage de Périgée et
Coiffe

3ᵉᵐᵉ Etage

2ᵉᵐᵉ Etage

1ᵉʳ Etage

Figure 67. ELDO A or Europa.

There was initially considerable technical discussion – and controversy – as to the best way to design the tanks, which were the major structural element, and the one where most weight could be saved. It was realised that the high internal pressure – about two atmospheres – required to avoid cavitation at the propellant turbopump inlets could be used to stiffen up a very light structure so that it could take heavy longitudinal and bending loads. A pressure of 30 psi over a 10 ft diameter circle represents a total force of 340,000 lb, which was quite adequate to accelerate upper stages of around 35,000 lb to an acceleration of some 7–8 *g*.

The liquid oxygen tank was put forward of the kerosene tank in order to bring the centre of gravity as far forward as possible with respect to the centre of pressure, so as to ease the task of the control system. A lighter kerosene tank structure could be obtained if some of the end loading were taken by stringers and frames, since the mass of the liquid oxygen tank under acceleration would need excessively high pressures in the tank.

The choice of tank material, too, was the subject of considerable study, both light alloy and stainless steel being close contenders. It was essential for the tank material to have good strength at liquid oxygen temperature (–183 °C), to be able to withstand kinetic heating, to be readily and efficiently weldable, and be available in uniform sheets of closely controlled thickness. Stainless steel was eventually chosen, made by Firth

Vickers for railway coaches, and with very good physical properties over the range of temperatures required. Individual cylindrical sections were prefabricated from 36 inch wide strips of material.

The two Rolls Royce RZ 2 rocket engines each had a rated thrust at launch of 150,000 lb, giving a take-off acceleration of 1.3 g. The original thrust rating of the RZ 2 engine was 137,000 lb, and this was used for the single stage flights. The higher rating, obtained by an increase in combustion chamber pressure, was used in the later flights.

The turbines were driven by hot gases produced by fuel rich combustion of a small proportion of the main propellants in separate gas generators, and two turbopump units supplied the propellants to the regeneratively cooled thrust chambers. The bright yellow plumes produced from the gas generators can be seen very clearly as the vehicle lifts off. The total propellant flow rate was 1,210 lb/s for the two motors. At high altitudes, the thrust of an engine increases, due both to an improvement in specific impulse brought about by a more efficient expansion of the exhaust gases, and to increases in propellant flow rates as the hydrostatic heads at the propellant pumps rise with vehicle acceleration. Sea level S.I. was around 248, vacuum S.I. was around 285.

The main function of the pressurisation system was to provide pressure in the liquid oxygen tank to preserve its structural integrity, to provide sufficient pressure at the pump inlets to suppress cavitation, and to control boiling of the liquid oxygen. The pressurisation system operated in two principal modes: 'standby', with the tanks full of propellants, the pressure in the liquid oxygen tank was maintained at about 9 lb/in² and about 5 lb/in² in the kerosene tank; and 'flight', when the pressures were raised to 30 and 15 lb/in² respectively, starting about one minute before lift-off. The 'standby' pressures were chosen to provide adequate stability of the vehicle under conditions of ground loading, winds, etc., and the liquid oxygen in its tank would be in a state of equilibrium boiling. For 'flight', a low-pressure vent valve on the liquid oxygen tank was closed, and the combined effects of atmospheric heat input and of nitrogen supply from the pneumatic control unit raised the pressure to flight level in about 30 seconds. During the raising of pressure in the liquid oxygen tank, the pressure in the kerosene tank was raised so that the differential pressure across the intertank diaphragm was maintained at a safe value. The engines were then started. During flight, internal supplies of gas were needed so that the pumps did not cavitate or the tanks collapse. This was done with two heat exchangers or evaporators, one of which used the waste heat in the exhaust gases from one turbine to provide gaseous oxygen for pressurising the liquid oxygen tank, using a feed from the liquid oxygen pump discharge. The other, associated with the other engine, evaporated liquid nitrogen supplied from a storage bottle on the vehicle to pressurise the kerosene tank.

One distinguishing feature of the flight models of Blue Streak (and, later, Black Arrow) was the spiral painted around the vehicle, which can be seen in the photograph on the right. This was for the benefit of the cameras filming the launch, and was the brainchild of an Australian mathematician, Mary Whitehead. In an interview some years later, she had this to say of the pattern:

> The zig-zag pattern on the Blue Streak – that was at my request, because we were required to know whether the missile rolled as it went off [the launcher]. There were, I think, three or four cameras – they were called launcher high speed cameras – around [the launch site], so with that pattern, if the rocket rolled you could measure it really easily, depending on where that diagonal was relevant to the top and bottom stripes.
>
> I had seen the black and white checks that were used on other missiles. I think that might have been in America, and this was the same sort of thing. But making a continuous line like that, you could measure to a degree whether or not it had rolled: as far as I know, I don't think they did.

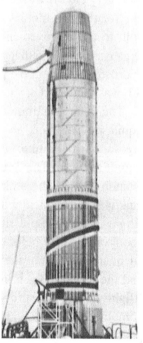

The interstage skirt was a stainless steel corrugated truncated cone, 1.23 m high, tapering from 2.8 m to 2.0 m, which weighed 137 kg. A small solid fuel motor was used to separate the two stages: it provided a thrust of 40 kN for 2 seconds.

The design of the second stage, also called Coralie, was constrained by several factors. Its size was set by the weight that Blue Streak could carry, and Blue Streak had been designed as a ballistic missile, not as the first stage of a satellite launcher. The 2 m (approximately 78 inches) diameter was certainly wider than the planned 54-inch Black Knight stage, which had the advantage that the vehicle was less flexible. The original brochure design had only one motor: the four motor design made control easier, particularly in the roll mode, and also meant that the interstage structure could be shorter and thus lighter.

Figure 68. The characteristic spiral pattern used on flight models of Blue Streak and Black Arrow.

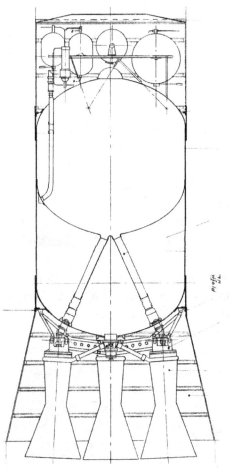

Figure 69. The French second stage of Europa, also known as Coralie.

The stage had a mass of 1,850 kg, of which 9,850 kg were fuel (100 kg of this was left at burn out) – 6,440 kg of dinitrogen tetroxide and 3,410 kg of UDMH. The chambers were pressure-fed from a gas generator, which used 41 kg of dinitrogen tetroxide together with 18.5 kg of UDMH and 98 kg of water, the water being used to lower the temperature of the gases to 350 °C. The pressure in the tanks was maintained at 18.6 bar (the combustion chamber pressure was 13.7 bar). Each chamber had a thrust of 68.1 kN (15,300lb) and a burn time of 75 seconds.

There is no doubt that choice of pressurised tanks did make the stage structurally rather inefficient. The choice was made in part to save time – the French had had little experience with turbopumps. A gas generator was lighter than simply using compressed helium gas, and a similar system had been in use on the Veronique sounding rocket. In addition, the same gases were used to drive the steering mechanism for the chambers and to generate the on-board electrical power supply.

Attempts were made to test the stage separately, with a vehicle called Cora[1]. The Cora 1 version used only the French Coralie stage, while the Cora 2 version added the German Astris stage. The Italian Europa nose fairing was also tested. For these tests the rocket nozzles were shortened to allow sea-level operation and four fins were added for stability. In the event, only the Cora 1 version was tested, and this failed on two out of three attempts.

The interstage structure was a cylinder 1.23 m long with a mass of 137 kg. Separation of the third stage was by explosive bolts, which caused problems in some of the early flights.

The German third stage, also known as Astris, was powered by dinitrogen tetroxide and Aerozine 50, an equal mixture of hydrazine and UDMH. This was stored in a spherical titanium tank 1.2 mm thick. The helium gas was stored in two bottles made of glass re-inforced plastic pressurised to 300 bar. The main motor had a thrust of 22.5 kN and two small control chambers of 500 N thrust (in vacuum). Since the chambers were to run in vacuum, the pressure in the chambers was a relatively low 11 bar. For a small vacuum stage where a restart capability could be useful, a pressure feed had advantages over turbopumps. The all-up mass was 3,370 kg (7,420 lb) and the empty mass was 610 kg (1,340 lb).

Germany had, of course, been the world leader in rocketry up to 1945, but the diaspora of rocket scientists and technicians, coupled with the perceived militaristic nature of rocketry development, meant that no significant work had been done since. On the other hand, there are advantages in starting again from scratch. Some of the initial German ideas for the third stage were very ambitious, and design studies for a high energy stage were outlined by Dietrich Koelle of Bölkow at a European Spaceflight Symposium in May 1963[2]. The paper mentioned that 'Since 1961 intensive studies have been carried out at Bölkow Entwicklungen concerning an optimised high energy stage OPMOS'.

The three designs were 'based on the propellant combinations H2/O2 and H2/F2 pressure-fed, and H2/O2 pump-fed. The results of the studies indicate that the payload capacity of the ELDO launch vehicle can be increased by the introduction of a high energy upper stage from the present 220 lb to some 1,550 lb for escape missions.' The studies were not merely theoretical: 'Small hydrogen cooled engines have been on test for some time at the firm of Bölkow'.

Figure 70. German proposal for a high energy third stage.

The motor chamber was also distinctly unusual. It would provide 8,650 lb thrust, and used an Expansion–Deflection nozzle, where the flow is expanded radially and then turned axially. It has the advantages that it is a good deal shorter than the conventional rocket chamber, as seen in the artist's representation above, and also allows for a greater expansion of the exhaust gases.

Calling the PAS a fourth stage was perhaps overdoing things slightly, but it was designed to enable Europa to put a satellite into geostationary orbit. It was described thus by J. Nouaille, who was the Project Management Director, in 1968:

> ... A lower metallic skirt attaches the stage to the top of the Europa I third stage. At the upper part of this skirt a separation mechanism releases the PAS system from the third stage when appropriate.
>
> Four small rockets fixed on top of the perigee motor, approximately in the plane of the centre of gravity of the PAS system, spin up the vehicle, as soon as separated from the third stage, to a velocity of 120 rpm.
>
> The perigee motor is then ignited, giving the vehicle an increment in velocity of approximately 2,450 m/s.
>
> The main characteristics of this motor are:

empty mass	70 kg	mass of propellant	685 kg
specific impulse	278 s	maximum pressure	61 bar
maximum acceleration	10.3g		

> After burnout of the motor, a second separation occurs between the empty spinning perigee stage and the spinning satellite.
>
> During the whole of its operation, the perigee stage is controlled and monitored by electronic equipments attached inside the upper skirt of the stage (equipment bay).
>
> Apogee Motor.
>
> The apogee motor is part of the satellite. It is ignited by a telecommand order at the most appropriate moment, taking into account the actual orbit and attitude data of the satellite.
>
> Its main characteristics are:

empty mass	36 kg	propellant mass	156 kg
specific impulse	270 s	maximum pressure	42 bars
velocity increment	1470 m/s	approximately.	

> The launching procedure of the ELDO-PAS satellite ... includes
>
> (a) an injection from Guiana into a parking orbit by the three lower stages of the launcher: after burn-out of the third stage, the attitude of the vehicle remains controlled by a cold gas jet system installed on the third stage. The platform of the Inertial Guidance system is used as a reference during this phase. Ground stations in Kourou (Guiana), Fortalezza (Brazil) and Brazzaville (Congo) are used to monitor the vehicle.

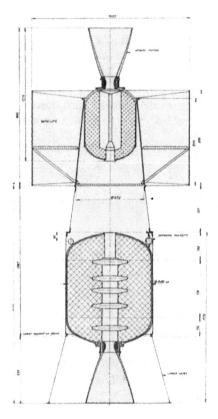

Figure 71. The Perigee Apogee System, or PAS.

(b) When crossing the Equator, approximately at the longitude of Brazzaville, an accurate orientation of the PAS system is achieved, the PAS assembly is separated from the third stage, and then, immediately, the perigee motor is ignited, burnt out and separated.

(c) The satellite is spinning on a transfer orbit about 300 km perigee and 35,000 km apogee. An accurate tracking is obtained from the Gove station (Australia) and the main data on this orbit and on the attitude of the satellite are collected.

These data are transmitted to the main Control Centre in Darmstadt. The optimum time of ignition of the apogee motor is then calculated. Ignition will take place somewhere over the Atlantic, either at the second apogee of the transfer orbit or more probably at the fourth apogee.

The Launches

Each flight model of Blue Streak was numbered F1, F2, F3, and so on. The first three launches – F1, F2 and F3 – were to be just of Blue Streak itself, with a dummy nosecone weighted to simulate a payload. F4 and F5 had all dummy upper stages; F6/1 and F6/2 had live first and second stages; from F7 onwards all stages were live.

F1

F1 reached Australia on 18 January 1964. It was set up in the launch gantry, and was static fired (that is, the tanks filled and the engines ignited, but the vehicle remained tethered to the ground and not released) on 30 April. The weather caused delays to the launch throughout May, and after other delays, the launch date was set for Friday, 5 June.

A report prepared by HGR Robinson of RAE states:

> The vehicle was successfully launched at 9.11 a.m. after an extremely smooth and efficient final count down, both as regards vehicle and range… The vehicle lifted off

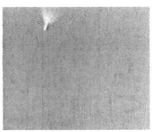

Figure 72. Here the rocket can be seen to be tumbling, as the fuel sloshing in the nearly empty tanks caused the autopilot to lose control.

and programmed downrange according to plan, its flight path and walking impact point following closely to nominal. At about 130 seconds, however, telemetry records indicated the commencement of incipient instability. This became marked at 140 seconds, developing into an uncontrolled corkscrew at 145 seconds. At 147.5 seconds the engine ceased thrusting, some six seconds before the planned time for engine cut. The termination of powered flight has been diagnosed as arising from fuel starvation caused by the manoeuvres of the vehicle during its final period of instability.[3]

The problem lay in what was called 'fuel sloshing' – that is, the vibrations of the vehicle caused the remaining fuel in the tanks to slosh from side to side. As the 'slosh' built up, the control system was unable to cope, and the vehicle corkscrewed then tumbled. It was not a difficult problem to solve – the control system could be adapted to cope, and in any case, the vibrations would be different when the upper stages were added. Figure 72 shows three frames from the film of the flight showing the last few seconds.

Velocity at engine cut:	9625 ft/second
Height at engine cut:	38.9 n.m.
Impact range:	548 n.m.
Impact time:	850 seconds
Apogee height:	85 n.m.
(n.m. = nautical miles)	

F2 and F3

Repeats of F1, F2 was launched on 20 October 1964 and F3 on 22 March 1965. Both flights were extremely successful, meeting all objectives.

Modification of the autopilot reduced the sloshing on F2 to a low and stable value; additional anti-sloshing baffles were installed in the liquid oxygen tank for F3.

F4

This was a simulation of the complete vehicle, but with dummy upper stages, and the first launch to have the motors uprated to the full 150,000 lb thrust. F4 was launched on 24 May 1966 with a planned first stage boost duration of about 144.3 seconds and with the cut-off was to have been by exhaustion of the liquid oxygen. The flight was terminated after 135 seconds by the Range Safety Officer when it appeared that the vehicle was straying outside the range boundaries. This was a somewhat controversial decision, particularly when it was found that the vehicle had been inside limits; the range officer had acted on false tracking data caused by large cross-polarisation of the tracking transponder signal.

F5

F5 was a repeat of F4, and was launched on 15 Nov 1966. The flight was a success.

F6/1

Launched on 4 August 1967, the first and second stages were live, with a dummy third stage and satellite. The first stage performance was as planned, but the explosive bolts of the first/second stage separation system fired prematurely and the second stage failed to ignite. This was thought to be caused by an electrical fault which caused the second stage electronic sequencer to be reset. This meant the command to open the main valves was not given and the motors did not fire even though the main tanks had been pressurised by the gas generator.[4]

F6/2

This was a repeat of F6/1, launched on 5 December 1967. Again, the first stage performance was as planned, but this time the first and second stages failed to separate.

F7

The RAE report on F7 was as follows:

> This vehicle was launched on the 30th November, 1968 … The most important defect during the trial was the complete failure of the 3rd stage immediately after separation from the 2nd stage. Final assessment has been unable to establish the cause of the failure but it has highlighted three areas which may have been either singularly or in combination responsible for the failure. These are firstly the

pressurisation pipes which were of rigid construction. These may have fractured and for F8 a flexible element is included. Secondly, unscheduled operation of the break up system due to spikes appearing in the signal from the WREBUS system in the second stage; filters are being fitted to F8. Thirdly, the failure could have been occasioned by a rupture of the tank diaphragm which separates the two propellant liquids. This diaphragm may have been weakened during the preparation phase, and it appears that this is the most likely cause of the failure.

F8

A paper in the ELDO archives has this to say about F8:

> Following the F7 trial, the Secretariat tried to inculcate a greater awareness of the need for better technical discipline and control of operations during a trial. Meetings and discussions took place with Member States on the subject of inspection and defect reporting in particular. During the F8 trial, some improvement was obvious, but it is still apparent that these disciplines are not accepted as having the importance attached to them which the Secretariat would wish. The supply and control of spares was also still far from satisfactory in the upper stage areas.[5]

F8 was launched on 3 July 1969. Both the first and second stages functioned correctly, but after the signal was sent to separate the third stage, it appeared to explode. The RAE report suggests that the failure was identical to that of F7, and was not a mechanical malfunction but an electrical malfunction.

F9

The subsequent RAE report describes the flight thus:

> On the 12 June 1970 the vehicle was launched at 10.40 am local time ... The first stage functioned correctly as predicted in the flight plan, and the second stage separated and performed as predicted. The third stage separated from the second stage and its engines ignited correctly. After engine ignition occurred the third stage helium tanks lost pressure progressively which caused the third stage engines to lose thrust and also to give intermittent thrusting. These factors gave rise to uncovering of the fuel depletion sensors and a premature engine shut down before all the propellants were used up and before orbital velocity was achieved. The satellite did in fact separate correctly from the third stage when the engine cut off signal was given.
>
> A second major fault which occurred during the flight of the vehicle was the non-jettisoning of the satellite fairings during second stage thrusting. This fault occurred due to the unscheduled separation of a plug and socket connection between the third stage and the satellite. This plug and socket was in the circuit which should have carried the command signal to ignite the fairing jettisoning device; the continuity of this plug and socket was monitored and a disconnect was registered at +78 seconds.

The failure to achieve orbit was a combination of these two faults. Post flight calculations show that an orbit would have been achieved by the satellite even with the under-performance of the third stage had the fairings been jettisoned. On the other hand had the third stage performed correctly the complete third stage and satellite with fairings attached should have acquired orbital velocity.

A later report pinpoints the cause of the plug failure:

Investigation has shown that upon assembly of the connector, air was sealed into the cylinder at 1 atmosphere by two toroidal seals on the piston. Upon reaching a less dense atmosphere during flight, the differential pressure was sufficient to operate the piston and to separate the plug and socket. The device operated correctly in F7 and F8 because the cylinder and piston were dismantled several times before final assembly for flight. This had the effect of slightly damaging the toroidal seals and allowing a slight air leakage to occur.

There was also a problem with the pressurisation of the third stages tanks, meaning that the thrust in the last part of the flight became irregular.

F10

For budgetary reasons, there was no F10.

F11

F11 was the first and only flight from Kourou in South America. It was launched on 5 November 1971.

During the first stage burn, the vehicle went out of control and broke up due to a failure of the electronic guidance mounted at the top of the third stage. As the vehicle accelerated, air resistance caused the temperature of the fairings to rise, and at the same time, an electrostatic charge built up on the fairings. Air at low pressure and a high temperature can conduct relatively easily; there was a discharge from the fairings to the main third stage body which disrupted the electronic systems, leading to a loss in control.

F12 and F13 were never launched: the Europa programme was abandoned on 27 April 1973. Blue Streak never flew again.

Eleven Blue Streaks were launched: F1 to F9 (there were two F6s, F6/1 and F6/2) and F11.

F12 is at French Guiana, or parts of it are. The stainless steel tanks (which would not corrode in the equatorial heat and rain) are being used as a chicken coop.

F13 is at the Deutsches Museum, Munich.

F14 is at the Aircraft Museum at East Lothian, outside Edinburgh.

F15 is at the Euro Space Centre, Redu, Belgium.

F16 was not finally completed (and is now on display at the Space Museum at Leicester).

F17 and F18: by the time of the final cancellation these vehicles were only parts, and not fully assembled.

In addition to these vehicles, several non-flight prototypes were built. These included D1 to D4, some of which were for trials at Hatfield, others were taken to Spadeadam for static firings. Another, designated DA, was shipped to Australia before the flight vehicles, and set up on the launch site for static testing. This was to test the Woomera site and give experience to the Australian team. DB was static fired at Spadeadam to check the engines. In addition, there was a further prototype vehicle, DG, used to prove the Blue Streak launch site at Kourou, in French Guiana.

Improving Europa

The first geosynchronous communications satellite was Syncom 2, in July 1963 (since the orbit was inclined to the equator, it was not, strictly speaking, geostationary). The first geostationary communications satellite was Syncom 3, launched in August 1964, and used to transmit the 1964 Olympics in Tokyo to the United States.

It was the French who, as early as 1964, realised the limitations of Europa and proposed that it should be dropped in favour of a more powerful design. To be fair, the design had been put together without much of a rationale behind the overall concept other than to produce a satellite launcher, and there had been no clear idea of what satellites it might be launching. Although Europa might be able to put a respectable payload into low earth orbit, there was simply no demand in Europe for such a capability. On the other hand, the French were quick to see the possibilities in communications satellites (unlike the British, who remained sceptical for a long time). ELDO A's capacity for geostationary orbit was minimal or non-existent.

The French Government's proposal was to replace the planned upper stages of ELDO A with what were usually referred to as 'high energy stages' – in other words, liquid hydrogen. An ELDO review paper of 1964 shows the thinking behind these designs [translated from the original French by the author]:

> By 1963, the Secretariat had requested proposals from the member states for the design of launchers using upper stages with high energy propellants.
>
> The feasibility studies were placed by the Secretariat in November 1963. They consisted essentially of two types of launchers:

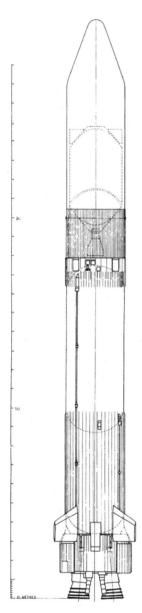

Figure 73. Original
proposal for ELDO B1.
Note the extra fins for
aerodynamic stability.

– a launcher ELDO-B whose design used as first stage the Blue Streak of the Initial Programme. This launcher should be able to place at least 2 tonnes in low orbit and, with an apogee motor, at least 500kg in a synchronous orbit.

– a launcher ELDO-C which had been defined by its performance (6 to 10 tonnes in low orbit – 1.7 tonnes to escape velocity).

The reports corresponding to these studies have been studied by the Secretariat. Work complementary to these studies on these launchers has also been undertaken.

As a consequence, a general line of conduct can now be shown, which is based on the following principles:

1) using liquid hydrogen/oxygen gives advantages which can be seen from many points of view (high performance at a relatively modest cost, experience already acquired, function proved by launchers in the USA etc.)

2) an economic way of using this propulsion in different high energy stages by using one motor which can be grouped together and most importantly can follow the dimensions of the stages in question ...

Whatever the choice of definitive designs for the launcher ELDO B, it appears that it is now possible to envisage beginning development work by the start of 1965.

This would involve ... the development of a motor: it seems that a motor of thrust of 6 to 7 tonnes would be optimal ... It is therefore the intention of the Secretariat to place in very general terms, within the body of the budget of the Future Programme 1964, three contracts for the study of such a motor; the development could begin, after the choice of the best design, at the start of 1965.[6]

One of the contracts issued was to Rolls Royce, who then developed the RZ 20 motor, intended for the upper stages of ELDO B.[7]

There were in effect two proposed designs which were closely linked: B1 and B2. The third design, ELDO C, was much sketchier.

B1 would have one upper stage, B2 would have two. Each of the stages would use the same rocket motor – as mentioned above, with a thrust of 6–7 tonnes (60–70 kN, or around 14,000 lb). By comparison, the American Centaur stage, first attempted launch 1962, had two motors each of 15,000 lb thrust. The B1 stage would have one of these motors, the B2 stage would have four such motors, and the B1 stage would be the third stage of the B2 design.

One problem was the large volume which liquid hydrogen occupies. This meant that the relatively light B1 stage was almost the same length of the first stage, and to assist aerodynamic stability at low speeds just after lift-off, it was proposed to fit four fins onto the kerosene tank of Blue Streak, although this idea only appears in early sketches for ELDO B. Another was that the maximum weight Blue Streak could carry in the form of upper stages was limited by the first stage thrust to 16,100 kg, and the extra weight on top of the vehicle meant that the stresses were greater so that thicker gauge stainless steel would be needed, making the first stage heavier.

ELDO B2 meant pushing Blue Streak to its limit – and perhaps beyond! The weight of the upper stages meant that the launcher would only be possible if the RZ 2 motor were uprated from 150,000 lb thrust to 165,000 lb, and the resultant vehicle would have a lift off acceleration of only 0.25 *g*. This implies a lift-off weight of the order of 265,000 lb. It would also mean strengthening the liquid oxygen tank by increasing the skin thickness from 0.6 mm to 1.5 mm, with a weight penalty of 0.85 tons. (Another solution was, of course, the strap-on boosters already described.)

The French failed to gather enough support from the other member states to switch the programme from Europa to ELDO B: at the Intergovernmental Conference in July 1966, a five-year revised programme for ELDO A and PAS vehicles was agreed; the B1 and B2 launchers were effectively abandoned at the same conference.

A little money was still available for 'studies'. In the following years, different firms produced a plethora of designs for an ELDO B launcher, some not much more than sketches, others fully worked out designs, but as a consequence of the intransigence of the British Government, none of these were to come to fruition. However, they were to appear once again when studies began for a Europa III.

For a variety of reasons, opposition to ELDO B was led by the UK. Firstly, the new Wilson Government were against ELDO in itself; the last thing they and the Treasury wanted was an open-ended commitment to further spending. Secondly, it did not share the French view that the future lay in communications satellites, and moreover, even if it did, it was not convinced that there would be sufficient demand for a second, competing system. Thirdly, it did not believe that it would be 'economic' in the sense that the United States could always undercut ELDO in price. (There is a further discussion about this in the section about the Treasury.)

There was a fourth factor too: the UK had no inherent objection to using American launchers (the early military Skynet satellites were launched using Delta and Titan launchers, although later ones have used Ariane 4 and Ariane 5). The French, on the other hand, were very resistant to the idea of relying on the United States. In particular, they were wary of the restrictions that might be placed on competing commercial systems.

There was one other requirement if Europa was to be used to launch communications satellites: a new launch site. Firstly, whatever site was chosen would have to be as close to the equator as possible – unless the satellite is in an equatorial orbit, it will appear to 'wander' north and south in the sky. Secondly, communications satellites need to be launched in an easterly direction to take advantage of the spin of the Earth. The difference this makes is quite considerable: the speed of rotation of the Earth at the equator is close to half a kilometre a second. A clear launch corridor is also needed: the first stage of the satellite launcher will fall to the ground not that far from the launch site, and if the flight has to be terminated for any reason, or the launcher explodes, then the debris cannot be allowed to fall on populated areas. Woomera had very restricted launch corridors, which were mainly in a northerly direction. As a range for testing rockets and missiles, Woomera was ideal, but it had considerable geographical limitations as a satellite launching base.

Effectively then the site had to be on the eastern seaboard of a large ocean, where spent stages could fall without hazard (except to passing shipping!). ELDO considered several sites, although only two were serious contenders. One was the Australian proposal of Darwin in northern Australia, and the other was the French proposal of Kourou in French Guiana, an overseas *région* and *département* of France located in South America. The choice fell on Kourou, which is now used for the Ariane launchers. Darwin was ruled out on various counts. One was that it was technically inadequate (this seems rather odd – the infrastructure at Kourou cannot have been very advanced at that stage), from the safety angle as being too heavily populated, and cost – which again seems a little odd. Woomera had cost ELDO nothing as it was being provided by the Australians, and it might well be expected that Australia would pay for a large part of the new site. One valid technical point was that it was further from the equator (12°S as opposed to 5°N).

The impression here and elsewhere is that the French are in the driving seat, and that choices were often made for political reasons, with technical reasons coming second (and the reasoning often questionable: would northern Australia be 'technically inadequate' compared with the South American jungle?). The new vehicle was given the title of Europa II, although the changes were fairly

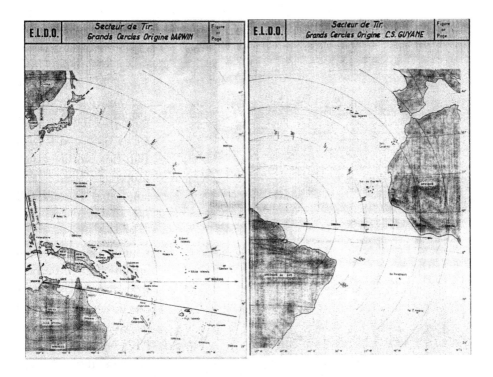

Figure 74. 'Arcs of fire' from Darwin and from Kourou.

minimal. The British objected to even this fairly modest proposal on the grounds that the PAS system was not part of the 'original programme'. This was a mantra that the British Government would produce time and again.

Even so, 170 kg was not a particularly substantial payload for a communications satellite. To be of much use, this would have to be increased, and one cheap and easy way of doing this was to add strap-on boosters.

Indeed, as early as January 1963 the Italians came up with a proposal for 10 ft diameter solid fuel motors to be strapped on to Blue Streak, rather in the style of Titan III. The proposal was passed through to RAE, where Harold Robinson thought some of the assumptions rather optimistic.[8] The Italians thought the combination could put 10 tons into orbit; Robinson was more pessimistic at 6.25 tons. Making motors this size with no prior experience would have been a considerable challenge, and in addition, Blue Streak would have to be strengthened quite considerably to take such a load.[9]

Since the French were now the leading lights in ELDO, it was inevitable that most of the proposals for uprating the launcher would come from them. Of the

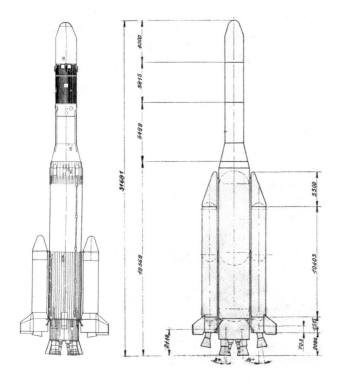

many proposals made to improve Europa, the most significant from French companies such as SEREB (Société pour l'Étude et la Réalisation d'Engins Balistiques). One such proposal used P16 solid fuel boosters (P for French *poudre*, early solid fuel rockets being made from gunpowder) derived from the first stage of the French silo-based ballistic missile. The L17 liquid fuel boosters derived from the Diamant satellite launcher.

Figure 75. On the left is Europa with the French P16 boosters. Drawn on the right is the version with French L17 boosters.

	Blue Streak	P16	L17
Diameter/m	3.1	1.5	1.4
Propellants/kg	88,500	16,440	33,500
Thrust/kN	1335	550*	668
Burn time/second	160	75	111
Total impulse/MNs	213	41 × 2 = 82	74 × 2 = 148
Payload**	170	300	290

* This is the mean thrust (in the original French 'la pousée moyenne – 550 kN' as opposed to 'la pousée maximum – 630 kN').
** to geostationary orbit

As can be seen from the figures for the estimated performance, the boosters produce a useful payload increase – but even 300 kg is distinctly inadequate for a modern communications satellite. On the other hand, only two boosters are being considered. Ariane had a 'pick and mix' approach to boosters, with two or four solid or liquid boosters. Four boosters would certainly increase the payload a good deal further – and give the opportunity to stretch the upper stages at the same time.

Even larger boosters were being considered, as this talk by Dr K Iserland of ELDO in 1968, given to the Royal Aeronautical Society, shows:

… on the other hand, the use of liquid boosters is also being studied, and, in particular, two practical solutions – one based on Blue Streak, in other words using two Blue Streaks as additional boosters, and the other based on the technique of Diamant B, using the first stage of the improved version of the French Diamant B launcher … in the first case three Blue Streaks are attached together side by side and they would all be lit up together from the start. Propellant would be pumped continuously from the outer boosters to the engine of the central Blue Streak so that, at booster jettisoning, the central or core stage is still full and continues to burn for a full burning time. The version deriving from Diamant B type uses four boosters clustered around the central Blue Streak. They each have a diameter of 2.4 m, as compared with the 3 m of Blue Streak, and have 4 engines of 36 tonnes thrust each. The outer four boosters are functioning as a zero stage … i.e., the core stage is lit up at the end of the booster phase only.

It might be said that having to resort to these rather extreme measures points up the inadequacy of Blue Streak as the main core. A better approach might well have been to design a bigger core – which is what Europa III was all about.

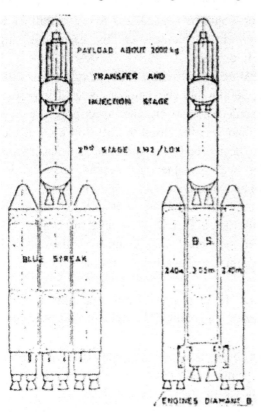

Figure 76. On the left: three Blue Streaks strapped together, and on the right, Blue Streak with French Diamant strap on boosters. Both versions have a liquid hydrogen upper stage.

Europa III

In December 1969, an ELDO report outlined various feasibility studies for a Europa III design[10]. It was stipulated that the new vehicle should have the capacity to launch 400–700 kg into geostationary orbit, and a brief history of previous proposals was outlined – the most significant of which had been made in 1969, when there was a proposal to replace Blue Streak with a French designed first stage designated the L135, which burned N_2O_4/UDMH. It would have a diameter greater than 3 m and the capacity to put a payload of 600 kg into geostationary orbit.

The report also thought that the proposal to uprate Europa II with boosters for relatively low cost to give around a 280 kg geostationary payload would not meet the payload criterion. A new booster would be needed, and four possibilities for the first stage were then considered (in the context of this discussion, 'high energy' can be read to mean 'liquid hydrogen').

A: This would be based around Blue Streak with a high energy stage, which would have been capable of putting 500 kg into geostationary orbit. The problem was that this design lay at the lower end of the performance spectrum.

With strap-on boosters, the payload was increased to 750–800 kg, but there were two drawbacks: firstly, that the design had been stretched as far as it would go and so no further development was possible; and second, that the 3 m diameter of the stage was thought to be a disadvantage.

There was also the consideration of the availability of Blue Streak after 1976 given the British Government's attitude to ELDO – although no doubt Rolls Royce and HSD could have produced them easily enough on a commercial basis. The running costs of Spadeadam might also have been a problem, since it was currently being funded by Britain.

B: this was based on the French L135 design already mentioned, with a diameter of 3.6 m, and a capability of launching 700–750 kg to geostationary orbit.

C: the first stage would have been powered by four RZ 2 chambers, downrated back to 137,000 lb (it is not made clear why the downrating was thought to have been a good idea). It was thought that this vehicle was capable of 850–900kg to geostationary orbit. The report shows a distinct lack of enthusiasm for the design, with the comment that Europe was 'more experienced in N_2O_4/UDMH technology than in lox/kerosene'. This is an odd comment, since liquid oxygen/kerosene is hardly the most sophisticated of technologies, and the motors were to be provided by Rolls Royce, which did have a fair amount of experience in the technology. It was probably the political dimension which ruled this design out – the main motors for the vehicle would have been produced in a country which was not a member, and which had done its best to wreck the organisation! The tank structure would have been light alloy rather than the stainless steel balloon design of Blue Streak.

The timescale for each of these three designs would be set by the high energy upper stage – that is, the liquid hydrogen upper stage.

D: this was an all hydrogen design, which looked distinctly aspirational – in other words, whereas engineers might relish the challenge of producing such a design; politically and economically it represented a distinct challenge.

The estimated cost of each design (in MMU) was:

A	B	C	D
641*	770	747	852
629**			

* with solid boosters   ** with liquid boosters

MMU = Million Monetary Units, the notional currency of ELDO, effectively equivalent to one American dollar. 1 MU = £0.413

On the other hand, given the speculative nature of most of these designs, any attempt to cost them seems distinctly futile. They are probably in the right order, but that is about as much as can be said for them..

The conclusion of the report ran roughly as follows. Option A was viable only if cost were a consideration. B represents the best compromise – although, again, it was not made clear what the compromise was. C matched the performance criterion but 'represents from the engineering standpoint a compromise with the EUROPA I and II vehicle system'. This is another fairly political point, since the only part of Europa that had worked without fault on every flight was the RZ 2 motor, and that was the only point of commonality. As for D – it was acknowledged that the design was not very realistic at the present time.

Despite these figures, the report of EUROPA II AD HOC group in May 1972 came to very different conclusions. (The 'low cost configurations' being referred to are upgrades of Europa, or option A in the previous report.)

> The EUROPA II ad hoc group concludes that, even without making provision for additional development launchings, the low cost configurations will most probably not result in a significant saving with respect to EUROPA II; if provision is made for additional firings – the necessity of which is accepted, except by the German delegation – all low cost launchers will be more expensive than EUROPA III.

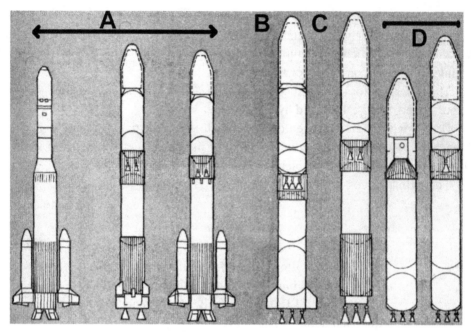

Figure 77. Some of the various proposals for Europa III.

The ad hoc group notes that the low cost launchers do not have the same growth potential as the EUROPA III launcher whose performance increase would be more cost effective.

Under the circumstances, the majority of the EUROPA II ad hoc group cannot recommend further study of the so-called low cost configurations.[11]

Whether or not Blue Streak-based options were technically adequate or financially viable, there is a very strong likelihood that they would be politically unacceptable, given the past attitude of the UK Government.

As is well known, option B was chosen, and after further evolution of the design, went on to become Ariane I – but that is another story.

More interesting from the British point of view was option C. Unlike Blue Streak, it would have light alloy tanks rather than stainless steel, but still with stringers to the kerosene tank (it was not made clear why stringers would be needed with conventional tanks). The main stage motor would have four RZ 2 Mk III engines each of 62.2 tonnes (137,000 lb) thrust, with the ability to swivel in the tangential plane only (this feature might have been borrowed from the Gamma 201 of Black Knight). Quite why the thrust had been down rated from 150,000 lb back to the original 137,000 lb was also not made clear.

The estimated cost of the new motor was £7,536,500 (18.1 MMU), and the annual cost of running Spadeadam was put at 2.25 MMU. There would be a completely open engine bay, 3.6 m diameter, with each motor having a separate pump and turbine.[12]

Val Cleaver of Rolls Royce produced the brochure, which he sent with a covering letter. The letter did not display a great deal of enthusiasm for the proposition, describing the RZ 2 as '1950s technology', and hinting strongly that a motor using high energy propellants would be far more interesting. Cleaver might have been a good engineer, but he was obviously not so good as a salesman! Whatever his feelings about being asked to produce a design of that sort, he might have done better to keep them to himself.

The ironic part of the story is that a four engined RZ 2 first stage would have almost certainly have been cheaper to develop than the L135 design which became Ariane and would have kept the UK in the launcher business. With any luck, it could have been as successful as Ariane was. Unfortunately, the British Government was actively hostile to the idea of a new launcher, and the new design was effectively a French vehicle. Thus ELDO died, and Ariane was born.

---

[1] HAEU, Florence, ELDO 80.

[2] *Design Studies On A High Energy Third Stage For The European Launch Vehicle.* Dietrich Koelle, May 1963. (Private collection.)

[3] TNA: PRO DSIR 23/31973. Flight trial of F1 of EUROPA 1 (ELDO satellite launcher system - first stage). RAE TM Space 45, August 1964.

[4] HAEU, Florence, ELDO. F6/1 launch, September 1967.

[5] HAEU, Florence, ELDO archives.

[6] Ibid.

[7] HAEU, Florence, ELDO 3569. *E.L.D.O. B1 & B2 Vehicles Oxygen/Hydrogen Upper Stages. Rolls-Royce Thrust Chamber Design Study to E.L.D.O. Contract CTR/17/7/10.* February 1967.

[8] TNA: PRO AVIA 65/1567. ELDO satellite launcher system.

[9] HAEU, Florence, ELDO 3515.

[10] HAEU, Florence, ELDO 3035.

[11] HAEU, Florence, ELDO 1561. *Low cost launchers - conclusions of the EUROPA II AD HOC group.* 30 May 72.

[12] HAEU, Florence, ELDO 4287. *Europa IIIC with 4 RZ 2 engines.*

Chapter 11

Black Knight

What was Black Knight, and why was it necessary?

One of the many unknowns when work began on Blue Streak was what would happen when the vehicle carrying the warhead re-entered the atmosphere. To cover a range of 2,000 miles or more, the vehicle would have to be travelling at very high speed – around four kilometres a second. The time of flight would be of the order of 20 minutes, most of it spent in the vacuum of space. Various problems arise when the vehicle re-enters the atmosphere. Firstly, will it be aerodynamically stable? Secondly, will it be able to withstand the forces imposed by the deceleration? Thirdly, will it simply burn up, as a meteorite does?

The point at which all the aerodynamic forces of the vehicle act is called the 'centre of pressure'. It is important that the centre of pressure is well behind the centre of gravity, otherwise the vehicle might 'flip over'. Whichever shape is chosen, it must be stable, else its flight path cannot be predicted with any accuracy.

A sphere can be ruled out immediately: it is aerodynamically unstable and will tumble as it falls (Eiffel demonstrated this by dropping cannonballs off his tower). An alternative shape is a cone with a rounded base, and this can be oriented so it enters the atmosphere 'blunt end first' or 'sharp end first'. The 'blunt end first' configuration is familiar from the Mercury, Gemini and Apollo capsules. The 'blunt end first' also has a much higher drag, or air resistance.

The 'sharp end first', having a lower drag, is not slowed down so much in the thinner upper layers of the atmosphere. Instead, it is still moving fast when it meets the denser air lower down. The result is that the deceleration is then more abrupt, the forces greater and the peak heating greater. The advantage from a military point of view is that it would be much more difficult to stop, and it would also be rather more accurate.

The heating effect was another unknown. Effectively, the kinetic energy of the vehicle is almost all converted to heat in a very short period of time. This does not happen by 'friction', as is commonly stated, but by a different mechanism. The air in front of the vehicle is being compressed, and when gases are compressed they heat up. To compress them, work has to be done; this work appears as heat. If the heat is given no chance to escape, then the heating is described as 'adiabatic'. Heat is then transferred from the hot gases to the vehicle.

RAE decided to go for the 'sharp end first' design, and carried out calculations on the effects of re-entry.[1] Calculations are one thing; what happens in reality is another. There was only one way to find out: fire a model re-entry head vertically upwards, and see what happens when it comes back down again. To achieve a speed of 4 km/s on re-entry meant sending the body 800 km high. In addition, RAE wanted to carry instruments in the head to measure temperatures and accelerations, so it had to be a reasonable size. A re-entry head weight of 200 lb was chosen.

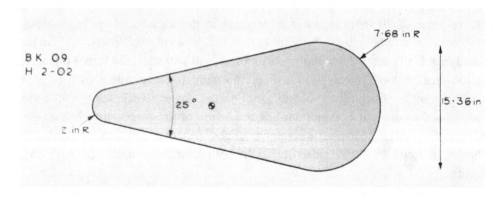

Figure 78. The BK09 re-entry head, representative of the 'sharp end' first design.

An example of the shape chosen is shown above – this is a drawing of the re-entry head flown on BK09.[2]

The next question was what sort of vehicle could launch the model re-entry vehicle? Could a solid fuelled vehicle do the job? The advantage of using solid fuel motors is that they can be clustered together 'off the shelf'. A further advantage is that of staging: it is quite easy to arrange a stack of solid fuel motors into a three or four stage vehicle.

One crude method of estimating the effectiveness of a rocket motor is to estimate its total impulse – that is, thrust × burn time. Black Knight as originally designed gave a total impulse of around 2,300,000 lbf.seconds, whereas one of the larger solid fuel motors of the time, the Raven VI, gave a total impulse of

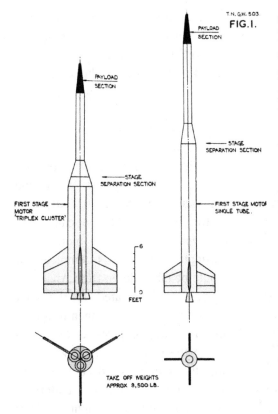

T.N. G.W. 503.
FIG.I.

PAYLOAD
SECTION

STAGE
SEPARATION SECTION

FIRST STAGE MOTOR
SINGLE TUBE.

PAYLOAD
SECTION

STAGE
SEPARATION SECTION

FIRST STAGE
MOTOR
'TRIPLEX CLUSTER'

6

0
FEET

TAKE OFF WEIGHTS
APPROX 9,500 LB.

Figure 79. Sketches for a solid fuelled test vehicle. These were only speculative and were not fully worked.

around 450,000 lbf.seconds. Hence five Ravens would give the same impulse. Using solid fuel motors also made staging easier, producing a more efficient vehicle. The sketches on the left show some possible two stage vehicles.[3]

On the other hand, another way of estimating the performance of rocket vehicles, as mentioned earlier, is the mass fraction or mass ratio, which is defined as (initial mass)/(final mass). Here Black Knight would have won hands down: any British solid fuelled vehicle of the time would have had a very poor mass ratio, whereas that of Black Knight was extremely good.

The RAE estimated that the mass ratio for a solid fuel motor of around one million lbf.seconds impulse would be in the region of five. For Black Knight, the ratio turns out to be 18. In performance terms, this means that, all other factors being equal, Black Knight would have a final velocity nearly double that of a solid fuel motor design (to be precise, 1.8 times greater). In addition, a large solid fuel vehicle would not have been so easy to steer, whereas Black Knight was steered by swivelling the liquid fuel chambers.

But there was also another reason, never quite spelled out, which was that Britain had never built a ballistic rocket before, let alone one on the scale of Blue Streak. Black Knight would be a chance to gain experience not only in the building of such rockets, but also other matters such as launch technique, and so on. Indeed, de Havilland was given the responsibility for the launches at Woomera, in anticipation of Blue Streak, and somewhat to the chagrin of Saunders Roe.

Since the specification[4] was for a quick and inexpensive solution, then it was obvious HTP motors would be chosen, and since RAE was running the programme, then it was also fairly obvious that RPE's motor, the Gamma, would be chosen. The Gamma had a thrust of around 4,100 lb; four of them clustered together gave a thrust of 16,400 lb. This set the weight of the vehicle: a lift off acceleration of 0.3 *g* meant an all-up weight of (16,400/1.3) lb or around 12,600 lb.

Armstrong Siddeley (as they then were) at Anstey were given the contract to build and test the new motor, which would be called the Gamma 201. When the motor for a particular vehicle was completed, it would be taken down from Anstey to Cowes, to be mated to the tank section. Saunders Roe then needed a site to carry out static testing – where the vehicle is filled with fuel and the motors fired, but without releasing it. The Needles battery at the end of the Isle of Wight was War Office property, but was now surplus to requirements. Saunders Roe proposed building the site close to the battery, just round to the south, in a natural amphitheatre formed in the chalk.[5] The area was known as High Down, and Figure 81 below shows the site soon after completion.

Figure 80. The Gamma 201, designed and built by Armstrong Siddeley using the Gamma chamber developed by the RPE.

The two gantries were where the vehicles were held down for static testing. The site has not yet been landscaped, as can be seen from the areas of raw chalk.

Some of the site is still in existence, principally the curving concrete walkway in the amphitheatre. Figure 81 (bottom) shows the site as it was in 2010, more than 50 years after it was constructed.

In the centre was the control room: test firings were carried out from here and the various instrumentation read outs were recorded. In the 1950s, the main method of recording data was by using a pen recorder with a long strip of paper. All the instrumentation would have been analogue. At each end of the walkway

Figure 81. Top: the High Down test site under construction in 1957, and below, the site as it is today, now owned by the National Trust.

was a gantry, the bases of which remain today. The vehicle would have been mounted inside the gantry, and below was a large iron blast deflector, which could be water-cooled during engine tests. The gantries and control rooms were designed to be as identical to their Australian counterparts as possible so as to reduce the chance of problems on countdown or launch. There would also have been tanks behind the main site for the kerosene and HTP, as well as workshops and other facilities. According to the architect, the part of the site that gave him greatest difficulty was the water reservoir for cooling the blast deflectors: the water had to be pumped up from nearby Yarmouth through a very inadequate water main.[6]

In these days of Green Belt, planning permission, NIMBYism, and so on, it seems incredible that a rocket test site could be built on one of the most scenic sites in Britain without objection, but it was also obvious from the files that the idea of anyone objecting did not occur to the official mind. The site belonged to the War Office, was now surplus to requirements, and a test site was needed. Planning permission as we now know it was not necessary in those days.

It is indeed a scenic spot, now owned by the National Trust. Cars are not allowed along the cliff road up from Yarmouth, but there is a bus which runs up to the Needles Battery for those who do not care for the walk. The site overlooks the Needles rocks and lighthouse. Beyond the lighthouse is the Shingle Bank, and even on the calmest days, broken water can be seen here. In the far distance are the chalk cliffs of Studland Bay and Swanage, with the Purbeck Hills in the background. Equally, it must have been a bleak place in a southwesterly gale in winter.

Figure 82. An early Black Knight at Woomera. This is an early 'proving round', with a non-separating head.

The Saunders Roe brochure for the proposed test site stated: 'The cost of renovating existing buildings, cutting the road and installing the complete test facility is estimated to be no more than £45,000.'[7] This turned out to be somewhat optimistic, but to build a static test stand for that figure is good value for money.

Black Knight was a slim cylinder, 3 ft in diameter. Its exact height would vary depending on the payload, but it was 32 ft 3 inches to the top of the separation bay. Initially it was a single stage vehicle; later versions used a solid fuel Cuckoo motor as a second stage. Some also used some small Imp motors to push the re-entry head away from the main body, but to describe this as a three stage version, as some RAE reports did, is being a little optimistic.

Starting at the bottom, there was the propulsion bay, 36 inches diameter and 56 inches high, which held the four rocket chambers, the turbine to drive the pumps, hydraulics and other equipment, and was built by Armstrong Siddeley at Anstey, before being taken down to Cowes to be attached to the tank section. It was easy to dismount for transport, and could be tested and calibrated before being assembled into the vehicle.

The thrust chambers were 'toed in' slightly so that the thrust line passed through the centre of gravity. Four fins were fastened to the engine bay; these were for aerodynamic stability during flight. Attached to two of the fins were pods. One held a radar transponder (a transponder receives the radar signals and rebroadcasts them to give a stronger signal), the other a telemetry transmitter to return data, and an electronic flashgun. This was set to flash every four seconds, and since the launch was at night, the vehicle's progress could be followed optically.[8]

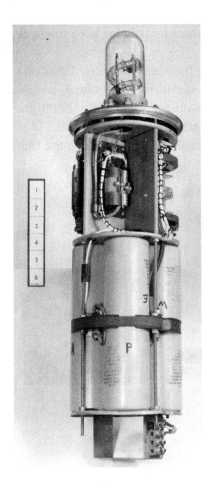

Figure 83. Electronic flash gun installed in one of the pods attached to the fins.

Above the engine bay was the HTP tank, 208 inches long along the cylindrical section. Two tubes of manganese dioxide were attached to the tank, and if the destruct signal was sent, explosive would drive the black powder into the tank. Manganese dioxide is a catalyst for the decomposition of hydrogen peroxide, and the resultant steam and oxygen gas would blow the tank apart.

The bay between the HTP and kerosene tanks contained air bottles which were used to pressurise the HTP tank. The kerosene tank was not directly pressurised, as the hydrostatic head was sufficient to prime the pumps. One striking feature is the size of the HTP tank compared with the kerosene tank: 208 inches compared with 48 inches. The mass of HTP was over eight times that of kerosene, although the greater density meant the volume ratio was smaller.

Above the kerosene tank was the pressurised electronics bay. This housed instrumentation, telemetry, control system electrics, autopilot gyroscopes, a command receiver and electrical power supplies well away from the worst effects of engine induced vibration and heating. Above this would be the payload, which would differ from flight to flight. The tubes running down the side of the vehicle were fibreglass fairings covering the kerosene feed pipe and the various electrical cables.

Initial ideas for the tank structure was to make it from very thin aluminium (26 Standard Wire Gauge (SWG), which is 0.018 inches or 0.46 mm) supported by 16 stringers.[9] This may well have been by analogy with Blue Streak, although the stringers in the case for Blue Streak were there for a different reason. The RAE was uneasy about the structure, mainly because the stringers would experience aerodynamic heating on ascent, which would lead to a variety of problems with them, the main being expansion. If the stringers expanded and the tank did not, then unpredictable stresses would be set up. At least one such tank was constructed; photographs of the setting up of the gantries at High Down and at Woomera show a white painted stringered missile body being used to check the facilities.

The change to 20 SWG (0.036 inches or 0.914 mm) tank thickness did incur a slight weight penalty, which was estimated at 50 lb. The initial weight breakdown for the vehicle was given as 1,100 lb vehicle weight plus 200 lb for the head, or 1,300 lb. 11,250 lb fuel gave a total weight of 12,550 lb. As is always the case, weights crept up during the detailed design process; the first vehicle, BK01, came in at 1,480 lb.[10]

Figure 84. An early non flight vehicle being used to test out the gantry at Woomera. This was a thin walled stringered version which was not put into production.

Close inspection shows a series of holes cut in the side of the motor bay. These were designed to reduce what is called 'base drag'. When the vehicle is climbing up through the atmosphere, it encounters air resistance. If the rocket has a blunt end, the air will swirl around behind it, producing extra drag. The purpose of these 'bleed holes' was to let air flow in, reducing this base drag.

To test the idea, a small solid fuelled vehicle was prepared, which had four small rocket motors and a take-off weight of 140 lb, the rear section being a scaled version of the Black Knight propulsion bay. Several of these were test fired in Wales, with the conclusion that the bleed holes gave an estimated improvement in the final velocity of up to 200 ft per second. Further tests were needed to check on what is called 'recirculation' where hot exhaust gases get swept back into the engine bay.[11]

Figure 85. Base drag test vehicle in flight.

One effect which was not appreciated at the time was the expansion of the exhaust plumes in the near vacuum of high altitude. This did cause problems in some of the early launches, when the kerosene supply to the motors was cut, leading to a period of what was called 'cold thrusting' – that is, the motors were running on HTP only. This works, but is much less effective. The problem was investigated by firing a Gamma motor in a vacuum chamber at Westcott. The solution was to fit spring loaded flaps over the holes, so that they would open in the atmosphere, but as the air got thinner, the springs would force the flaps shut.

To improve performance further, a small solid fuel motor was added as a second stage. This was the Cuckoo – originally intended as a launch boost for the Skylark (the name apparently derived from the idea that the Cuckoo kicked the Skylark from out of its nest!). The original version was the Cuckoo IA, first flown in April 1960; the Cuckoo IB was identical apart from a changed interface from Skylark to Black Knight. It was replaced by the Cuckoo II, which had an improved performance. The Cuckoo would later go on to be developed further in extended versions of Skylark. The oddest feature is that the second stage is mounted pointing downwards. At the very top, under the jettisonable fairing, is a gas bottle with jets. This was used to spin up the Cuckoo motor after separation from the main stage, which then coasted up to apogee then fell back to earth. An ionisation gauge was used to detect the start of the atmosphere, at which point the motor was fired. After motor cut off, the head was pushed away from the empty casing so that they would re-enter separately.

An even more elaborate system was used in later flights. This involved a cradle which was referred to as a 'sabot' (French for 'clog', but in this context it has the artillery meaning of a device used in a firearm or cannon to fire a projectile which must be held in a precise position). The re-entry head was held in the sabot, which was attached to the Cuckoo motor by a long nylon rope. The aim was to propel the head well clear of the spent motor case.

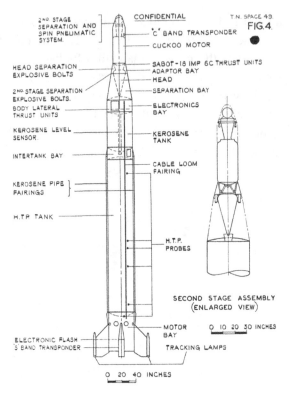

Figure 86. BK16 layout.

The sabot system was necessary to keep the second stage re-entry out of the field of view of the instruments tracking the re-entry head. Separation took place at high altitude so as to give the head sufficient time to be at least 20,000 ft ahead of the second stage at 300,000 ft altitude. It was built on a magnesium alloy plate, on which there was an aluminium seat shaped to the re-entry head. It gave the head a velocity increment of the order of 400 ft/second using four Imp X motors. The thrust of the Imps blew off four panels on the separation bay, and in the case of BK20 it is possible that there was some interference at separation. Despite exhaustive ground testing on the range at Larkhill (Salisbury Plain) it was not possible to simulate the effects of vacuum and low temperature on the nylon.[12]

The lanyard as originally conceived consisted of 8 ft of steel wire followed by 120 ft of undrawn nylon. It was changed to 310 ft of slightly thinner nylon. The steel wire and first 20 ft of the nylon were covered with a woven asbestos sheath to protect it against the flame of the Imp motors, which had a burn time of a third of a second.

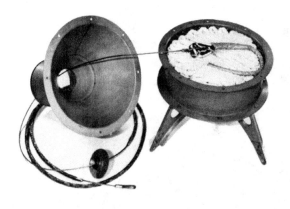

The final arrangement was sometimes and somewhat optimistically referred to as a three stage version: the Black Knight first stage, the Cuckoo second stage, and the Imp motors making the third stage. Starting at the top, below the nose cap is a gas bottle, which with attached jets, was used to spin the stage up. Below that is the Cuckoo motor, pointing downwards, then the sabot and

Figure 87. The lanyard and sabot arrangement.

lanyard arrangement, with the re-entry head itself at the very bottom.[13]

As we shall see in the next chapter, the lanyard system was not always a success for a variety of reasons.

Changes to Black Knight were made incrementally throughout its life. The other major improvement was to replace the original rocket chambers with ones derived from the Stentor motor – the Stentor being developed by Armstrong Siddeley for the Blue Steel missile. The new chamber was lighter, but equally importantly, it was much easier to control the mixture ratio of HTP to kerosene in the new motor, making it more efficient. The maximum thrust was also greater – potentially 25,000 lb, but in practice an intermediate thrust level of 21,600 lb was used. The new motor was named the Gamma 301. A further refinement was proposed for the 54-inch Black Knight, the Gamma 304, which would have had only one turbine and pump feeding each chamber, rather than the four separate ones in the Gamma 301.

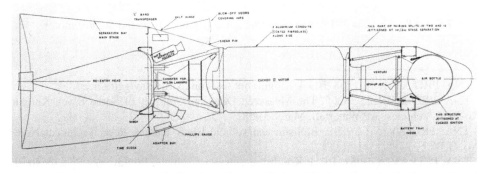

Figure 88. The upper stages for the later Dazzle flights. The head was pushed away from the Cuckoo second stage with small Imp motors, using a lanyard/sabot arrangement.

Figure 89. Gamma 301 motor using the small chamber from the Stentor motor.

---

[1] TNA: PRO AVIA 13/1268. Black Knight: hot tests on inertia navigator head.

[2] TNA: PRO DSIR 23/36265. The Measured Drag and Stability of Some Black Knight Re-entry Vehicles. AP Waterfall. March 1968.

[3] TNA: PRO AVIA 6/19890. Ballistic test vehicle Black Knight. RAE Technical Note No. G.W. 503.

[4] TNA: PRO AVIA 48/57. Black Knight: re-entry test vehicle. Preliminary Specification for Re-Entry Test Vehicle. Guided Weapons Department RAE, 27 April 1955.

[5] TNA: PRO AVIA 48/57. Black Knight: re-entry test vehicle. Saunders Roe Technical Publication 144.

[6] Personal communication.

[7] TNA: PRO AVIA 48/57. Black Knight: re-entry test vehicle.

[8] TNA: PRO AVIA 6/21517. Black Knight electronic flash installation for optical tracking.

[9] TNA: PRO AVIA 13/1269. 'Black Knight': body tests.

[10] Ibid.

[11] TNA: PRO AVIA 6/19890. Ballistic test vehicle Black Knight. (Technical Note Number G.W. 503, HGR Robinson.)

[12] TNA: PRO AVIA 6/21754. Black Knight vehicles of project DAZZLE and their flight behaviour. RAE Technical Report 68076.

[13] TNA: PRO DSIR 23/36396. The Black Knight vehicles of project DAZZLE and their flight behaviour.

Chapter 12

Black Knight and the Re-entry Experiments

Of the 22 Black Knight launches, two were proving flights and one was for ELDO, testing the range instrumentation. The remaining 19 were all for re-entry experiments. Initially, these were to test out the design of the re-entry, but soon they broadened out into a more general study of re-entry phenomena.

The first two flights were the proving flights; it was the third launch, BK04, which was the first to carry a separating re-entry head. This had thermocouples on the head to measure the temperatures at re-entry, the data being radioed back to the ground. Later re-entry vehicles would have a tape recorder to store the data – which was another good reason to ensure that the re-entry head was found after the flight. The data showed that the peak heating was similar to that predicted, and thus the design was now proved experimentally. There were other issues which could be explored in later flights.

After the early flights had verified the re-entry body design, the direction of the programme began to shift. Defence against ballistic missile attack seemed almost impossible, but there was now an opportunity to investigate whether such a defence might be possible. There was also another objective – to discover how best to make Britain's missiles safe from an anti-ballistic missile defence.

The first few flights had shown some interesting phenomena. Firstly, that the exhaust plume from an ascending rocket gave a very strong radar response.[1] There was the possibility of using this to detect enemy launches, although this would mean some form of over-the-horizon radar. Secondly, that the re-entry vehicle gave a very weak radar response – what today would be called 'stealthy'. This tied in with work being done at RAE by the mathematician Grant Dawson, who was studying the radar response of the V bombers.

Ballistic missile defences have been divided into exo-atmospheric and endo-atmospheric – or, to put it more simply, intercepting the re-entry vehicle outside the atmosphere or once it had re-entered. In order to intercept the vehicle, it first had to be tracked. Radar was the only way of tracking the vehicle outside the atmosphere – and, as mentioned, the re-entry vehicle shape had a low radar cross section – particularly viewed from head on. It was also relatively easy to hide the re-entry vehicle within a host of decoys which gave similar radar responses.

Interception within the atmosphere has to be done within a very short space of time – certainly less than a minute. Again, one problem is how to discriminate between the re-entry vehicle and decoys. Thus further flights were planned, using optical instruments to observe the re-entry. These were the 'Gaslight' series of experiments. The results were sufficiently promising to lead on to a further set of experiments, Dazzle, with American participation. For these flights, the range at Woomera would be much more heavily instrumented.

Roy Dommett, who was involved in the Dazzle experiments, describes them thus:

> The DAZZLE programme was sold to the Governments on the basis of exploiting the Blue Streak technology of a low radar observable profile, which was then an advanced concept for the west.
>
> Chosen was the simple conical GW20 shape for which the UK already had derived an extensive experimental aerodynamic data base. The intention was to observe the re-entries of bodies with heat shields made in simple, reasonably well understood materials. Our agreed choices were fused silica, copper, PTFE (Teflon) and loaded durestos (an asbestos-phenolic composite).
>
> For comparison there were to be two reference copper spheres flown. Their manufacture proved surprisingly difficult. The copper shapes were turned to shape by hand held tools in a workshop behind a garage in North London by men wearing armoured vests, and then had to be kept spotlessly clean to avoid sodium contamination. The PTFE was moulded from powder in large sections under pressure, which bulked down about 30% more than expected. ICI, the supplier, was very helpful as no one had ever made such big pieces before, and its final profile varied noticeably with the room temperature. Being PTFE, it was very difficult to machine. The silica glass sections were made in rough by a glass blowing firm in the north near Newcastle using sand moulds, and we had change from £100. The crud had to be machined off with diamond tools in F1E workshop at RAE where we discovered that glass stress relieved itself hours after it was touched. It could only be assembled by having layers of asbestos felt mat between every glass-glass and glass-metal interface.
>
> The conical copper bodies were expected to fail at some point during re-entry by softening and distortion of the nose, but up to that time they would be clean and the observables would be entirely due to the interactions with the atmosphere unaffected by contamination from ablation products. It was thought that the PTFE would massively sublime at a much lower surface temperature than the silica and the products probably suppressing the flow observables, and that the durestos would ablate messily, enhancing them.

He also has this to say about the sabot system described in the previous chapter:

> To be absolutely sure of the quality of the data, it was decided that the re-entry experiments should be pushed into the atmosphere ahead of the upper boost stage using a sabot that was firmly restrained by a lanyard to the upper boost stage, so that the experiment should have been several thousands of feet ahead of it during re-entry. The sabot was driven by four Imp solid propellent motors. In vacuum the plumes spread enormously, and the section of the tether near the sabot had to be in steel. Playing out the tether could exceed the critical speed for the undrawn nylon rope, chosen to avoid elastic bouncing around, and a very careful packing technique had to be found. Finally the rope still broke in flight quite early, despite extensive ground based testing, then it was eventually realised that nylon type plastics have "attached water molecules" which boil off in vacuum, cooling the rope and making it brittle. We also found it near impossible to get the re-entry vehicles to come in without a coning motion of the order of 20 degrees generated by the separation disturbances.

Black Knight Launches

Note: these are listed in the order in which they were fired, not the vehicle number.

BK01

Single stage. Launched 7 September 1958 at 20:03. Apogee 140 miles. No re-entry head.

The preliminary post firing meeting stated that:

> ... the vehicle appeared to follow the anticipated velocity and acceleration programme until about 132.7 secs from time zero when the motor flame went out. At the same instant, transmission from the body telemetry sender ceased, and there was failure of all information channels on the head sender ... some seconds later, a large bright flash was observed which seemed to travel laterally.
>
> The vehicle was recovered in two main portions, one consisting of the engine bay and part of the HTP tank and the other the head, electronics bay and approximately half the kerosene tank.
>
> Fragments of the engine bay, and HTP tank were scattered over a wide area, and the state of the engine bay contents suggested that an explosion had occurred on impact. Extensive burning of cable looms in the lower half of the bay had occurred ...[2]

The vehicle was fitted with a destruct mechanism in case of failure. If the appropriate signal was sent, small explosive charges would blow manganese dioxide powder into the HTP tank. Manganese dioxide is a fine black powder, and a very effective catalyst for the decomposition of hydrogen peroxide. Post flight analysis showed that the aerial had picked up a stray signal and had inadvertently triggered the destruct mechanism. This would cause the HTP to explode, destroying the vehicle – hence the bright flash, which was probably the remainder of the kerosene burning in the hot steam and oxygen produced by the decomposition. In all other respects, the vehicle had behaved as designed, and from most points of view, the flight could be called a success.

BK03

Single stage. Launched 12 March 1959 at 20:20. Apogee 334 miles. No re-entry head.

BK03 was the second proving trial, and was successful except for an engine malfunction late in flight resulting in a long period of 'cold' thrusting (that is, decomposition of HTP in the absence of kerosene). The fault was subsequently traced to excessive heating of the propulsion bay, in which temperatures were measured during flight.

Control of the vehicle was satisfactory both during 'hot' burning and 'cold' burning. In this trial, the guidance telescope tracking was made the primary source of information and radar tracking was retained as the stand-by; this proved very successful. Very good tracking information was received until engine flameout, after which radar information was used during 'cold' burning.

BK04

Single stage. Launched 11 June 1959 at 22:33. Apogee 499 miles.

BK04, the first re-entry experiment with a separating spin-stabilised head, was very successful. The main stage reached its designed burn-out velocity and both guidance and control of the vehicle were satisfactory. Telemetry on the main stage worked throughout the flight to apogee and re-entry, and valuable information on systems performance was obtained. The head separation, turnover and spin-stabilisation system was successful. The head re-entered (200,000 ft) at 11,740 ft/second, telemetry worked throughout to impact, except for a short period during re-entry, and some information on re-entry dynamics and heating was obtained. Recovery of the head with the patches of materials under test attached to it yielded valuable information on ablation during re-entry.

The photograph in Figure 90 is a time lapse picture of the re-entry of the head and of the main rocket body. (The stars in the background appear as streaks due to the long exposure time.) This was one of the reasons Woomera was chosen for

Figure 90. Re-entry photograph for the first separating head, BK04.

the trials: clear skies with no cloud or smog (the flights were also scheduled for moonless nights, for obvious reasons). A further advantage of the range is that the remains of both the head and the rocket body could be collected and examined later (provided they could be found!).

Data was sent from the re-entry head by radio (later heads would have a tape recorder) for analysis. The graph in Figure 91 below shows the temperature at the head of the re-entry body.[3] The results from the flight vindicated the choice of design: the heating was well within acceptable limits.

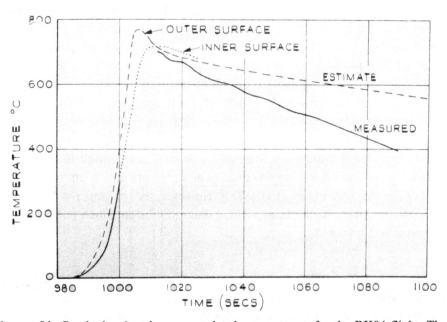

Figure 91. Graph showing the re-entry head temperatures for the BK04 flight. These data showed that the peak temperatures were close to the predicted value, and that any re-entry head based on this design would not burn up on re-entry.

Although Black Knight had fulfilled its original purpose, RAE was interested in some of the other phenomena observed during re-entry, and the range at Woomera was equipped with better optical and radar equipment for the gaslight and Dazzle series of flights. The US was also interested in the results, which meant that the programme became a UK/US/Australian collaboration.

BK04 also set up the altitude record for a single stage vehicle until January 1962, when the launch of a Thor lifted an Echo-2-prototype balloon to a height of more than 900 miles.

BK05

Single stage. Launched 29 June 1959 at 21:03. Apogee 275 miles.

BK05, a re-entry experiment using a double cone eroding head, was designed for greater penetration at high speed into the atmosphere with the object of obtaining much greater heating, particularly in the nose cone which was made of doorstops. A complicated parachute recovery system was built into the head in an attempt to prevent damage to the nose cone on impact.

Overheating in the propulsion bay, as in BK03 (unfortunately not confirmed until after BK05) again caused premature engine cut-out resulting in a reduced re-entry velocity.

A hitherto unsuspected long decay time of thrust at engine shut-down resulted in collision of body and head at separation. The body telemetry continued to function to re-entry but head telemetry ceased just after head separation. The head aerial was probably broken by the impact with the body. The head was recovered and it was found that the parachute had torn out the inner core of the head and the base dome had been pulled off. Early deployment of the parachute would have resulted in excessive drag loads and it can only be assumed that this happened.

Some supporting evidence is that the barometric switch used to deploy the parachute was found on recovery to operate at a pressure equivalent to 22,000 ft instead of the expected height of 10,000 ft. However, the trial was not a complete failure since recovery of the head yielded data on erosion, albeit at a lower re-entry speed than intended.

BK06

Single stage. Launched 30 October at 1959. Apogee 455 miles.

BK06 was a repeat of BK05 with a similar head but using a tape recorder to record separation and re-entry data. Vehicle performance was good, a re-entry velocity of 11,220 ft/second being achieved at 200,000 ft. There was some thrust even after the eight seconds allowed between burn-out and head separation; this caused collision between main stage and head, initiated the ejection of the pyrotechnic flashes and deployed the parachute on the ascent instead of later, as intended, during descent. The tape recorder in the head was switched on correctly and covered the separation phase and later part of the re-entry. The tape cassette, with recordings intact, was recovered together with the eroded durestos nose cone.

BK08

Two stage. Launched 24 May 1960 at 21:00. Apogee 350 miles.

BK08, the first two-stage vehicle to be fired, was intended to obtain re-entry of the head at a higher speed. Main stage performance was good, but the second stage did not separate from the main stage and so was not ignited. The failure of explosive bolts or inertia switch circuitry was the probable cause. The trial, however, proved the aerodynamics of a new configuration, the control stability with the heavier vehicle, the stressing with greatly increased forward weight and the necessarily modified guidance arrangements.

Figure 92. The BK08 re-entry head being set up at Woomera.

BK09

Two stage. Launched 21 June 1960 at 19:35. Apogee 301 miles.

BK09, the second two-stage vehicle, was very successful. Separation of the second stage, initiation of the second stage boost and separation of the head from the second stage boost was satisfactory. The second stage boost ignited at the correct height on the downward trajectory prior to re-entry, and a re-entry velocity of 15,000 ft/second was achieved at 200,000 ft. The tape recorder in the head recorded data during re-entry down to 80,000 ft. Just prior to this an abnormal and completely unexpected increase in head oscillation occurred. The head broke up shortly after this and unfortunately the last inch or so of tape which had passed through the tape head was lost. This corresponded to the period immediately prior to head break-up. The break-up of the head at a low height during re-entry indicated that either the plank construction of the head was unsatisfactory or abnormally high loading was applied during re-entry, maybe resulting from an unstable oscillation.

The attempt to observe the re-entry with the Gaslight system showed that the instruments were not sensitive enough, that a better acquisition system was necessary, and that increased Black Knight performance was needed to raise the level of observables.

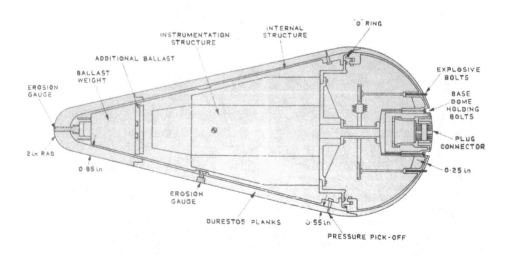

Figure 93. The BK09 re-entry head.

BK07

Single stage. Launched 25 July 1960 at 21:24. Apogee 330 miles.

BK07 was a single-stage vehicle with a high drag heat shield head equipped to give data on heat transfer and re-entry dynamics. Extensive instrumentation was put in the motor bay to investigate base heating and pressure distribution. In addition, lightly loaded spring flaps were fitted to the pressure bleed holes in the propulsion bay to check the direction of flow through these holes. Their movement was monitored by telemetry.

Propulsion was satisfactory except that towards the end of the burning phase one of the four motors reverted to 'cold' thrusting and this resulted in a reduced re-entry velocity. This motor fault was subsequently attributed to a failure of a kerosene feed pipe.

The head separated from the main body but the additional thrust units in the head, provided to give increased separation, did not operate, nor did the turn over and spin thrust units. However, the head did re-enter nose first, but at a large initial incidence. The recovered head shows that impact was on the nose and that there was no re-entry burning on the afterbody. Head telemetry was extremely good and re-entry data was obtained. Complete dynamic analysis of the re-entry head was possible and head temperatures during re-entry were obtained.

The tracking lamps which were fitted for the first time to this vehicle were seen clearly by the guidance telescope operator and the kinetheodolite operators after engine flame-out until about 200 seconds. The electronic flash unit failed to function.

BK13

Single stage. Launched 7 February 1961 at 22:16. Apogee 427 miles.

BK13 was a single-stage vehicle with a double cone eroding head similar to BK05 and BK06 except for the substitution of a low-power telemetry beacon plus a tape recorder in place of the normal telemetry sender (BK05) or tape recorder (BK06), the use of a cine camera for filming the head wake during re-entry, and the deletion of the recovery parachute system. As in BK07, further measurements were made in the motor bay to investigate base heating and pressures.

Propulsion was good and a re-entry velocity of 10,870 ft/second was achieved at 200,000 ft. The flaps over the base bleed holes opened before take-off and closed later as in BK07. The kerosene level sensor and HTP probes worked and propellant usage in flight was determined. The head separation, turnover and spin systems were faultless, as were the pyrotechnic flashes on re-entry.

Unfortunately, due to an incorrect setting of the switch, the tape recorder in the head started too soon and the tape had run out before re-entry; for the same reason the camera in the head did not record the re-entry wake. The durestos nose cone and materials specimens were recovered and erosion measurements were made. The electronic flashes were observed clearly but the tracking lamps were not seen.

BK14

Two stage. Launched 9 May 1961 at 21:37. Apogee 258 miles.

BK14 was another two-stage vehicle with second stage and head similar to that of BK08. The initiation of the second stage light-up was to be by means of a Phillips ionisation gauge as on BK09. Upper atmosphere experiments carried out were a cosmic ray scintillation counter and electron temperature measurement.

The supply of kerosene ran out early, at 128 seconds, and this was followed by 14 seconds of 'cold' burning (i.e. HTP only). The final shut-down occurred at the correct time. Subsequent analysis of records indicated that a leak had developed in the kerosene supply system which accounted for the excessive kerosene flow rate.

Initiation of second stage separation was dependent on the operation of an inertia switch, and as a result of a drop in first stage performance, associated with cold burning, the acceleration was not high enough to operate the switch. Because of this, events following burnout, such as second stage separation, spin, ignition, head separation and recording did not take

Figure 94. BK14 in its gantry – a night time photograph at Woomera.

place. Had the second stage operated, the resultant re-entry velocity would have been adequate for a satisfactory experiment.

In view of this, alternative methods for arming second stage separation, not dependent on first stage performance, were subsequently employed.

BK17

Two stage. Launched 7 June 1961 20:50. Apogee 362 miles.

BK17 was another two-stage vehicle but with a lighter low-drag eroding head to give higher re-entry speed. The first stage performance was very good. The kerosene level sensor and HTP probes worked well and propellant usage in flight was determined. Visual observation and camera records of re-entry were confusing, but it soon became clear that the second stage had not functioned correctly. The head and tape recorder were recovered, and subsequent analysis of head tape record and body telemetry indicated that second stage separation occurred at re-entry and not at the end of first stage burning. It was possible to deduce from the records that the failure of second stage separation at first stage burn-out was due to failure of one of the two explosive bolts. In later vehicles explosive bolts and associated circuits were duplicated. At re-entry the second stage was torn off, followed by second stage burning.

The tape of the recorder in the head ran out before re-entry. In subsequent heads, the speed of the tape in the tape recorder was reduced so as to run for a longer period and ensure recording re-entry.

BK15

Single stage. Launched 1 May 1962 at 22:43. Apogee 494 miles.

BK15 was a re-entry physics experiment but limited by the availability of ground instrumentation on the range at the time, i.e. the 'Gaslight' project equipment and not the more sophisticated 'Dazzle' project equipment.

A single-stage vehicle was fitted with a separating uninstrumented 36-inch diameter copper sphere (the first pure metal head used). The object was to achieve re-entry of the sphere in advance of and well separated from the main body, to provide spatial resolution for ground instruments. This was to be done by turning the vehicle over in the yaw plane after engine burn-out, then separating and pushing the head vertically downwards away from the body when it had turned through 180°.

Figure 95. BK15 prior to launch.

A sabot containing thrust units was used to push the head away; the sabot itself was to have remained attached to the body by a lanyard. Subsidiary upper atmosphere experiments were also carried out and further data obtained on Gamma 201 engine performance and propellant usage. It was also intended to test for the first time an 'automatic pilot' in the ground guidance system. A head re-entry velocity of 11,600 ft/second was achieved at 200,000 ft.

The vehicle turnover and head separation devices worked, but the timing of the latter was incorrect; the head was separated before the vehicle had turned through 180°. The lanyard failed to hold the sabot to the body and the sabot therefore accompanied the head. The re-entry of the head was not recorded by any ground instrument nor was it seen by any observer.

This in itself is a significant result since it confirms the prediction that, because of the absence of ablation products and other contaminants in its wake, the re-entry into the atmosphere of a pure copper head should be a target difficult to detect by optical means. The sphere was recovered and, as expected, there was no heat discolouration of the surface; the maximum surface temperature did not exceed 350 °C during re-entry.

Figure 96. The BK15 copper sphere after re-entry.

BK16

Two stage. Launched 24 August 1962 at 21:08. Apogee 356 miles.

BK16 was the proving trial of the Black Knight vehicle for the further re-entry physics experiments, Project Dazzle. It was a two-stage vehicle powered for the first time by a Gamma 301 engine, and a transistor control system was also tested for the first time. Ignition of the second stage was timed to occur at about 1,000,000 ft, and the head was separated from the second stage at an increased velocity in order to achieve re-entry of the head well separated (15,000 ft) from the rest of the vehicle. The head was a 15° semi-angle copper cone, shape GW20, a type to be flown later in the Dazzle programme. As with BK15, limited ground instrumentation was available to obtain some re-entry data. Dynamics and head temperature measurements during re-entry were also included in this trial.

Figure 97. The new Gamma 301 engine, which could give higher thrust and better mixture control.

Propulsion was very good, a re-entry velocity of 14,600 ft/second being achieved at 200,000 ft. Telemetry was successful: all engine pressures, control system parameters and guidance data were successfully recorded

Once again, as with BK15, difficulties with guidance telescope tracking necessitated a change back to radar information for guidance until telescope tracking was resumed.

All the aims of the trial were achieved. The Gamma 301 engine and the transistor control system were both proved in flight. Separation and ignition of the second stage and separation of the head were achieved according to plan. The FPS.16 radars successfully tracked the transponder and excellent records were obtained from which the trajectory and all events (second stage separation, spin and ignition and head separation) were determined. The range at Woomera was provided with an excellent opportunity of rehearsing for the re-entry physics experiments to follow. The head tape recorder was recovered and data on re-entry dynamics and temperatures was obtained. The trial confirmed the expected re-entry characteristics of an uncontaminated low-drag head.

BK18

Two stage. Launched 30 November 1962 at 02:03. Apogee 358 miles.

BK18 was the second proving vehicle for the Gamma 301 engine and the transistor control system. The head was a 12.5° semi-angle doorstops cone, 2 inch nose radius and with a semi-elliptical base, and was fitted with accelerometers and rate gyroscopes for investigation of re-entry dynamics. No provision was made to measure re-entry heating.

Propulsion was once again excellent; a re-entry velocity of 15,750 ft/second was attained at 200,000 ft. The transistor control system was proved for the second time. The guidance telescope tracking problem was again evident, and the guidance radar information was used throughout flight for guidance and proved most satisfactory. All the vehicle systems were successful. The second stage flare was ignited *in vacuo* and proved to be a useful acquisition aid for sighting ground instruments. Ground instrumentation was successfully operated and re-entry instruments data was obtained. The head tape recorder was recovered and all the data was successfully recorded, from which the dynamic behaviour of the head during re-entry was determined.

BK11

Single stage. Launched 17 October 1963. Apogee 322 miles.

This was launched in support of the ELDO programme, and was designed to test the safety systems and the broadband telemetry. The interim report issued after the flight had this to say:

> Two WREBUS [Weapons Research Establishment Break Up System] command destruct systems were tested with a comprehensive programme during the whole of the flight. One WREBUS system was commanded by a transmitter located at Red Lake, and the other by a temporary low-power installation at the rangehead. The test programme featured both manual and automatic command functions.
>
> ... During the whole of the flight the broadband telemetry system ... functioned well and records of equipment monitoring were obtained.[4]

A second vehicle, BK10, was in reserve in case the tests had to be repeated. Since the flight had met all its objectives, BK10 was never fired and is now in the World Museum, Liverpool.

Project Dazzle

Figure 98. These radars give some idea of the scale of investment in range instrumentation.

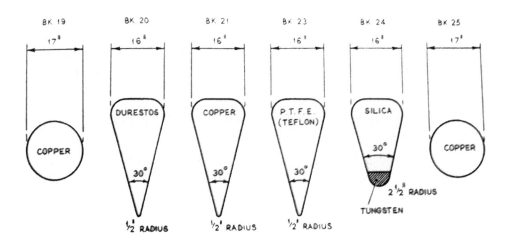

Figure 99. The re-entry heads of Project Dazzle.

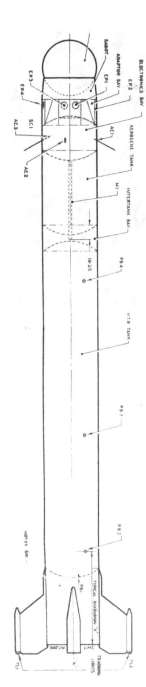

Figure 100. BK19 with a copper sphere re-entry head.

The final six Black Knight launches were part of Project Dazzle, and the various re-entry heads can be seen in Figure 99 above. The range instrumentation was greatly improved, as can be seen in the pictures above.[5] The purpose of these flights was to study re-entry phenomena more closely. Dazzle was a joint UK/US/Australian project – the UK providing the vehicles and re-entry heads, Australia supplying the range facilities, and the US supplying much of the instrumentation.

BK19

Two stage. Launched 6 August 1964 at 02:45. Apogee 374 miles.

This was the first launch with the Gamma 301 motor rated at the full thrust of 21,600 lb, and all vehicle systems were successful. The re-entry head for this flight was a copper sphere, which was recovered the next morning. Although it had split into two and had been scored and dented by the impact with the ground, it was in surprisingly good condition.

BK20

Two stage. Launched 6 November 1964 at 23:15. Apogee 391 miles.

The launch was successful with the first stage performance exceeding expectations. Problems arose with the Imp motors and the sabot: the head was fired at an angle meaning that re-entry did not occur where predicted. The lanyard broke before the sabot had been significantly retarded, which meant it followed close behind the head. In addition, the clutch on the tape recorder inside the head began to stick and the tape ran erratically. No useful data was obtained from it.

Figure 101. The picture shows the launch of BK20: since all but one of the firings took place on dark moonless nights, such pictures are rare.

BK21

Two stage. Launched 24 April 1965 at 20:32. Apogee 404.6 miles.

Both the main stage and second stage performance was satisfactory. The re-entry head was a GW 20 shape made from copper. A thicker rope was used to retard the sabot, but again it broke without retarding the sabot significantly. The copper cone disintegrated on re-entry, and the tape recorder was not found until two months after the trial.

BK23

Two stage. Launched 27 July 1965 at 03:30. Apogee 306.1 miles. Re-entry head: GW 20 coated with PTFE.

Shut down of the first stage occurred about 3½ seconds early, and as a result it under-performed by about 700 ft/second. The head appeared to disintegrate on re-entry at a height of about 28,600 ft.

BK24

Two stage. Launched 29 September 1965 at 00:05. Apogee 376 miles. Re-entry head: GW 20 made of silica.

There was a misalignment of the radar beam (about 0.8°) tracking the vehicle which sent the vehicle further downrange than expected, but the second stage velocity increment was fortunately at 10° to the vertical, and uprange, which helped compensate for the error.

The head survived re-entry down to 100,000 ft, when flakes of silica were observed flaking off for several seconds before the break up at 45,000 ft. Only the transponder and the stainless steel base plate were recovered.

BK25

Two stage. Launched 25 November 1965 at 22:50. Apogee 393 miles. Re-entry head: Copper sphere.

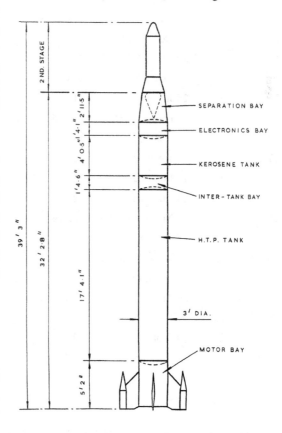

Figure 102. The layout of the later two stage Black Knight vehicles.

The performance of the first stage was excellent. A flare was fitted to the second stage to highlight ignition. The sabot and lanyard system seemed to be a success. The head was recovered the following morning. It had survived the impact with the ground but the halves had flattened on impact and flew apart.

Summary of Black Knight Launches

Vehicle	Stages	Launch date	Empty weight (lb)	All up weight (lb)	Apogee (miles)
1	1	7 September 1958	1,424	13,072	140
3	1	12 March 1959	1,474	12,660	334
4	1	11 June 1959	1,474	13,194	499
5	1	29 June 1959	1,500	12,645	275
6	1	30 October 1959	1,541	13,095	455
8	2	24 May 1960	2,003	13,719	350
9	2	21 June 1960	2,022	13,739	301
7	1	25 July 1960	1, 600	13,371	330
13	1	7 February 1960	1,555	12,813	427
14	2	9 May 1961	2,062	13,230	258
17	2	7 June 1961	2,015	13,235	362
15	1	1 May 1962	1,554	13,448	494
16	2	24 August 1962	2,183	13,787	356
18	2	30 November 1962	2,174	13,941	358
11	1	17 October 1963	-	-	322
19	2	6 August 1964	2,228	14,187	374
20	2	6 November 1964	2,181	14,122	391
21	2	24 April 1965	2,173	14,106	404.6
23	2	27 July 1965	2,191	14,150	306.1
24	2	29 September 1965	2,246	14,182	376
25	2	25 November 1965	2,149	14,021	393

Crusade and the 54-inch Black Knight

RAE had plans for yet another set of re-entry experiments after Dazzle, which were code named CRUSADE (derived, apparently, from 'Co-operative Re-entry Undertaking, Signature and Discrimination Evaluation'). RAE also wanted to improve the performance of Black Knight still further. The Gamma 301 motor had been used at thrust levels of 21,600 lb, but up to 25,000 lb was quite feasible. Increasing the thrust meant that a bigger vehicle could be designed. Thus was born the 54-inch Black Knight[6].

The original 36-inch Black Knight had been a long and slender vehicle – increasing its weight without increasing the diameter would have made it rather too long. The new vehicle would have tanks of 54 inches diameter, which made it shorter than the original vehicle, despite being heavier.

Saunders Roe produced a variety of schemes (Figure 103) presenting to them to the RAE at a meeting in July 1962. The first ideas were to keep the engine bay at 36 inches, and taper the HTP tank, although the result looked distinctly inelegant.

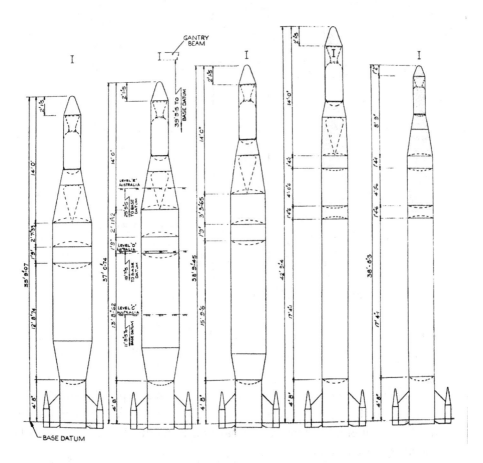

Figure 103. Various proposals from Saunders Roe for an enlarged Black Knight. On the far right is Black Knight 16, included for comparison purposes. Second right is a standard 36-inch Black Knight with a Kestrel second stage.

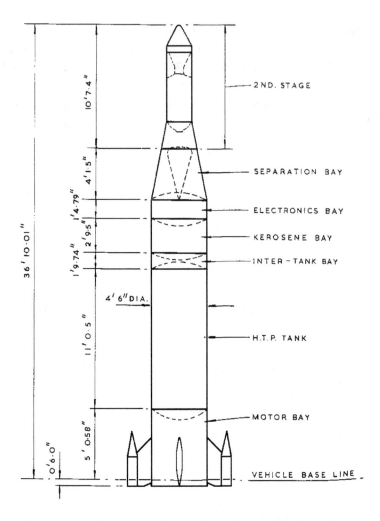

Figure 104. The final version of the 54-inch Black Knight. Work had begun on BK26 when the project was dropped in favour of the Black Arrow satellite launcher.

It was decided to change to a 54-inch bay and thus a parallel sided vehicle relatively late in the design process, and for aerodynamic rather than structural reasons. The parallel sided tank would have been marginally heavier, according to Saunders Roe's calculations.

It was not only the main stage that could be enlarged: keeping the Cuckoo motor meant that despite the increase in size, performance was not that much better. Instead, a new motor was designed – the Kestrel. This was 24 inches in diameter – the first Westcott motor to exceed 17 inches diameter.

	Thrust	Burn time	S.I.	Weight full	Weight empty
Cuckoo II	8,200 lb	10 seconds	213	500 lb	83 lb
Kestrel	27,000 lb	10 seconds	230	1164 lb	84 lb

(NB – different sources give different data)

And comparing the two first stage configurations:

	Weight empty	Fuel Capacity	Total
36 inch	1,380 lb	11,600 lb	12,980 lb
54 inch	1,480 lb	15,600 lb	17,080 lb

The last Dazzle flight took place in November 1965. The first flight of the 54-inch Black Knight was still some way away – perhaps 12 months or so – but the intention was to use BK22 for the first experiment in the Crusade programme. This was one of the last of the Gamma 201 engined vehicles, and had been an ELDO experiment back-up, upgraded to a two stage vehicle. Seven firings had been pencilled in, which included much heavier (250–300 lb) re-entry vehicles, and some decoys. Construction of a new engine bay and tank structure for BK26 had begun when RAE was given the choice between Crusade and Black Arrow, and Black Arrow won. Further re-entry experiments were dropped, although the US carried out further flights using old Redstone missiles, in a programme codenamed Sparta. One Redstone was left surplus to requirements at the end of the programme, and adapted to launch an Australian satellite.

How Successful was Black Knight?

The rocket itself had a relatively simple task to perform, which was to boost its payload as high as possible, from where the re-entry vehicles would fall with as great a velocity as possible. Some launches were completely successful in that the rocket and the experiments yielded all the data required. Sometimes the launches were successful, but the experiments failed, yielding little data. However, even when the vehicle's performance was below optimum, the experiments could still yield good results.

Failures that would have jeopardised orbital attempts had less impact on re-entry studies. The very first flight of all, BK01, ended prematurely when the destruct system operated inadvertently. There were also problems with engine overheating leading to kerosene starvation and resultant 'cold thrusting', particularly in the second (BK03) and fourth flights (BK05), but again, these were solved relatively early in the programme. 'Cold thrusting' occurs when the engine consumes HTP in the absence of kerosene: decomposition still takes place, but the thrust is very sharply reduced. In addition, on many flights the kerosene was exhausted before the HTP, resulting in a few seconds of cold thrust after 'all burnt'. The discrepancy between the calibration during test firing and the actual launch was never pinned down.

The first stage of the vehicle performed exactly as intended on 15 of the 22 Black Knight flights. Other launches had problems in one way or another:

BK01: the self-destruct mechanism was accidentally triggered near the end of the flight.

BK03 and BK05: overheating in the engine bay lead to a fuel lock in the kerosene pipes, resulting in a long period of cold thrusting.

BK07: one chamber reverted to cold thrust after 100 seconds. Over 80% nominal velocity achieved.

BK14: pipe failure caused loss of kerosene: cold thrusting after 130 seconds. 85% nominal velocity achieved.

BK12: 6.8% difference in mixture ratio between flight and calibration.

BK23: premature shut down of engine due to gearbox failure 3 seconds before expected flame out.

Of the 22 flights listed, seven would not have made it into orbit if they had been satellite launch vehicles. Most of these problems could be considered as developmental difficulties that occur with any new technology. As a very first attempt at a modern liquid fuel ballistic vehicle, this is a fairly good record, with no major failures at all. It is also a tribute to the engineers at Saunders Roe and at Armstrong Siddeley Motors.

However, Black Knight was a success in a different direction. It gave RAE the confidence in the basic design, and as a consequence, many further projects were proposed using Black Knight as a basis. One, of course, was the Black Prince launcher and its various derivatives discussed in the BSSLV chapter. Another was Black Arrow – the subject of the next chapter.

Was Black Knight Good Value for Money?

Certainly the individual vehicles were cheap enough. A letter from Saunders Roe to the Ministry of Aviation in October 1958 has this to say:

> In answer to the enquiry you made concerning the cost of additional Black Knight rounds, I quote below a Memorandum from our Commercial Manager which I hope provides you with the information you wanted:-
>
> The approximate cost of the production of the vehicle is £41,000 as it leaves our factory at East Cowes. This figure includes the Armstrong Siddeley engine at a cost

of £15,000 each but excludes the cost of items of normal Embodiment Loan Equipment.

In addition there is the cost of testing and setting up the vehicle at High Down and later in Australia. As you will appreciate this is a very difficult figure to assess but I would suggest it is about £7,000 per vehicle. It is possible that this would be reduced if series production was underway but it would appear that we shall be constantly modifying each individual Equipment in the next year or two, and £7,000, therefore, would be a fairly safe figure to use.[7]

To modern eyes, these figures are astonishing. The idea that the Gamma 201 rocket motor cost only £15,000 tells us a great deal about inflation between then and now – but also that Black Knight was not exactly overpriced. This contrasts with a Treasury memo of January 1961, talking about Black Knight.

The memo begins with some general comments, which show some misconceptions: 'In 1956, we agreed to the expenditure of £5m., for which it was expected to get 12 firings a year...' The mistake seems to lie with the Treasury – there was never any suggestion of 12 firings a year. The memo then goes on to say, 'The first ten firings will work out at a cost of about £½m. each!' There is a certain note of incredulity in the author's tone, which makes one wonder what sum of money the Treasury would think reasonable for a programme in which the re-entry heads are shot 500 miles out into space. The total expenditure of the programme to date had been around £6 million, and the Ministry of Aviation were now asking for more funds. The author goes on to say:

On balance I think I recommend approval of this proposal – just. Any doubts I have are stilled by one further consideration which may appear cowardly but is, I believe, realistic: I do not think we have any hope at the present moment of killing the Black Knight series of experiments: and even if we had, to persuade Ministers to do so now would ruin our chances of killing the Blue Streak launcher project, for we could not hope to persuade Ministers to face the political odium of two further cancellations close together. Black Knight, although pretty expensive (and I would expect the C.&A.G.* at some time to get on to it) is at least working successfully. It has had a good press. It provides a useful vehicle for a certain amount of incidental upper atmosphere research of the kind Universities can share in. Its cancellation would be very strongly opposed in the Ministry, would draw a great deal of adverse criticism in public–after all, we have now got over the most expensive early stages – and would only save less than £1m. a year. Far better, I think, to keep our sights on the larger fish, Blue Streak, than to spoil Ministerial appetites with this smaller fry.[8]
[*C.&A.G.: Comptroller and Auditor General – part of the National Audit Office]

The author is proposing that the Treasury should let the Black Knight programme carry on simply because it gives it a greater chance to cancel Blue

Streak. For cynicism and superciliousness ('a certain amount of incidental upper atmosphere research …') this is difficult to beat.

The author noted that since the next series of flights were to be done in co-operation with the Americans, it would be difficult to reduce the programme further. After all, 'Given that we indulge in this hobby at all, co-operation with the US is surely sensible and desirable'[9].

One of the functions of the Treasury is to keep a close eye on government expenditure: it does help if they get their facts right, and are rather more careful when it comes to making comment on technical matters which are somewhat beyond their grasp.

---

[1] TNA: PRO AVIA 92/128. Policy and financial control of Westland Aircraft Ltd on Black Knight project,

[2] TNA: PRO AVIA 48/58. Black Knight: trials. Preliminary Post Firing Meeting Friday October 3rd, 1958. Guided Weapons Department, RAE.

[3] TNA: PRO AVIA 13/1268. 'Black Knight': hot tests on inertia navigator head. *Summary of Direct Kinetic Heating Measurements on Black Knight*, Dommett RL and Peattie IW. Guided Weapons Department RAE, 3 November 1961.

[4] *Interim Report on the Range Safety Proving Trial*. Space Department, RAE. 2 January 1964. (Science Museum Library and Archive)

[5] TNA: PRO DSIR 23/35658. Project dazzle Pt 5.

[6] Minutes of a meeting at Osborne House 19 July 1962. (Science Museum Library and Archive.)

[7] TNA: PRO AVIA 65/667. 'Black Knight' project: policy and financial control. PD Irons of Saunders Roe to WG Downey at the Ministry of Supply, 30 October 1958.

[8] TNA: PRO T 225/2124. Space and general guided projectile research: Black Knight test vehicle; Ministry of Aviation firing programme. Memo by JA Marshall, 2 January 1961.

[9] Ibid.

Chapter 13

Black Arrow

Some of the most interesting moments in the history of a project may never be recorded on paper, since they happen as a result of casual conversation, or, at least, not in meetings with formal minutes. It is not recorded when it was first suggested to develop Black Knight into a satellite launcher, but judging from the various sketches and proposals that began appearing at the end of the 1950s, the idea took root fairly quickly.

The original Black Knight was not well suited as a satellite launcher. It was simply too small, and could not carry any significant weight in the form of upper stages. Here, for comparison, is some data for Black Knight and various small satellite launchers:

	Thrust/lb	Burn time/ seconds	Total impulse/ lb.seconds
36-inch Black Knight*	21,600	120	2,600,000
54-inch Black Knight	25,000	140	3,500,000
Vanguard	30,300	145	4,400,000
Diamant A	68,000	110	7,500,000
Scout A	58,000	47	2,700,000
Black Arrow	50,000	127	6,350,000

*Using the later Gamma 301 motor.

These figures are only approximate, and total impulse is not the only measure of how effective a rocket stage may be, but it gives a good indication. Scout does not appear to be that powerful, but it was a four stage vehicle, which improved its efficiency as a launcher. Vanguard was not the first American launch vehicle (although it was intended as such); Diamant A was the first French satellite launcher.

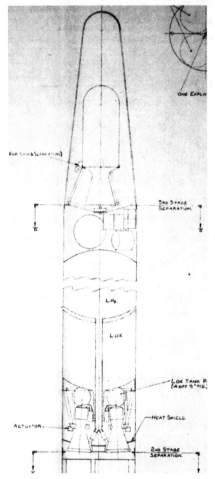

Black Knight had originally been designed with the Gamma 201 motor in mind, which had a thrust of only 16,400 lb, limiting its all-up weight to around 14,000 lb. The Gamma 301 had a greater thrust, which meant that in theory Black Knight could now carry more in the way of upper stages. The Gamma motor could have been further uprated to 25,000 lb thrust, but increasing the thrust without increasing the fuel load means reducing the burn time – the total impulse will stay very much the same. The proposed 54-inch Black Knight is getting closer to Vanguard, although with some way to go.

At around this time, RAE were also becoming seriously interested in liquid hydrogen as a fuel. There was considerable experimental work being carried out at RPE, and also the proposed liquid hydrogen third stage for Blue Streak. Black Knight offered a way of building and testing a liquid hydrogen stage, and

Figure 105. A liquid hydrogen stage for Black Knight. Unfortunately, no room has been left for a satellite!

producing a satellite launcher at the same time. There are various sketches of such vehicles, but most of the work was summarised in a publication from Space Department of RAE entitled 'The Orbital Capabilities of Black Knight with a Hydrogen-Fuelled Second Stage' dated April 1963[1]. Four vehicles are studied, with first stage thrusts of 21,600 lb, 25,000 lb, 40,000 lb and 50,000 lb. Version 1 is 36 inches diameter, all the others are 54 inches. It is not made very clear how the thrusts of 40,000 and 50,000lb are to be achieved. Various combinations of upper stage are modelled – a pump-fed or pressure-fed liquid hydrogen second stage with a solid fuel third stage. Again, what emerges fairly clearly is the inadequacy of Black Knight as a first stage. The most that could be put into orbit using a pump fed second stage is 88 lb, 102 lb, 324 lb and 377 lb respectively. Given that only versions 1 and 2 relate directly to Black Knight, then this is not very impressive.

The conclusions of the report were that:

> This study of the possible use of enlarged versions of Black Knight together with a second stage using liquid oxygen and liquid hydrogen as propellants indicates that such a vehicle can place a payload of about 80 lb into a circular polar orbit at an altitude of 300 nautical miles. Addition of a third stage using a solid propellant increases the payload to about 375 lb.
>
> The advantages of turbo-pump feed for the second stage engine compared to the use of pressurised tanks is quite clear at the stage sizes considered and the payloads just given are for the turbo-pump version. A pressure-fed version would not be capable even of placing the second stage in orbit and in the three stage version, would substantially reduce the payload.
>
> The results of the trajectory calculations for this class of vehicles show that the second stage thrust should be at least twice the total initial mass of the upper stages and payload. This means that the engine developed for this second stage might well have applications for larger upper stages on other vehicles.

There is, of course, another option: using strap-on boosters. This was not considered in the context of the liquid hydrogen stages, but there is no doubt that it would have improved performance very considerably. On the other hand, the idea of Black Knight with added boosters did arise in an entirely different context – RAE was asked by an official at the Ministry of Aviation in October 1961: 'What has the RAE done on solid propulsion and what has been done on smaller rockets? (Opinion and any possible objections)'

The context is not entirely clear (the document seems to be the transcript of a telephone call), but Dr Schirrmacher gave a fairly lengthy reply, and part of this reads:

> For smaller satellites two solutions are worth mentioning:
>
> Black Knight boosted by two Ravens and carrying a Rook as 2nd stage with a Cuckoo as 3rd stage. One will get 100lb into a 200mile orbit. Present development at Westcott for an improvement of the mass fracture [*sic* – probably fraction] indicates that the payload could probably be increased by a factor 2 [*sic*].
>
> It is at the moment not clear whether the Black Knight has to be structurally altered. In this case it is estimated that there are altogether two years of development time with Saunders Roe at £650,000 per year and the probable alteration to the gantry. £1.5M would be a likely sum.[2]

The 'two solutions' mentioned is probably a reference to an RAE report written in January 1961 headed 'Black Knight as a Satellite Launcher'. It takes as its starting point the 36-inch Black Knight 'Dazzle' vehicle, with a thrust of 21,600 lb. It mentions that extra fuel could be carried in the main stage (i.e., Black Knight) but without specifying exactly how much. In true RAE style, the report also tends to the theoretical when it uses solid motors 'similar to' the Rook, Raven or Cuckoo. It is also not clear why only two Raven strap-on boosters would be used: using four would be almost as easy and would improve the performance considerably (near the end of the Black Arrow programme, RAE did begin investigating an uprated Black Arrow with four such boosters). The conclusion to the report states: 'It is clear from these results that, unless exotic fuels are used [a reference to liquid hydrogen], the potentialities of Black Knight as a satellite launcher are very limited.'[3]

The later 54-inch Black Knight is rather a different, and quite a respectable launcher could be made from it using contemporary solid motors. Such a possibility is explored further in Appendix B.

There was another contender for an all-British small satellite launcher, one with a somewhat curious history. In October 1958, David Andrews of Bristol Siddeley Engines (BSE) produced a brochure for an IRBM fuelled by kerosene and HTP, using four of the large Stentor chambers clustered together in a configuration similar to that of the Gamma 301[4]. Using HTP rather than liquid oxygen as an oxidant in a missile has very clear and definite advantages, as the HTP can be stored in the vehicle for long periods, and there is no problem with ice etc. As a single stage vehicle, its payload over the 2,000 miles range required was small if not minimal. The brochure does not give any breakdown for the weight of the vehicle, but it does give some very detailed drawings of the engine. On the other hand, lightweight warheads were now a possibility after the negotiations with the US concerning atomic weapons, and, failing that, a small solid or HTP stage could have been added. Two stages had been avoided in the design of Blue Streak owing to lack of experience in staging, but by late 1958 this was hardly a novel technique.

The brochure seems to have produced no official reaction at all, even during the Skybolt crisis when it looked as though Britain might be thrown back onto its own resources. On the other hand, dropping the development of Blue Streak for an entirely new design would have been difficult for many reasons: writing off the work done so far would have been difficult politically, and by then Blue Streak had a strong grip on both industry and the Government. In these circumstances a momentum develops: the new design did not have enough to offer to make the change worthwhile.

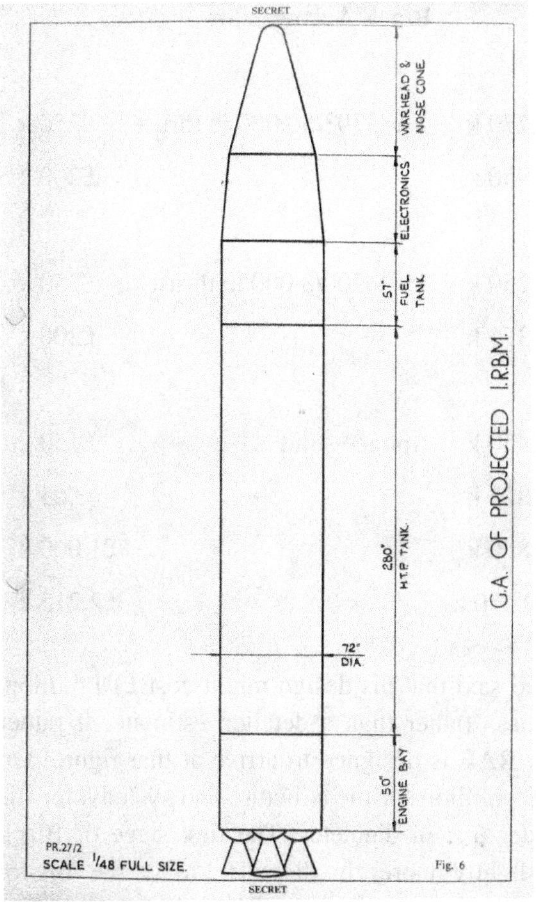
SECRET
WARHEAD & NOSE CONE
ELECTRONICS
57" FUEL TANK
280" H.T.P. TANK
72" DIA.
50" ENGINE BAY
G.A. OF PROJECTED I.R.B.M.
PR.27/2
SCALE 1/48 FULL SIZE.
Fig. 6
SECRET

Figure 106. Proposed IRBM design using HTP/kerosene as fuels.

The vehicle was a good deal smaller than Blue Streak. The design sketch indicates a diameter of 6 ft and a height rather less than 50 ft – effectively an enlarged Black Knight. This would have made the silo smaller and cheaper. The Stentor motor was already under development for Blue Steel: it might have been quicker and cheaper than finishing the development of the RZ 2. Most important of all, using a non-cryogenic fuel would have increased its credibility immensely. The American Titan II missile, which used storable liquid propellants, and was housed in a silo whose design owed something to the Blue Streak silo design, remained in service until the mid-1980s.

However, it was not to be. Andrews did not forget the idea, however, and later issued a brochure entitled 'A Three Stage Satellite Launcher Based on Black Knight and its Technology'[5]. The brochure is undated, but probably originates from 1962 or 1963. He was an engine designer, and so whilst that part of the brochure is again covered in great detail, there is no General Arrangement drawing, merely an artist's impression of what the vehicle might look like. A payload of 650 lb to a 300 nautical mile orbit is given, but this is more of an educated guess rather than a figure derived from detailed calculation.

The Bristol Siddeley proposal and the Black Arrow design were costed against each other by RAE with the following breakdown:[6]

Bristol Siddeley Proposal		**Black Arrow**	
First stage:			
PR.27/90,000 lb thrust	£1,770 k	2 × G303/50,000 lb thrust	£350 k
Structure & systems	£5,950 k		£200 k
Second stage:			
G303/25,000 lb thrust	£250 k	2 × G200/8,000 lb thrust	£250 k
Structure & systems	£375 k		£200 k
Third stage:			
PR.38/2,000 – 3,000 lb thrust	£450 k	Apogee solid	£150 k
Structure & systems	£825 k		£65 k
Guidance	£850 k		£1,000 k
Total so far:	£10,470 k		£2,215 k

Andrews at Bristol Siddeley had said that his design might cost £10 million, but this was almost certainly a guess rather than a detailed estimate. It rather looks as though the breakdown by RAE is designed to arrive at that figure! On the other hand, a figure of nearly £6 million for the structure and systems for the first stage is absurd. It is a cylinder 6 ft in diameter. The first stage of Black Arrow is a cylinder 2 metres (slightly more) in diameter. Why the Bristol Siddeley first stage should cost 30 times as much as the Black Arrow stage is a mystery.

Then some additional items are thrown in – launch sites, range additions and so on – which brought the grand total thus:

Bristol Siddeley:	£11,470,000
Black Arrow:	£2,915,000

The costing was done at RAE, and Black Arrow was the design produced by Saunders Roe in collaboration with RAE. If BSE were to have seen these figures, they might have had legitimate grounds for complaint!

Plans for enlarging Black Knight had been under consideration for some time. The liquid hydrogen proposals had sketched out vehicles with larger thrust, but without any detailed plans for how this might be done. During 1962, Saunders

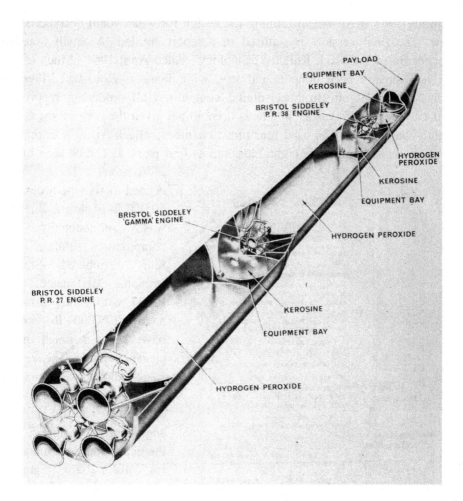

Figure 107. The three stage HTP/kerosene launcher proposed by BSE.

Roe turned their attention to a variety of configurations for six and eight chamber motors. The six chamber arrangement was awkward; an eight chamber version was both more logical and more powerful. The next question was how the chambers should be arranged. One possibility was to have four fixed chambers and four swivelling, but control of the vehicle became more difficult. The question was whether the existing arrangement in the Gamma 301 was powerful enough to swivel two chambers rather than one, and tests showed that it was. The motor bay size had not been fixed, and Saunders Roe gave different configurations with diameters of 57.5 inches, 63 inches and 64 inches. In the end, other considerations fixed the size of the bay, as we shall see.

During 1963, RAE began refining the design for what would become Black Arrow. The final version crystallised in a report entitled 'A Small Satellite Launcher Based on Black Knight Technology', dated April 1964[7]. Much of the work had already been done in industry, with Saunders Roe, BSE, Ferranti (guidance) and Elliott Brothers (digital computing) all producing reports in February. The design was as simple as it could be, given the constraints. Black Knight's Gamma 301 had used four thrust chambers. Black Arrow's Gamma 8 first stage engine would use eight, clustered as four pairs, as can be seen in the illustrations. These were needed to give the necessary 50,000 lb of thrust. But this was undoubtedly a compromise solution: the four chambered Stentor version, designated the PR.27, with a total thrust of nearly 100,000 lb, would have been a much more elegant solution. The second stage engine, Gamma 2, would use two Gamma chambers, with the turbine and pumps redesigned, and the expansion cone enlarged to make for greater efficiency in the vacuum of space. The third stage would be a spin stabilised solid fuel motor, which would be named Waxwing.

If any design was going to succeed this would be it – the technology was all tried and tested, and involved nothing new or exotic. As a result, it should also be cheap. From there on, the vehicle almost designed itself, although there is one oddity in the dimensions. All weights and dimensions are marked on the General

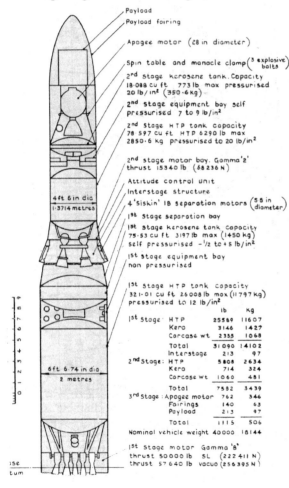

Figure 108. Black Arrow: the final design.

Arrangement drawing in imperial units – pounds, ft, inches, except one. The diameter of the first stage is 2.0 metres. There is an odd rationale behind this.

Europa 1, the ELDO vehicle, was a combination of Blue Streak plus the French and German upper stages. The French stage, Coralie, had a diameter of 2.0 metres, and so the first stage of Black Arrow was given this diameter as well, so that there was still the possibility of mating Black Arrow to the pre-existing interstage adapter if the opportunity arose – effectively a new version of Black Prince.[8] Sadly, the occasion never did arise.

The first stage consisted of engine bay, HTP tank, intertank bay for electronics, kerosene tank and the interstage separation bay. There would be some weight saving if the two tanks shared the same bulkhead, but the decision was taken to separate them to prevent problems if one tank leaked into the other. The HTP tank but not the kerosene tank was pressurised with nitrogen from bottles in the intertank section.

The first stage engine bay had the eight rocket chambers arranged in four pairs of two. Each of the pairs was mounted radially, and were gimballed for vehicle control. Two turbopumps were used to feed the engines. The tanks contained 28,700 lb of propellants, which were consumed in 130 seconds giving a thrust of 50,000 lb at sea level. There was a mixture ratio control keeping the HTP to kerosene ratio to 8.2 : 1.

Figure 109. The Gamma 8 motor (left) powered the first stage of Black Arrow, and the Gamma 2 (right) powered the second stage.

The separation sequence began as the first stage engines ran short of HTP, and soon after the first stage acceleration fell below 3 *g*, eight explosive bolts were to blow the two stages apart. At the same time four Siskin solid fuel rockets ignited on the second stage – at that point it would be in free fall, and so the propellants would be literally floating in their tanks. The motors provided the force needed to settle the propellants in their tanks (these are sometimes referred to as 'ullage rockets'). HTP to start the second stage ignition was held in bottles in the interstage section, being forced in under the pressure of nitrogen gas. When the Siskin boosters had burned out, eight more explosive bolts blow off the interstage section.

The second stage used rocket motors which were almost identical to those in the first stage, except that they had extended expansion nozzles for high altitude operation, where there was a much lower external pressure, and there were, of course, only two motors instead of eight. They consumed 6,725 lb of propellant in 125 seconds for a thrust of 15,300 lb. The engines were fully gimballed for control. There was a fuel ratio control system as in the first stage, and, like the first stage, it shut down when the HTP was exhausted. Both tanks were pressurised with nitrogen.

The second stage diameter of 54 inches was a hangover from the 54-inch Black Knight: 'the second stage, of 54" diameter, is derived from the vehicle being developed for the defence penetration programme'.[9]

So the second stage had the motor bay, the HTP tank, an intertank bay, the kerosene tank and the separation bay. The electronics were contained in the sealed intertank bay, and included the flight sequence programmer, the attitude reference unit, telemetry, the tracking beacon, the command destruct receivers, the control system servo amplifiers and power supplies.

After the second stage had exhausted its propellants and cut out, it separated, and the third stage plus payload would coast for around 300 seconds towards the apogee of the transfer orbit ellipse. During this time the vehicle would be stabilised by a nitrogen gas attitude control system. Then six solid fuel motors were fired to spin-stabilise the third stage and satellite to around 20 radians per second or about 200 rpm. Five seconds later the third stage and satellite were released.

The vehicle's third stage was a specially designed solid propellant motor called Waxwing, with 695 lb of propellant, burning for around 55 seconds and producing an average thrust of 3,500 lb. The motor was very efficient, having a ratio of fuel/total mass close to 0.9. The satellite separated from the empty casing of the Waxwing 120 seconds after its ignition by firing gas-generating cartridges. (It turned out that not even this was a sufficient time: it is thought that some residual thrust from the Waxwing caused it to collide with Prospero after it had separated.) This meant three separate objects were put into orbit: the empty motor casing of mass 77 lb, the motor/payload separation bay of mass 31 lb, and the satellite itself.

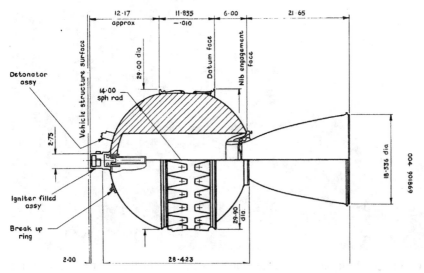

Figure 110. The Waxwing solid fuel motor was the third stage of Black Arrow.

The Waxwing was sometimes referred to as an apogee motor since it was not ignited until it was near the top of the sub orbital trajectory provided by the first two stages. At this point the third stage and payload would have fallen back to Earth, but the Waxwing gave sufficient extra velocity to put into the payload and itself into a near circular orbit.

The payload shroud or nose cone was made of magnesium alloy. Designed in two parts, it had an overall length of 11 ft 6 inches and a maximum diameter of 54 inches with a mass of 150 lb. Attached to the second stage, it was jettisoned 185 seconds into the flight, during the second stage burn.

When designing small satellite launchers, it is important to get the design right first time. The payload of Black Arrow was only 132 lb from a total lift-off weight of 40,000 lb. If the vehicle has to be redesigned or modified for whatever reason, and ends up not very much heavier (132 lb on the third stage obviously reduces the payload to zero, but there is more latitude on the lower stages), then the payload becomes zero! 132 lb is only 0.33% of the total vehicle weight. Losing 132 lb from a payload of a ton or more has a lot less impact.

Figure 111. A Black Arrow vehicle can be seen here in the static test stand at High Down, on the Isle of Wight. The Needles lighthouse can be seen in the background.

The design having been finalised, the project was put to the Cabinet by Julian Amery, then Minister of Aviation. Again, there was Treasury resistance. A compromise was eventually reached whereby Black Arrow could only go ahead if Crusade and the 54-inch Black Knight were cancelled, with a saving of £4.8 million,[10] and Black Arrow was chosen in preference to further re-entry studies.[11] In addition, Westland Aerospace, who had taken over Saunders Roe, and Bristol Siddeley, who were to build the new engine, offered to pay a substantial sum, £1 million, towards the development costs. Not surprisingly, Amery greeted this offer with delight. A letter from Westland in February 1965 noted: 'We expect you to make Management arrangements comparable with the excellent ones you have achieved on the present Black Knight programme. This has been RAE based.'[12]

However, in retrospect it seems odd that the two projects should conflict: the first stage of Black Arrow would have suited Crusade well. This, however, needs to be seen within the context of British deterrent. By 1964, the UK was committed to the Polaris programme, and Polaris in its original form was considered too small for an extensive system of decoys, and so the later Chevaline system sacrificed one of the three warheads to carry the decoys. When the idea of flight trials for the Chevaline system was considered, Black Arrow was deemed too expensive for the purpose.

Figure 112. Announcement of the go-ahead for Black Arrow.

The decision to go ahead with Black Arrow was taken in the last days of the Douglas Home Government, and this was to have serious consequences for the project. The incoming Labour Government, not surprisingly, wanted to review many of the major military and technical aerospace projects, and Black Arrow was put on hold. For a Government whose leader won an election with speeches about 'the white heat of the technological revolution', the attitude to technology was ambivalent. To put Frank Cousins, a trade union leader with few apparent qualifications for the position, as Minister of Technology was a breath-taking piece of political cynicism on Harold Wilson's part. To be fair, part of Wilson's policy was to reduce the number of highly skilled or qualified people working on these projects, so that they could become available for civil work, helping the export drive and so on.

The Treasury got their chance to oppose Black Arrow once again when they prepared a brief for the incoming Chancellor, Jim Callaghan. The brief would have been one of many prepared for him, and this one concerned the space programme and ELDO. (A full version of the brief can be found in Appendix A. It is a very good summary of the Treasury's position throughout the 1950s and 1960s.) The paragraph on Black Arrow reads as follows:

Apart from some effort within Ministry of Aviation establishments, and a limited amount of research by industry (at government expense) into the basic technology of building satellites, the principal commitment of the UK 'national civil space programme' is the development [handwritten: 'at an estimated cost of £10m.'] of the small satellite launcher based on the military research rocket Black Knight. The decision to go ahead with this development was taken at the beginning of September by the late government. It was justified [handwritten: 'by the protagonists'] on the grounds that, if Britain is to go in with Europe in collaborative launcher and satellite development programmes, it was necessary for Britain to have her own 'research tool' which would provide her with the necessary technological know-how and an opportunity to test in flight bits of the communications satellites which she will be contributing to the European programme. But no decision has yet been taken on whether the UK should in fact participate in any further European launcher or communications satellite programmes–as stated above, the Treasury would be opposed since to such participation. Until the decisions on this have been taken, it is in the Treasury's view wholly premature to embark on the development of the United Kingdom small satellite launcher. We recommend therefore that until Ministers have had the opportunity to consider collectively the future course of UK space policy, no work should proceed on a small satellite launcher, and that the project should be regarded as in abeyance. If the Chancellor agrees, we will submit a draft minute for him to send to those of his colleagues who are concerned with these matters.[13]

Figure 113. One of the two Black Knight gantries was rebuilt for Black Arrow. In the distance is the equipment centre, from where the launches were controlled.

To be fair to the Treasury, there was an economic crisis at the time (when is there not?), and Wilson had made great play with the trade deficit in the election campaign – something which rather backfired on him, as it put sterling under further pressure (those were the days of fixed currency rates, with the pound being worth $2.80).

The firms that were to be involved in Black Arrow, Saunders Roe (who were now part of Westland) and Bristol Siddeley Engines, were given three monthly holding contracts merely to keep the project ticking over. The bigger problem was that as Black Knight wound down, then the teams of engineers and designs involved would disperse unless there was something to keep them usefully employed – which a holding contract did not really do. The offer the two firms had made to contribute towards the development costs faded away as the delays dragged out, and whilst more and more defence work was being cancelled.

Review after review took place[14], and each time the project was opposed by the Treasury, which did make some valid points. These points were tied up with general British space policy, which was in a muddle. The new Government wished to withdraw from ELDO but to do so with the least political damage. In particular, they were worried not to give any impression of lack of enthusiasm for joint European projects, and so withdrawal from ELDO at the same time as approving Black Arrow was felt to be giving the wrong signals. On the other hand, if Britain did withdraw from ELDO, Black Arrow could be used to counter arguments that Britain was withdrawing from space research altogether. As usual with such contrary policies, inertia ruled.

There were economic reviews too, to determine whether Black Arrow would be 'profitable' or not. It is surprising that no one made the analogy with the likes of the Gloster Whittle aircraft: that was certainly not profitable, yet without it, or other test vehicles like it, there would be no jumbo jets. Was the Wright Brothers' aircraft 'profitable'? And the papers are not helped by the jargon used by the economists: one phrase to be found in the papers is 'exogenous stochastic error' when talking of the development costs of the vehicle, which in the context appears to refer to unpredicted cost overruns with external origins – in other words, it might end up costing more. The reviews also involved RAE in a good deal of work, having to put forward justifications (usually the same ones) for the vehicle almost every few months.

The arguments for Black Arrow were that it gave British scientists experience in space research at minimal cost, to which the Treasury reply was why did the scientists need such experience? The scope for research satellites was felt by the Treasury to be negligible, Black Arrow was far too small to be of any use in launching communications satellites, and there seemed to be few other uses which would benefit the UK. In many ways the Treasury was right, but the project was fought for by the Ministry of Aviation since it was seen as the last chance for British rocketry: if Black Arrow did not go ahead, the British space effort, or at least rocketry effort, would die.

The Treasury's attitude to the project was summed up in a letter from FR Barratt in Treasury Chambers to Walter Abson at the Ministry of Aviation in April 1965:

> It is claimed that the launcher would be of value in giving UK scientists and technicians experience of injecting satellites into orbit and controlling them, and would also enable us to test satellites and components in an actual space environment. But why should we in fact wish to give UK scientists and technicians this experience? What satellites is the UK going to be putting up? And for what reasons? No decision has yet been taken in regard to a programme of communication satellite and development. Ministers have not yet even been invited to consider the possibility of such a programme. Nor has any military requirement for a UK satellite launching capability been stated. The indications are that no such requirement will arise.
>
> It is suggested that possession of a small satellite launcher will enable us the better 'to compete for contracts for space projects'. I must remind you that the Working Party … concluded that export prospects in the space field were relatively quite small … It might be as well if you were to specify what contracts we are more likely to get if we have a small satellite launcher. I should myself be very surprised if there was anything significant. If there isn't, I suggest that you would do better to drop this particular argument.
>
> … I do not myself understand how the development of a small launcher on the basis of 'proven techniques' will have much relevance to ELDO activities on, say, high energy upper stages. Doubtless you can explain this. In any event, however, Ministers have yet to reach a final decision on UK policy towards ELDO: it is possible, to say the least, that they will ultimately take the view that we should aim to withdraw from ELDO activities. It is also possible that the attitude taken by the Italians and others at the recent ELDO conference will lead before long to the collapse of the organisation. It cannot therefore be assumed at this stage either that ELDO will continue to exist or that, if it does, the UK will continue to participate in it, or will wish to influence its activities.[15]

Apart from anything else, this is a marvellous example of the Civil Service at work. The sentence 'Ministers have not yet even been invited to consider the possibility of such a programme' tends to suggest how much policy was actually made by the senior civil servants rather than by ministers, and there is also a slight implicit suggestion that options are put up to Ministers to be rubber stamped. And then there is the put down in the third paragraph: 'Doubtless you can explain this.'

It is also worth noting the attitude to ELDO. But, more importantly, this is a very neat summary of the official attitude to most of these projects. When the Ministry of Technology again tried to get Cabinet approval in 1966, the then Chancellor of the Exchequer, Jim Callaghan, wrote a three page letter of protest reminiscent of those written by Selwyn Lloyd on Blue Streak. Again, the project had to be referred back for further economic studies before it was resubmitted to Cabinet. And the idea of economic reviews, cost benefit analysis, and the like was a new phenomenon. In the 1950s, projects were given a go-ahead mainly on their technical merit. Now the economists were creeping in. This is not intended to be an entirely pejorative comment. The costings of a good number of earlier projects had been completely unrealistic, and a sharper eye on estimates would be welcome. But that is not what the economists were looking at. How do you try to estimate an economic return from basic research? But it was probably the earlier gross under-estimates of earlier projects that made the Treasury so sceptical.

In addition, there were even those in the Space Department at Farnborough who felt that a satellite launcher took up too much of the space budget, to the detriment of other projects. Others in the Ministry of Defence felt that the offer of free launches from the Americans should be used instead, or, even if they did have to pay for them, then it was still cheaper than developing a British launcher. Needless to say, once Britain's last vehicle was cancelled, the offer of free launches evaporated with it.

There were further financial problems: the Ministry of Defence wanted nothing to do with the funding of the programme. This meant that there might have been potential problems with the use of Woomera, as Black Arrow was, strictly speaking, a civilian programme, and so did not fall under the Joint Agreement between Britain and Australia. And indeed at this time the future of Woomera was being reconsidered – ELDO was moving to South America and thus there would be no more Blue Streak launches, and the number of new missiles under test was shrinking fast. Thoughts turned to alternative Black Arrow launch sites. Barbados was high on the list, but the RAE also looked at launch sites in the UK

Britain is not well situated geographically for satellite launching (but neither was Woomera). Payloads can be boosted by firing the rocket in the same direction of the spin of the Earth: eastwards. But this would mean that the spent rocket stages would fall on Europe, and, not surprisingly, people in Europe might object to this. The only clear path would involve launches almost due north, into polar orbit, and for this, two sites were considered. One was the existing rocket test site on the islands of North and South Uist in the Hebrides; the other was on the north coast of Norfolk in the region of Brancaster. South Uist was ruled out on grounds both of accessibility and of infrastructure – indeed, it was considered by the Ministry of Aviation to be less accessible than Woomera. Certainly the weather in the Outer Hebrides would have posed a problem. The planned flight paths are shown in Figure 114. The first stage impact points are shown, and the second stage would impact close to the North Pole, north of the 85° parallel.

A team went from RAE to look at the area, and concluded:

No particular site has been chosen at this time. There are several areas north of the coastal road which

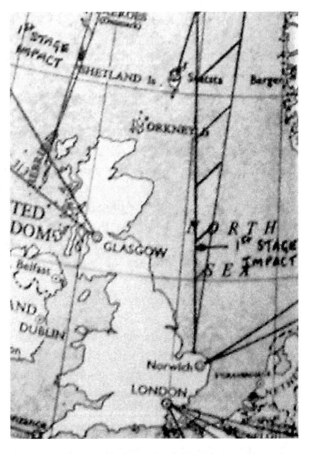

Figure 114. The flight path of Black Arrow launched from Norfolk.

appear adequate. Ideally a suitable site would be within about five miles of a disused airfield or W.D.-owned site with existing roads and buildings. All that would be required in these circumstances would be the construction of a launch site on the coast line and all the other operations could be conducted at the other site, the two being linked together by microwave and hard wire links.[16]

A further memo in October 1966 noted that:

For Black Arrow, a site in East Anglia for polar launching has been studied in considerable detail by Space Department, RAE, and found to be technically very attractive. Their itemised estimate for equipment gives a figure of around £1M, to which must be added the complementary civil engineering costs to cover installation and general services. These latter costs are very dependent on the site chosen and at a first broad estimate might amount to as much as £2M, making a grand total of perhaps £3M; a considerable reduction should be available if a suitable disused camp or airfield became available. Such a site would supersede Highdown for vehicle check-out and static firing purposes, and would have the simplest (and cheapest) supply and maintenance problems.

Polar launching of Black Arrow from N. Uist appears quite feasible, but such a site would not supersede Highdown, and would have substantial technical and logistical disadvantages compared with East Anglia; clearly the remoteness of the area would increase considerably installation, operation and maintenance costs.

The Norfolk launch site looked quite promising until someone realised that the North Sea was beginning to fill up with oil rigs. A chart was rapidly fetched from the Ministry of Power:

> …discussions have been held with the Ministry of Power, and a chart obtained showing the position of existing and proposed oil rigs. Although it would be possible to show statistically that the chances of hitting an oil rig would be acceptably low, it seems probable that political considerations would inhibit the establishment of a launch site in Norfolk.

The chances of a spent first stage falling on an oil rig were minute, but given the consequences if one did, the proposal was dropped. The other launch sites being considered were outside the UK (although it is slightly surprising that the Lossiemouth/Inverness area was not considered) and were too expensive to set up. It was back to Woomera.

The Flights

R0 – 28 June 1968

R0 was launched on a trajectory to the north west, towards Talgarno in Western Australia. The first two stages were live; the third stage was inert.

The launch procedure was that the vehicle was held down on the launch pad by a ball and claw mechanism, the engines were started, and when full thrust was reached, which took around 4 seconds, the vehicle was released. But as soon as it cleared the pad, the vehicle immediately developed a large rolling oscillation. The cause of the fault was an open circuit in the feedback loop that controlled one of the pairs of motors: a wire had probably broken.

At about 64 seconds into the flight, the control system could not cope and the vehicle tumbled. One of the payload fairings broke away, followed by the payload, then the Gamma 8 first stage motor stopped working. The vehicle was destroyed by ground command when it was at an altitude of 9,000 ft on its descent.

The response of the motor pair concerned was normal up to about 4.1 seconds after opening the first stage engine start valve, 0.5 seconds before the release jack was opened. Loss of the signal meant that the pair of motors concerned would swing to their full extent and back again on receiving a correction, instead of the small deflection needed to put the vehicle back on to its correct course.

This can be seen quite clearly in the film of the launch, where one of the set of rocket exhausts can be seen swinging from side to side. Figure 115 shows the vehicle a few seconds after lift-off, and one of the sets of exhausts can be clearly seen pointing to one side.

Eyewitness statements were taken from those who had been watching the flight. One, by Ken Smith of RAE, reads as follows:

> Ignition was seen to occur just after zero time, and the vehicle lifted off as expected a few seconds later. The ascent appeared normal, and the downrange pitch programme was clearly occurring. Just after the time count of 60 seconds, two or three glowing fragments were detached and fell away from the ascending vehicle. Immediately afterwards, the clean flame

Figure 115. The exhaust jets swinging from side to side.

pattern changed to an intense white smoky trail. The vehicle pitched violently, then appeared to recover to its original path, but the rate of ascent diminished. Then tumbling began; the vehicle turning slowly nose over tail several times and beginning to descend. The intercom call was made 'vehicle descending'. The FSO announced that break-up would be initiate when the vehicle reached 9,000 ft on the descent. After a short interval, the vehicle still falling and tumbling, he announced break-up 'NOW'. An instant later the vehicle disintegrated in a bright luminescent explosion. The loud sharp detonation was hard several seconds later.[17]

R1 – 4 March 1970.

The cause of the failure of R0 was relatively simple, but it was decided to repeat the flight, so that the R1 launch was an exact repeat of R0. This time the vehicle behaved exactly as intended.

R2 – 2 September 1970

R2 was launched on 2 September 1970, carrying a spherical satellite, christened Orba, as its payload. This was intended to be a very simple satellite (there was no money in the budget for anything more complicated) to measure the atmospheric drag in low orbits by observing its orbital decay. Figure 116 shows the

Figure 116. The Orba satellite on top of the Waxwing motor.

satellite on top of the Waxwing motor, whilst the payload shrouds are being fitted around it.

The first stage was completely successful, but the second stage shut down 15 seconds early, leaving 30% of the HTP unburned. This turned out to be due to a leak in the HTP tank pressurisation system, with the result that the nitrogen gas ran out early and so there was no pressure in the tank to help feed the propellants. With insufficient pressure the turbopumps cavitate and their effectiveness is much reduced. Hence the second stage thrust dropped to almost nothing. The third stage separated correctly, and fired, but the velocity was insufficient to reach orbit, and the payload crashed into the Gulf of Carpentaria. There were other problems which the subsequent RAE report describes:

> Two other defects were recorded during the flight:
>
> The solenoid start switch in the attitude control system failed to latch open on first initiation.
>
> Only one of the two fairings separated correctly from the vehicle at the correct time – separation of the remaining section was delayed until third stage spin up.[18]

In addition to the drop in pressure in the HTP tank, either of these faults would have prevented the vehicle reaching orbit.

After this flight, an extensive review of the vehicle was set in motion, with eleven technical panels being set up. They began their work in December 1970, and submitted reports and recommendations by the end of June 1971. Relatively few deficiencies were found, and most of these related to the problems that had cropped up in the three development launches. Ian Peattie, who was the Project Officer at RAE for the launch vehicle, commented wryly that the review achieved its objective 'once certain panel members were persuaded that a fundamental re-design of the launch vehicle was not within the terms of reference.'

R3 – 28 October 1971

R3 was dispatched to Australia early in 1971, and the second stage arrived at Woomera on 26 July, followed by the first stage on 17 August. Static firing of the second stage occurred on 1 September, and the two stages and the back-up satellite had been assembled by 1 October. The complete vehicle was given a static firing test on 8 October, and the flight model satellite was fitted by 22 October. A decision was made to delay the launch until 26 October, but systems checking delayed the launch further.

Derek Mack, one of the Saunders Roe launch team (Saunders Roe had by then become the British Hovercraft Corporation), remembers the morning of 28 October as a cool, fresh Australian spring day, with clear skies. The overnight crew had filled the HTP tanks and adjusted the kerosene levels, as well as arming the many pyrotechnic systems on the vehicle. The gantry was wheeled back at 11:00, but there was some alarm when the Attitude Reference Unit, which steers the vehicle, began to give erratic signals. There was relief when it was realised that this was due to the vehicle swaying gently in the light breeze. The vehicle lifted off smoothly, and the various telemetry stations north of Woomera reported that all events had been successful. However, this did not yet mean that the launch had been successful: it was only when the global satellite station at Fairbanks reported an operational signal from a satellite on a frequency of 137 MHz that the team knew that they had an orbiting satellite. The party could begin, but there was a sour taste to it.

R3 launched the Prospero satellite (X3) into orbit on 28 October 1971, in a text book launch.[19] The programme had meanwhile been cancelled by an announcement in Parliament by the new Minister at the Department of Trade and Industry, Frederick Corfield, on 29 July 1971. The teams that had built Black Arrow and launched it were out of a job.

Prospero had a mass of 66 kg, and was launched into an orbit of perigee 557 km, apogee 1,598 km, and an inclination to the equator of 82°. It is still in orbit. It carried four experiments:

(a) To determine the thermal stability of a number of new surface finishes.
(b) To determine the behaviour of new silicon solar cells.
(c) An experiment in hybrid electronic assemblies.
(d) An experiment by Birmingham University to determine the flux of micro meteorites.

The satellite was formed from eight faces covered with 3,000 solar cells. Since the spacecraft would be in the earth's shadow for part of its orbit, rechargeable batteries were also carried.

The flight sequence for the Prospero satellite launch was:

Event	Time (seconds)
Lift-off	0
First stage engine shut down (HTP depleted)	126.9
Stage separation/second stage ignition	133.5
Inter stage bay separation	139.1
Payload fairing separation	180.0
Second stage shut down (HTP depleted)	256.9
Pressurise attitude control system	262.5
Spin-up rockets	575.0
Stage separation	577.0
Third stage ignition	590.0
Payload separation	710.1

The fifth vehicle, R4, was never fired, and is now on display in the Science Museum, London.

The Cancellation

Governments tend to make the announcements of the cancellation of a project as brief as possible. The Opposition and the Press will not follow up the cancellation of a project such as Black Arrow unless there is the whiff of a scandal. The Press was not interested; and in this case there was very little that the Opposition could use to attack the Government.

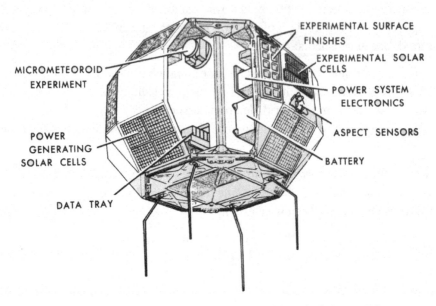

Figure 117. The Prospero satellite launched by Black Arrow R3.

One group of people that is rarely informed of the true reasons for the cancellation are those who work on the projects, and whose livelihood depends on them. Not surprisingly, urban legends or conspiracy theories begin to emerge. The cancellation of Black Arrow was no exception.

Engineers are usually conservative by nature, and often Conservative by political inclination. One of the 'bogey men' of the time was Tony Benn (known earlier in his career as Anthony Wedgwood Benn, but who had now adopted a more demotic name), and to those who worked on the project, part of the mythology was that it was cancelled by Benn and the socialists, when it was actually cancelled by a Tory Government!

In reality, virtually all the opposition came from the Treasury, as the following memo from February 1969 illustrates:

> I have no doubt that the Cabinet will give overwhelming approval to the Ministry of Technology's [Tony Benn] proposals for Black Arrow. At the S.T. meeting on Friday, 21st February, all Ministers from all Departments except the Treasury were not only in favour of the proposals, but emphatically so. The reservations of D.E.A. officials [Department of Economic Affairs] are not apparently shared by D.E.A. Ministers, including Peter Shore. The least enthusiasm was shown by Sir Solly [Zuckerman, Chief Scientific Adviser to the Government], but he gave qualified support and clearly did not brief the Lord President to oppose.

2. I should in fairness add that Tony Benn put his case very attractively. It clearly has some merit and while I suspect there may be a considerable degree of optimism influencing the supporters of Black Arrow, I doubt if the Treasury arguments however skilfully deployed, will sway other Cabinet members.

3. In the circumstances I would suggest that a last ditch fight by the Treasury against Black Arrow in Cabinet could be mistaken. It might undermine the Treasury's general position on more hopeful causes. I feel a tactically more rewarding line would be for you and the Chief Secretary to say that having looked at the proposals you feel that there is merit in them (including the proposal for a British launcher) and that you do not propose to object.[20]

The Ministry of Aviation had been subsumed into the Ministry of Technology in February 1967. It was further reorganised with the advent of the Conservative Government lead by Edward Heath in 1970, becoming part of the new Department of Trade and Industry (DTI). But after all the attacks on Black Arrow by the Treasury, it was actually the Aviation department of the Ministry of Technology which began the process by which Black Arrow would be cancelled. The memo in question is undated, but seems to have been written around October 1970:

As I mentioned the other day, I feel there might be considerable advantage in arranging an impartial examination of the National Space Technology Programme in the light of the recent Black Arrow launch failure [the R2 launch].

It is of the utmost importance that the next firing of Black Arrow, currently scheduled for May 1971, should be successful. We, RAE and Industry are already engaged in analysing the technical causes of the failure, but we have to recognise that there are also wider implications. Another failure, and our national technological competence as well as the future of the National Space Technology Programme would be in question. Examination of the programme as a whole by an outsider of suitable qualifications could be useful in ensuring that it develops from this point on in the best possible way.

The examination would have to take as a starting point an acceptance of our primary objective in space, which is to attain the capability in satellite technology enabling us to offer space hardware, internationally and on an industrial scale. The investigation should in addition not question the broad institutional framework of the Programme–in other words it would be accepted that the effort was a joint Government and Industry one. Within these constraints, however, the investigation should be given the widest possible remit to examine the means we have employed to reach our objective.

The formal terms of reference might be on the following lines:-

To assess the relevance and appropriateness of the Programme, in its present form, to the goal of establishing a significant national competence in satellite technology.

To study in particular the role of the national launcher (Black Arrow) in the programme; the level of effort needed to develop it into a dependable vehicle; and the cost of alternatives to it.

To report on the management of the Programme, with special reference to the launcher element.

And to make recommendations.

I believe that an examination along these lines could be of a very real help to us, especially in providing the answer to two questions -- is the level of spend on the Black Arrow launcher programme a sensible one, or ought it to be increased very substantially in order to achieve real gains; and, whatever the level of sensible expenditure on a national launcher programme might be, would it be preferable and more economic to use an American launcher?

The difficulty, of course, is to find a man of sufficient managerial and technical qualification within the UK who was not already involved in the space programme. We are already separately engaged in the discussion of suitable names. My present purpose is therefore to seek your approval of the terms of reference set out above, on the basis of which we might approach a suitable candidate.[21]

The question was who among what have been termed as 'the great and the good' would be willing and available, although one stumbling block was the requirement for some technical knowledge. In the event, an impeccably qualified candidate was identified and, on 1 October 1970, accepted the invitation to undertake the inquiry: William Penney.

William Penney is best known for his work on Britain's atomic weapons, although he had many other scientific accomplishments to his name. He won a scholarship to study at the University of London, winning the Governor's Prize for Mathematics and graduating with First Class Honours in 1929. In 1944 he joined the British mission to Los Alamos, working on the use of the atomic bomb and its effects. On his return to England, he was put in charge of the British atomic bomb project, and saw the project through to the test of the first bomb in 1952. At this point Penney was offered a Chair at the University of Oxford. Always more inclined toward the academic life, he was keen to accept this post, but he was persuaded that the 'national interest' required him to continue as director of the AWRE at Aldermaston until 1959. From 1954 Penney served on the Board of the Atomic Energy Authority, becoming Chairman in 1964. He retired as Chairman in 1967, and then became Rector of Imperial College.

Given these achievements, it was unlikely that his findings would be disputed, and given his expertise in running demanding experimental programmes, he would have seemed to be the ideal man for the job. Being retired from any form of Government research meant he had no axe to grind, and would be widely seen as an impartial observer.

A briefing note for the Minister after Penney had submitted his report noted that:

Lord Penney's approach to his inquiry was informal. By meeting the people in industry and government concerned with Black Arrow, and discussing the project with them, he aimed to make personal assessment of the management of the programme at the same time as briefing himself on the details of the project. He made two visits to industry, to see the major Black Arrow contractors – British Hovercraft Corporation on the Isle of Wight, and Rolls-Royce at Ansty, Coventry. He visited RAE Farnborough on two occasions, and had a number of discussions with the staff of space division and other headquarters divisions with an interest in Black Arrow. For details of the alternative launchers to Black Arrow he relied on information supplied by the department: he made no visits abroad in the course of the inquiry.

As might be expected, the report was thorough and comprehensive, stretching to 24 pages and 68 sections. He fulfilled his brief admirably, looking at Black Arrow and its alternatives, then considering the viability of Black Arrow within the larger framework of British space policy. His conclusions and recommendations are worth quoting:

My conclusions are as follows:

The disappointing performance of Black Arrow launcher R2 in September 1970 was not due to poor project management, bad fundamental design, or low grade effort. We know we were taking a gamble in trying to make do with so few test launches, and the gamble went against us.

The cost of launching of the X3 satellite on the R3 vehicle is almost fully incurred, and the best policy would therefore be to launch X3 in July 1971 as planned. But in spite of all the work being done to follow up the R2 failure, we cannot be sure that the gamble will not go against us again on R3. The Ministry has neither the time nor the resources to build up greater confidence in Black Arrow before X3 is ready for launch.

It is probable that with the present launch rate of one Black Arrow a year, we will still not be fully confident of its reliability by 1974 when we are planning to launch X4, the second major technological satellite. Even if the Ministry agreed to fund an increase in the launch rate, only one or two extra Black Arrows could be built and launched by 1974.

There is a three-year gap between X3 and X4, but Black Arrows are being built at the rate of one a year. This mismatch between the production rates of launchers and worthwhile satellites may well continue beyond X4, and cannot easily be remedied by adjustments to the launcher programme which is already running at about the minimum level for efficiency ...

The current programme gives us too few Black Arrows to establish the vehicle as a proven launcher in a reasonable timescale, and too many to meet our requirements for satellite launches. It is therefore not a viable programme at present, and there is no easy way out of the dilemma.

And on the subject of alternative launchers, he notes:

> Black Arrow has no alternative use, and the nation would have much to gain and little to lose if it were cancelled in favour of American launchers. We would be abandoning a certain political independence and a guarantee of commercial security payments, but on these two points satisfactory safeguards should be available from the US authorities.
>
> Unless a formal approach is made quickly to the inhabitants on the availability of Scouts and other launchers for our technological satellite programme, further commitments will have to be made on Black Arrow vehicles as an insurance move.
>
> As soon as we are satisfied that we can get the launchers we need from the Americans on acceptable terms, the Black Arrow programme be brought to a close as soon as possible. However, the launching of X3 on the R3 vehicle should proceed, and there may be a need for a further launch if problems arise with X3/R3.
>
> I therefore recommend that:
>
> The Ministry should make a formal approach to the US authorities as soon as possible about the availability of launchers for X4 and subsequent satellites in the National Space Technology Programme, and terms on which they can be provided.
>
> Commitments on R5 and subsequent Black Arrow vehicles should be kept the minimum possible level while the Americans are being approached, and all work on them should be stopped as soon as satisfactory arrangements have been made for the supply of US launchers.
>
> The X3 satellite should be launched as planned on the R3 Black Arrow vehicle in July 1971; the R4 vehicle should be completed in all major respects and used as a reserve for R3 up to the launch.
>
> If X3 goes into orbit successfully and functions as planned, the Black Arrow launcher programme should be brought to a close without further launches.
>
> If X3 fails to go into orbit successfully or fails to work in orbit, the Ministry will have to decide whether to bring the launcher programme to a close at that point or repeat the X3 experiments by launching the X3R on the R4 vehicle. Unless they are sure that the R4 vehicle has a better chance of success than the R3, and it is worth spending £1½ million to repeat the satellite experiment, a further launch should not be sanctioned.
>
> The X4 satellite should be launched on Scout; and Scout or Thor Deltas should be bought as necessary for later satellites in the series.
>
> The Ministry should determine at a high level the views of British industry on the value of a technological satellite programme. If no such value can be identified the programme should be brought to a stop. If it is established that the programme is worthwhile, a plan should be drawn up for a series of future satellites so organised as to give the maximum benefit to British firms in their attempts to win contracts in the international market.[22]

It is difficult to argue with his conclusions, or, indeed, with his recommendations. As he correctly points out, there was a 'mismatch between the production rates of launchers and worthwhile satellites'. Making fewer launchers was not economic; there were not the resources to make more satellites.

The report was submitted to the Minister in January 1971, and made its way up the government hierarchy, culminating in a meeting held in the Prime Minister's room at the House of Commons at 5:15 pm on Monday 6th July, 1971.

Those present were the Prime Minister (Edward Heath), the Chancellor of the Duchy of Lancaster (Geoffrey Rippon), the Lord Privy Seal (Earl Jellicoe), the Secretary of State for Trade and Industry (John Davies), the Minister for Aerospace (Frederick Corfield), the Chief Secretary to the Treasury (Maurice Macmillan), and Sir Alan Cottrell, Chief Scientific Adviser. An excerpt from the minutes of the meeting reads:

> The Lord Privy Seal recalled that on 24 May the Ministerial Committee on Science and Technology had approved the proposals by the Minister of Aerospace that the Black Arrow programme should be stopped, that we should support the full development of the X4 satellite and that for the launching of small satellites we should in the future rely on the American Scout launcher. The Prime Minister had been doubtful about the impact of this decision on future European collaboration in science and technology, particularly as the French were developing their own launcher the Diamant. Since then however the political difficulties have largely dissolved. The Diamant programme had now been deferred and the French were themselves using the Scout launcher this year.
>
> The Prime Minister and the other Ministers present agreed that the proposals originally approved by the Science and Technology Committee in May should now be implemented. The Chancellor of the Duchy of Lancaster said he did not think that the cancellation of Black Arrow would be deemed inconsistent with anything the government had said when in opposition; there had been no commitment to back any project which was not successful.
>
> The Minister for Aerospace said he thought that the announcement of the decision would not cause great surprise and could be done by an Answer to an arranged Written Question.
>
> The Prime Minister agreed to this and suggested that the announcement should be as late as possible.[23]

Three weeks later, the following exchange appears in Hansard for 29 July 1970:

National Space Technology Programme

Mr. Onslow asked the Secretary of State for Trade and Industry what progress has been made with the review of the National Space Technology Programme; and if he will make a statement about the future of the Black Arrow Launcher.

Mr. Corfield: The first phase in the review of the National Space Technology Programme has now been completed. Plans to launch the X3 satellite on a Black Arrow vehicle later this year have been confirmed, but it has been decided that the

Black Arrow launcher programme will be terminated once that launch has taken place.

We have come to this decision on Black Arrow mainly because the maintenance of a national programme for launchers of a comparatively limited capability both unduly limits the scope of the National Space Technology Programme and absorbs a disproportionate share of the resources available for that programme.

We hope to complete our review in the early months of 1972. Meanwhile work is continuing in industry on research into basic satellite technology and on the development of the X4 satellite. X4 is planned to be launched in 1974 on a Scout vehicle to be purchased from N.A.S.A.

The curious (or perhaps not so curious) feature of the announcement is how little attention it received. True, Black Arrow always had had a low profile, but neither in Parliament nor in the press was there any great comment. Perhaps the last word should be given to the New Scientist magazine, which had this to say in August 1971:

Despite considerable success with small launchers – notably Skylark – the modern sport of rocketry evidently rouses little excitement in British hearts. The now-promised demise of the Black Arrow programme, the erstwhile Black Knight venture, is unfortunate only because its death throes have been so prolonged. Announcing last week that, after a final launching later this year when it hopefully will put the X 3 satellite into orbit, Mr Frederick Corfield, Minister for Aerospace, said that henceforth Britain's need to pursue experiments in space technology would be met with US launch vehicles. The reason essentially is that nearly all foreseeable space applications are going to require satellites in high geostationary orbits; Black Arrow falls far short of this requirement, being able to lift some 260 lb only into a near-Earth orbit.

Black Arrow Improvements

Inevitably, there were a variety of improvements suggested for Black Arrow[24,25], some more realistic than others. As mentioned in the chapter on rocket motors, BSE carried out a programme to produce an improved HTP motor. The chamber that was test fired was both lighter and more efficient than the Gamma chamber used in Black Arrow, and had a higher thrust rating. This inevitably gave rise to the proposition that if the vehicle were fitted with the new chambers, then the tanks could be extended, improving performance.

The BSE report which detailed the 'Larch' programme suggested the following changes:

	Standard vehicle	Uprated Vehicle
All-up-weight	40,000 lb	48,710 lb
Overall length	43 ft	47½ ft

First Stage:

Total weight:	31,095 lb	37,514 lb
Propellant load:	28,735 lb	35,205 lb
Approx stage length:	22½ ft	24½ ft
(plus interstage)		

Second stage:

Total weight:	7,798 lb	10,221 lb
Propellant load:	6,618 lb	9,063 lb
Approx stage length:	9½ ft	11½ ft

Third stage:

Total weight:	1,107 lb	975lb

Payload to 300n.m. polar orbit	232 lb	375 lb

The proposal seems to have reduced the weight of the third stage: quite how is not obvious. It is easy enough to vary the amount of propellant in a liquid fuel stage, but to do so in a solid fuel stage would need some considerable redesign.

The payload increase is more than 50%, so this does seem to be a worthwhile improvement. This particular programme seems another example of the left hand not knowing what the right is doing: the RAE were making various attempts to upgrade Black Arrow yet never appear to refer to the Larch motor at all. Given that a lot of work had already been done on the chamber (but not on the full scale motors), it might seem a particularly cost-effective method of upgrading.[26]

The other proposal which RAE were working on at the time of cancellation was to attach four solid fuel Raven boosters to the first stage, with the empty cases being jettisoned after use. Again, the idea is similar – use the extra thrust to stretch the tanks.

	Original:	Proposed:
First Stage Burn time	127 seconds	200 seconds
Second Stage Burn time	117 seconds	180 seconds

In this case, both the first and the second stage tanks have been extended. The result is an increase of payload to 470 lb, which is effectively double that of the unboosted vehicle. The graph in Figure 118, showing the acceleration of the vehicle, is rather interesting.[27] At 34 seconds, the acceleration falls to zero! This is presumably when the Raven boosters have burned out, and the upwards kink in the graph after that is a consequence of the empty cases being jettisoned. It showed that the RAE had stretched the vehicle as far as it can possibly go, and any further weight increase would have been a considerable embarrassment!

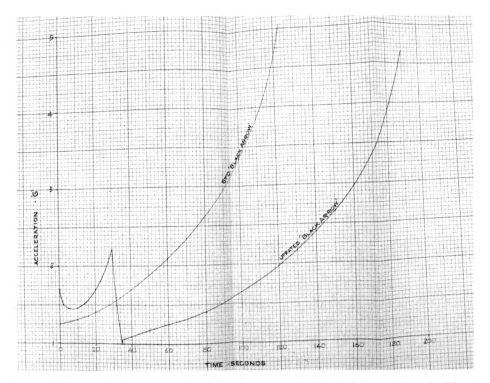

Figure 118. Graph showing acceleration versus time for Black Arrow with and without boosters.

One final suggestion on improving Black Arrow came from Saunders Roe: a vehicle called SLAVE.

Rather mundanely, this abbreviation stood for Satellite LAunch VEhicle. It was a Saunders Roe proposal dated around 1970, produced independently of RAE, and suggested using four of the large Stentor chambers for the first stage. The first stage could then be stretched correspondingly.

The idea was similar to BSE's proposal of some years previously, although there was no other connection between the two ideas. The second and third stages would have been unchanged.

This would have probably given the greatest payload of the three proposals outlined here, although developing the four chamber motor for the first stage would not have come cheap. It also had the advantage that there was still some 'stretch' left in the design.

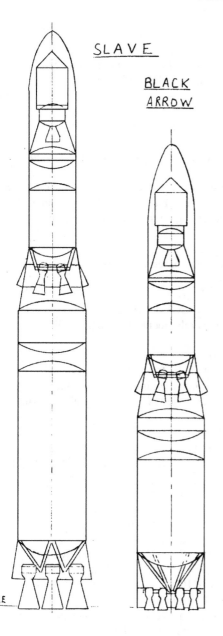

SLAVE

BLACK
ARROW

On the other hand, none of these ideas address the basic issue: what was the point? If there were no satellites to be launched, then there was little point in 'stretching' a vehicle which had run out of uses. The whole Black Arrow saga shows the confusion running through British space policy, and how a rather dubious decision taken in 1964 limped on for another seven years before being cancelled. Again one is tempted to say: do the thing properly, or don't do it at all.

Figure 119. A comparison of SLAVE and Black Arrow.

[1] TNA: PRO AVIA 6/19236. *The Orbital Capabilities of Black Knight with a Hydrogen-Fuelled Second Stage*. Technical Note Number Space 30. DJ Harper and HF Worsnop, April 1963.

[2] TNA: PRO AVIA 65/1567 ELDO satellite launcher system.

[3] Science Museum Library and Archive. Black Knight as a Satellite Launcher. AP Waterfall, Guided Weapons Department, RAE. 18 January 1961.

[4] TNA: PRO DEFE 7/2246. Development of Blue Streak. *A Four Chambered Engine Suitable for An IRBM P.R. 27/2*.

[5] Coventry Archives. PA1716/1/3/1. *A Three Stage Launcher Based on Black Knight and its Technology*. EGD Andrews, Chief Engineer (Rockets), Bristol Siddeley Engines Ltd. 1962. See also TNA: PRO DSIR 23/30957. Rocket engine research and development at Bristol Siddeley, 1963

[6] TNA: PRO AVIA 92/125. Policy on Black Arrow.

[7] TNA: PRO DSIR 23/32125. Small satellite launcher based on Black Knight technology (RAE).

[8] TNA: PRO AVIA 111/5. Project appraisal: Black Arrow.

[9] TNA: PRO AVIA 92/125. Policy on Black Arrow. CGWL, March 1964.

[10] TNA: PRO T 225/2125. Ministry of Aviation: development of a small satellite launcher based on Black Knight.

[11] TNA: PRO AVIA 92/125. Policy on Black Arrow.

[12] TNA: PRO AVIA 92/132. Black Arrow: contributions from industry.

[13] TNA: PRO T 225/2765. Ministry of Aviation space programme: future policy.

[14] TNA: PRO AVIA 92/133. Black Arrow launcher review: Director Space's Working Party (among others).

[15] TNA: PRO AVIA 92/126. Policy on Black Arrow.

[16] TNA: PRO AVIA 92/127. Policy on Black Arrow.

[17] TNA: PRO Science Museum Library and Archive. Black Arrow. Eye-Witness Account of R0 flight.

[18] Science Museum Library and Archive. Report on Black Arrow R2.

[19] TNA: PRO DSIR 23/40429. The Prospero satellite: first year in orbit

[20] TNA: PRO T 334/141. Ministry of Technology: expenditure on the development of small satellite launcher based on Black Night [*sic*] and Black Arrow

[21] TNA: PRO AVIA 92/282. Independent review of Black Arrow launcher by Lord Penny, Rector of Imperial College of Science and Technology, London.

[22] TNA: PRO AVIA 92/283. Independent review of Black Arrow launcher by Lord Penny, Rector of Imperial College of Science and Technology, London.

[23] TNA: PRO PREM 15/635. European space programme and proposed postponement of European Space Conference: Black Arrow national space technology.

[24] TNA: PRO DSIR 23/38157. Rockets Sub-Committee: possibilities of future versions of Black Arrow.

[25] TNA: PRO DSIR 23/32823. The Black Arrow Launching Vehicle – an Assessment of the Performance of Advanced Versions. RAE Technical Report No. 65072. JB Scott, April 1965.

[26] TNA: PRO AVIA6/22042. Black Arrow satellite launching vehicle.

[27] Rolls-Royce Ltd Rocket Design Office. Design Memorandum No. 33, Uprated 'Black Arrow' with 4 'Raven' Boosters. 16 June 1970.

Chapter 14

The 'Might Have Beens'

In any field like aerospace, there will always be projects on the drawing board which never make it through to hardware. Sometimes this is because they are simply bad designs, or sometimes because their rationale has disappeared. Quite often there is simply not the money for them. Thus, although the UK had the technical ability to produce an excellent large satellite launcher, it did not do so because it did not perceive a need for satellites at all, neither did it have the money to pay for them.

The historian must also be wary of the glossy brochures which are produced by the aerospace companies. Some are simply not technically feasible. One artist's impression in a BAC brochure had four Blue Steel missiles mounted under the wing of Concorde. Quite what the four missiles would do for the aerodynamics of the aircraft was not mentioned. Other brochures simply do not take into account political realities – whilst the Blue Streak/Centaur proposal was technically very interesting, it would have been a purely British design and would have needed an equatorial launch site. The prospect of the French agreeing to allow the British to use Kourou for their own benefit was remote.

There are other possible pitfalls for potential historians, particularly in regard to oral histories. People who worked on projects years ago might say, 'I remember a study being done on ...' without necessarily remembering why the study was being carried out, the outcome of the study, or, if it was rejected, why it was rejected. Similarly, there are many 'back of the envelope' ideas which never see the light of day, and rightly so.

Many of the engineers working on the projects at the time were not able to see the wider policy picture. This is partly because much of the work was secret and heavily compartmentalised, partly because commercial firms were not privy to the workings of government, and partly because engineers are sometimes prone to the mentality that because a thing can be done, it should be done. There is sometimes the feeling among them, unspoken, inchoate, but nevertheless real, that 'our toys were taken away'. What were the wider policy issues?

The UK undoubtedly had the capacity to develop large satellite launchers, and the saga related in this book is frustrating for the enthusiast because the drawings could have been converted to reality with two extra ingredients: money, and the will to go ahead. In the end, the projects did not go forward mainly because the governments of the time did not think them a worthwhile use of money and resources, and in hindsight, that decision can be argued to be the correct one.

First of all, what do you want a satellite launcher for? To launch satellites. Why do you want to launch satellites? Nowadays, that seems obvious, with telecommunications satellites and satellite TV and so on. But it was not obvious in the 1950s and early 1960s.

Interestingly, one of the very first UK studies on the possibilities of satellites was done as early as 1955, looking at their suitability for reconnaissance. In the 1950s much covert reconnaissance of Russia was being carried out by Britain and America using spy planes like the U2, or the high-altitude Canberra, partly because maps of the USSR were so poor. If you wanted to target a missile complex you had to know where it was and whether it even existed. Indeed, the Americans were to launch a vast number of reconnaissance satellites: there were 145 such launch attempts between February 1959 and May 1972. (This military involvement also makes launching other satellites very much cheaper: the launch facilities and radar tracking stations are already there, and the economies of scale are very considerable.) Information exchange between Britain and the US meant that the British requirement disappeared, although it was to surface much later in the 1980s as the proposed Zircon satellite, designed for signals intelligence.

Many other projects initiated by the US, although seemingly innocent, also had military origins. Next in usefulness were navigation satellites, forerunners of the ubiquitous GPS system so useful to yachtsmen and motorists. These had the Polaris submarines in mind: how does a submerged submarine know its position accurately enough to launch missiles? Raising an antenna to pick up a satellite is one answer.

Communication satellites were not taken seriously until the early 1960s. A paper prepared by the Post Office around December 1959 comes out against geosynchronous satellites as famously described by Arthur C. Clarke in 1945. They were considered too far in the future even for the US, and that the quarter second delay was too much for telephony – the users would not care for it.

A long Ministry of Aviation paper in April 1959 considered the uses then foreseen. First of all, it was reckoned that a Blue Streak/Black Knight launch vehicle and five satellites would cost at least £10 million to develop (would that were so!). This could not be justified on the Defence Budget because the foreseeable direct benefits were slender. Then it considered the various possible uses for satellites, including reconnaissance, communications, meteorology and navigation. The report noted that American claims that a manned satellite has important military value were difficult to appreciate on any reasonable time scale, and refers to them as belonging to a 'futuristic space platform age'. (In that respect they were right: no military uses for manned space stations have been found.) Weapons delivery posed severe technical problems on accuracy of delivery, reliability and security, and would raise serious political problems.

So although the report goes through the possibilities quite thoroughly, as with all these studies, the enthusiasm is lukewarm. Certainly, as time went by, defence interest in satellites, apart from communications, resulting in the Skynet satellites launched by the US, decreased steadily. Commercial applications were not even considered at this early stage.

The first transatlantic communications satellite was Telstar: launched by NASA aboard a Delta rocket from Cape Canaveral in 1962, it was the first privately sponsored space launch. Telstar was placed in an elliptical orbit with a period of 2 hours 37 minutes, but its availability for transatlantic signals was limited to 20 minutes in each orbit that passed over the Atlantic Ocean (which not all of them would). Unlike geostationary satellites, the transmitting and receiving dishes had to be steered continuously to track the satellite.

In 1962, a Commonwealth Conference on Communications Satellites was held in London. This might be looked upon as the last gasp of the old Imperial ideal, attempting to link the countries of the Commonwealth (still consisting, in the main, of the old 'white' Commonwealth). Much talk was made of the Black Prince concept, although the conference showed up one of the major weaknesses of Black Prince and Europa: whilst too big and expensive for the likes of scientific satellites, they were crucially that bit too small to put any appreciable payload in a geosynchronous orbit. Ingenious attempts were made to suggest elliptical 12-hour or 8-hour orbits, but these were at best an inadequate solution. Other papers and conferences suggested that such a system could be profitable, but even so, this was not sufficient incentive to produce a launcher that would be adequate.

The scientific community wanted to launch satellites for space research, but the UK science budget was already stretched, and whilst it might support a few research satellites, it certainly could not support a launcher programme. Even the number of purely British scientific satellites launched on American launchers is very small.

There was a final category of satellite put forward by RAE in the 1960s – the small technological satellite. These would be small and low cost, but would test out systems for more ambitious projects. This is quite a reasonable concept –

putting a variety of new technologies in one satellite has obvious weaknesses –
but even these were growing too large for Black Arrow. And the Treasury's
response? If these technologies need developing, then let commercial enterprises
develop them.

One new idea, and one that has not yet been fully exploited, is the ion motor.
In theory, this has an enormous S.I. Atoms are stripped of an electron and the
resultant ions are then accelerated through a very high voltage. This has the
potential to produce enormous exhaust velocities, the power necessary being
provided electrically from solar cells. The X5 satellite was intended to test this
concept. Its motor would accelerate the ions through 25 kV, with a beam current
of 25 mA, which would need 625W of power. The resultant thrust was only
0.015 N, but a small thrust over a very long period of time would have the same
effect as a large thrust for a short time.

But assuming that the policy makers in the UK decided that they did want to
launch satellites for whatever purpose, what were the designs that could have
pressed into service to launch them? The projects below are not arranged in any
great order of significance, but rather the intention is to take a somewhat
meandering walk through the 'might have beens' in terms of launcher design.

A Black Knight IRBM

One rather unusual proposal surfaced at the time when Skybolt was cancelled. At
the subsequent Nassau conference in December 1962, Macmillan was able to
persuade President Kennedy to provide Polaris, effectively without strings. This
seemed to be distinctly improbable at the outset, and there was obviously a short
period of almost panic in the British delegation. Without Skybolt and without
Polaris, Britain would have been thrown back entirely on its own resources,
which were then very few. Sir James Lighthill, at that time Director of RAE,
came up with a proposal for a missile based on Black Knight and wrote a note for
Macmillan headed 'A Possible British Deterrent'. It begins by saying:

> Advances in technology make it possible, today, to do Blue Streak's job with a
> rocket at one tenth of Blue Streak's weight. The main advances have been in
> reduction of weight of warhead, and the perfection of the technique of the two-stage
> rocket, whose first stage is shed after it has burnt itself out.[1]

He then put forward the idea that:

> … The very successful research vehicle BLACK KNIGHT (an entirely British
> development began in 1956 as a 'lead into' BLUE STREAK and first fired 2½ years
> later) could now do the job with a small second stage. Several two stage rockets with
> BLACK KNIGHT as first stage have actually been fired. In the last three firings, all

systems (including first and second stage separation, and complex data processing and transmission systems) worked as planned. In all fourteen firings of BLACK KNIGHT, the first stage flew successfully.

For delivery of nuclear warheads over 1500 nautical miles (the London Moscow distance), a suitable second stage would be one using the extremely well tried Gamma 201 engine (of 4000 pound thrust). Guidance would be based on the Ferranti miniature stable platform, of which the prototype has already been satisfactorily tested.

The fuels would be kerosene and hydrogen peroxide (which can be stored at room temperature for a whole month). The missiles can be fired from small hardened sites (a tenth of the size contemplated for BLUE STREAK), and would be vulnerable to a megaton explosion only if it took place less than half a mile away.

The system suggested would use proven components that started development in 1956. It would probably reach the first flight after two years of further development (at high priority), and begin to come into service after another two years (and some 30 test flights).

A tentative estimate of research and development cost includes £15m. spent on the missiles plus £20m. on the Woomera range, which I add together around up to £50m. The cost of operational missiles and launch sites is estimated at £200,000 and £500,000 respectively, which I add together around up to £1m. per missile. These figures suggest that for £100m. (essentially what remains to be spent by us on SKYBOLT) we could have 50 of these rockets. (They would be simpler and cheaper than the US rocket Minuteman because of the much shorter range required).[2]

In the end, Britain did acquire Polaris without strings, but Lighthill's suggestion was probably the most viable if Britain wanted to retain its deterrent yet 'go it alone'. Certainly some form of Black Knight would be the only ballistic missile which could have been developed in less than five or six years, and by far the cheapest.

Obviously RAE followed up this idea later, when the dust had settled:

An investigation was recently made into the possible use of Black Knight as an IRBM Two cases were considered, both based on the 54" Black Knight with a sea level thrust of 25,000 lb weight. The first assumes that HTP and kerosene were the propellants, with a vacuum specific impulse of 250 secs; the second that the specific impulse can be raised to 280 secs by using propellants such as NTO/UDMH. In both cases the second stage used the same propellants as the first stage. The maximum range of an 800 lb re-entry vehicle was found to be just over 1000 n. miles with the HTP/kerosene combination and 1500 n. miles with NTO/UDMH.[3]

And later in the report: 'It was found that a second stage weight of 3500 lb at a thrust of 6000 lb was about the optimum.'

Changing fuels would have negated one of the major advantages of the proposal: it was simple, cheap and made use of existing and proven hardware.

It is curious that a much more useful proposal was not pressed into service at this point. The 1958 proposal for an IRBM from Armstrong Siddeley, resurrected in 1960 (described in the Black Arrow chapter), would have been eminently suitable. The design is basically a much enlarged Black Knight, but instead of the small Gamma engine (4,000 lb thrust), it uses the large Stentor chamber: 24,000 lb thrust. Thus the proposed vehicle is six times more massive. It would have been easy enough to put a small second stage on top, and it would certainly have had the range to reach Moscow.

As an alternative missile to Blue Streak, it had the advantage that the propellants are storable, or reasonably so. Certainly, HTP would be easier to handle than liquid oxygen within a silo, and could be stored in the missile for long periods – months rather than weeks.

Certainly, an alternative history or counter-factual for Nassau would be interesting: what *would* Macmillan have done if Polaris was not available, or at least hampered by a sufficient number of strings as to make it political unpalatable? The only possibilities were some sort of air-launched missile (a ramjet cruise missile, the X-12 or Pandora, was being suggested at the time), or (with a large number of people having to eat quite a few words) a land-based missile, and here a Black Knight derivative is an obvious choice. Like Titan II and unlike Blue Streak, the fuels were storable in the missile, making for a much more credible weapon.

It is interesting, although coincidental, that such a missile would have looked very much like Black Arrow, whose development costs were a mere £10 million. The first flight of Black Arrow was in June 1968, and, given proper funding, this could have been reduced by at least two years or more. Lighthill had noted that it was now possible 'to do Blue Streak's job with a rocket at one tenth of Blue Streak's weight'. That was perhaps a touch optimistic, but Black Arrow at 40,000 lb was actually a fifth of the weight of Blue Streak's 200,000 lb.

Could Britain Have Launched an Astronaut into Orbit?

The science fiction author Stephen Baxter has written a short story, 'Prospero One', about the first and last launch of two British astronauts from Woomera. His 'Black Prince III' launch vehicle uses five Blue Streak boosters strapped together (this does seem a little over the top). But what would be needed for a manned capsule? Could it be done with British technology as of, for example, 1964 (and without five Blue Streaks)? The answer is yes – but only just.

America's first manned spacecraft was the Mercury capsule, which weighed 4,300 lb at launch and 3,000 lb once in orbit (the escape tower was jettisoned during ascent). Mercury was the smallest capsule into which a man could be squeezed; there was no room or weight spare for anything else, and it had very limited endurance. The more capable two man capsule, Gemini, weighed in at 8,500 lb in orbit. Gemini could stay in orbit for several days, and carry a reasonable amount of equipment.

Given the original Black Prince design had a payload of around 2,200 lb into low earth orbit, it would have to be uprated very considerably, but by this time RPE had developed liquid hydrogen motors of around 4,000 lb thrust, so making a motor of 8,000 lb thrust would not be that difficult. The American Centaur stage had a thrust of 30,000 lb; four of the RPE motors gives a thrust of 32,000 lb. Centaur was even built in stainless steel using the same balloon technique as Atlas and Blue Streak; de Havilland could no doubt have built a similar stage without much difficulty. Such a vehicle might be able to put a payload in the region of 4,000 lb into orbit, which is close, but not really good enough.

There is the possibility of strap-on boosters: four Black Knight boosters, stripped of all other equipment, might well serve here. They could be uprated to 25,000 lb thrust, giving an extra 100,000 lb thrust at lift-off. This would mean that there would be weight to spare, so that the liquid hydrogen stage could be considerably stretched. Now we might have a vehicle capable of putting 6,000lb or more into orbit – perhaps sufficient, but only just, and by stretching the design as far as it can go.

All this is achievable with the technology of 1964, although developing the liquid hydrogen stage would involve some years work. As to cost: Gemini cost around $1.25 billion, or roughly £500 million at the contemporary exchange rate. The question arises as to whether it would have been worthwhile putting a British astronaut into orbit … and, regrettably, the answer is no. Gemini was merely a stepping stone to Apollo. Now we have an International Space Station but the financial or scientific return from the ISS is negligible, whereas its costs are horrendous.

Alternative Histories

Writing alternative histories is a fruitless pastime: after all, why not go and write proper fiction? On the other hand, it can be interesting to speculate what *might* have happened…

• If RAE had decided to go ahead with Crusade and the 54-inch Black Knight, but adapted as a launcher using solid motors as described earlier. Now let us imagine a timeline thus:

The follow-on series of re-entry experiments after Dazzle, named Crusade, were cancelled in favour of developing Black Arrow. Let us postulate an alternative: Crusade goes ahead, and at the same time, the Black Knight launcher is developed. The 54-inch Black Knight would have been in production, as would have been the Kestrel. Waxwing development would not have been that much more expensive. One of the reasons for the cost of Black Arrow was not its development, but the cost of building one vehicle every 18 months or so – the production teams were proceeding at not more than 'tick over'. In addition, setting up the vehicle at the range at Woomera took around a month – which meant that for a further 17 months the range would have to be mothballed. The launch teams would have to be found alternative employment during those months – you cannot build a new team from scratch every time. If the range had been in use, fitting in an extra launch or two does not cost that much.

So the Crusade experiments carry on until 1968 or 1970. Does that mean that the Black Knight programme would have stopped there? Not necessarily. As part of the development of the Chevaline programme (improvements to the Polaris missile), a series of launches were carried out between 1975 and 1979 on a vehicle called Falstaff, which used the large solid fuel Stonechat motor (tested in 1969) – although Black Knight would have done the job admirably.

Hindsight is a wonderful thing, but it does seem that RAE might have been better off proceeding with the 54-inch Black Knight and basing a launcher on it. It would have had Crusade, Black Knight instead of Falstaff for the Chevaline testing, and a satellite launcher as good as or better than Black Arrow. But it was not to be.

• If the Government had cancelled Blue Streak completely in 1960. Black Knight still goes ahead as part of the re-entry experiments, and we get a better Black Arrow because there is more money available – it is not being spent on ELDO. On the other hand, would the money have been spent on Black Arrow?

• If the proposed launcher had remained an Anglo-French project, on the lines of Concorde. The British would have had to go along with the French whether they liked it or not, and the vehicle would probably have been a good deal more reliable than Europa. Whether or not the resultant launcher would have been worth the effort is another matter.

• If the French proposal for dropping ELDO A and going straight to ELDO B had won the day in 1965. One can sympathise to some extent with the Government's position on ELDO, but again it lacked the courage of its own convictions. Backing out of ELDO might have cost money and brought opprobrium down on British heads, but staying in and dragging their heels cost nearly as much money and was just as alienating. ELDO B might have had the prospect of being yet another money pit, but at least it was a project with some future, unlike ELDO A.

- If the proposal for the version of Europa III with four RZ 2 motors had gone ahead. It could well have been the success that Ariane was, but it would still have been unlikely that it would have changed the Government's attitude to launchers!

But none of these did actually happen.

---

[1] TNA: PRO AVIA 65/1851. WS-138A: Skybolt policy.
[2] Ibid.
[3] TNA: PRO DSIR 23/ 31835. Performance capability of Black Knight rocket in various roles (RAE TN Space 60).

Chapter 15

Conclusions

It can be argued that the Cold War became a war of resources, and one in which the UK effectively dropped out of at the start of the 1960s. It could be further argued that the collapse of Communism in Russia was also due to the collapse of a command economy directed to large military and technological programmes. However, let us concentrate on the UK.

As already mentioned, most of the projects discussed so far were military in origin, but many were never carried through to completion. This is not unique to the UK; similar cancellations happen in all countries developing new technology. One of the major factors contributing to the cancellation of such projects, not only in the UK but in the US, was the very rapid advance in technology during the 1950s.

In many ways, the major aerospace technologies, with the exception of computing and electronics, had become mature by the mid-1960s. Thus the jet engine, the rocket motor, supersonic aircraft and the rest had been successfully developed by this time. There have obviously been improvements, but they have been incremental rather than breakthroughs into new areas. It is also interesting that up until the 1970s, almost all technological advances came from government and military projects, whereas today the main driving force seems to be business and consumer interests, most notably in electronics and computing. Military spending is no longer the great driver of projects that it once was.

After the cancellation of Blue Streak there was virtually no further military interest in long-range missiles. The UK was left with the legacy of the work done so far on Blue Streak and Black Knight to pursue a rather half-hearted space programme. Considerable muddle in the subsequent policy left the UK disillusioned with space research – or, at least, with launchers – with the inevitable cancellations later in the 1960s.

The question then comes: why, when America, Russia and France were pursuing space exploration with vigour, and why, when countries such as China and India are launching satellites almost as a matter of routine, has the UK shown such little interest both at government level and among the people at large?

A useful German word can be used here: the *zeitgeist*, which might be translated as 'the spirit of the times', or the outlook characteristic of the period. America and Russia were pursuing their race in space as a way of fighting the Cold War at one remove, in an attempt to show the rest of the world who was technologically the more sophisticated. Britain had no such interest: at the end of the 1950s it was beginning the long retreat from Empire, and at the same time beginning to suffer from the economic and social ills which were to plague the country for the next 30 or 40 years. Another phrase has been used of the government at this time: 'managing the decline'. A country that feels itself to be in decline does not embark on new, challenging technological challenges.

As mentioned earlier, Macmillan's initial announcement in Parliament in 1959 was greeted with the response: '… is it just an attempt to keep up with the Joneses?' This was a fairly common attitude, as when Thorneycroft, then Minister of Aviation, was interviewed on television about the proposed Blue Streak launcher and ELDO in 1961. Ministers, when interviewed on television, have to expect difficult questions – but the tone of the questioning is interesting.

> Mr. Mackenzie: But couldn't it be argued that we, in Britain, have after all only a limited number of technologists available, even in any aspect of this area and that we might be better advised to get them off working, for example, in exploration of the problem of supersonic aircraft, or some more obviously commercial operation, rather than this rather exhibitionist activity of rocketeering?
>
> Minister: There's nothing exhibitionist about the brilliant Rolls Royce and de Havilland engineers who've, incidentally, done a great deal more than keep this in mothballs. We've just done two fully integrated static firings. The work is going well ahead and the Americans will tell you themselves that the payoff in other forms of industry – in metallurgy, electronics and the rest – have wide application to civil industry as a whole, is very great if we go into it.
>
> Mr. Mackenzie: But are we remotely in this competition? One knows how very far the Russians have gone, and the Americans and one has the awful feeling that this is the kind of feeble rearguard, final action to show the flag.
>
> Minister: Don't be so depressed, Robert. This is not a rearguard action at all. We are in this for eternity, all of us. It isn't just the question of doing it with the Atlas or the Blue Streak. We shall be making these rockets: I hope we shall be making them in Europe for a long time ahead, with great advantage to ourselves, to the world and to all the countries, including the smaller ones, that are in it.[1]

'Exhibitionist activity of rocketeering', 'feeble rear guard action'. And another quote from Mackenzie later in the interview: 'But I don't understand why, if the Americans are offering a launcher – which is presumably more advanced than the one we have – Blue Streak – why we may as well not write off Blue Streak and use their launcher for whatever purposes we've got in mind.'

And Mackenzie was wrong. Blue Streak was actually based on American technology, but it could be argued that in the process of anglicisation that a considerable number of improvements had been made.

Another example of the same frame of mind (and the frame of mind perpetually adopted by the Treasury) can be seen in a note from the Chief Secretary to the Treasury, John Boyd Carpenter, in July 1963:

MINISTER OF DEFENCE

I have seen the Minister of Aviation's minute to you of 16th July about military space.

I note that he does not believe that we shall be able to hold back over military space indefinitely. I must make it clear that I should find the utmost difficulty in agreeing to add to our programme what might well become yet another major defence role or commitment. I suggest we cannot start to build a vertical empire if our colleagues insist on our continuing to provide for the defence of a horizontal one. I am sure that, before we go any further, we need a cool appraisal of what our real military space requirements are, if any, and of the various ways in which they might be met, with full figures of probable costs and an analysis of the effect of such costs on the already horrible Costings. I understand that papers on all this are being prepared for the Defence Research Policy Committee and I hope that these, in particular that of the Ministry of Aviation, can be considered very soon.

These are examples of the *zeitgeist*, the feeling that space and rocketry are not Britain's concern, and more than that: that the UK does not have the resources to become involved, and that British projects will inevitably be inferior to American projects.

The apogee of enthusiasm for space in the UK was probably in 1964. This is the year when Black Knight had reached a total of more than 20 successful launches, when there were two successful Blue Streak launches, and when Black Arrow was given its go-ahead, being announced publicly at the Society of British Aircraft Companies (SBAC) dinner just before the 1964 election by the Minister of Aviation, Julian Amery. There was a feeling of optimism that ELDO might lead to a bright new future for Europe and for Woomera. Newspaper and magazine articles portrayed Woomera as a space port for the future. Even earlier in the 1950s, the hit BBC radio serial, Journey into Space, portrayed the launching of Commonwealth rockets to the Moon and to Mars from the Australian outback.

The last of the major aerospace projects were all initiated under Macmillan's Conservative Government. The Wilson Government in 1965 cancelled the TSR 2 and other major military aircraft projects. Concorde and Europa survived because of their international dimension: the UK was treaty-bound to these projects, the Foreign Office fought for them, the Government did not want to seem anti-European, and, most importantly of all, because the way the treaties were written, not a great deal of money would have been saved by cancellation.

The same was not true of Black Arrow, but by comparison with the likes of Concorde or TSR 2, it was a fairly insignificant affair. Spending was put on hold, to be doled out in three monthly offerings. Needless to say, this budgetary regime, the consequence of any lack of decision one way or the other, had the effect of both delaying the programme and increasing the cost, by preventing any long-term planning or ordering of materials.

Returning to the theme of the *zeitgeist*, it is interesting to look at the press view. There had been successes with the launches of Blue Streak in 1964 and 1965, and with the Black Knight launches. But the Black Knight programme had finished by 1964, and the ELDO launches were hardly good news, despite the fact that the Blue Streak stage had always performed as expected. There was always the Black Arrow programme, but this was deliberately (and by Treasury instruction) kept very low key at the outset. The R2 launch in September 1970 was a different matter. The failure drew widespread attention in the press.

The broadsheets kept their reporting quite factual, and there had obviously been some 'spin' from the Ministry of Aviation and from Farnborough. Almost all the papers refer to the 'seventeen seconds that cost success', obviously a reference to the drop in pressurisation. The tabloids were less forgiving.

Under the heading 'Broken Arrow,' the Daily Mail had the following to say:

> One Christmas, as a child, we got a train set called Golden Arrow which was gleaming, expensive, bursting with concealed power – and didn't work.
>
> So we can understand the chagrin of the boffins who get a space set called Black Arrow which was gleaming, expensive and … etc.
>
> The first all-British launch of a satellite to orbit Earth failed to lob into a space an object uncomfortably like a pawnbroker's ball.
>
> Its purpose, we are solemnly assured, was to tell us things we didn't know about the upper atmosphere. To this end, the Black Arrow project has been costing us £3 million a year.
>
> As the Americans are some years ahead of us in this sort of exploration, it is likely that we could get all the information we could possibly digest about the upper atmosphere simply by calling Washington at the cost of £1 per minute.
>
> If, however, we insist in going it – albeit late – alone, we would do well to mark the fact that NASA's budget is around £1,300 million a year. And even they are looking for European money to launch a recoverable, and therefore cheaper, space outfit.
>
> Our dilemma lies somewhere between the facts that even £3 million a year is too much to pay for a damp squib, while it would cost us many times that amount to buy

a share of the American Roman candle. Especially as they would want to light the blue touchpaper.

The worst part about the article is the tone of mockery. The Evening Standard, a few months later, under the headline 'WHAT A JOKE' was even more brutal:

> The French laugh at it. The rest of Europe ignores it.
>
> The Russians couldn't care less. Most Americans don't even know it exists.
>
> It is run on a budget that makes a shoestring look like a hawser. It depends on out of date equipment, and its future is in doubt.
>
> What is IT? The Black Arrow project.
>
> Britain's national space effort – the one intended to gain us admittance to the exclusive – and so far elusive – Space Club.
>
> But Black Arrow is a joke ... a joke on the British tax payer.

The article continues in this vein, but there is a more interesting passage at the end:

> The first one went haywire and had to be destroyed seconds after launch.
>
> The second performed perfectly, and the third, launched last September, failed to put a satellite into orbit and in so doing failed to gain Britain entry into the 'Space Club' ...
>
> Yet, ironically, the previous failures can be laid at the door of funds, or the lack of funds.
>
> This shortage of cash has led, in turn, to a shortage of time. For although there is only one firing a year, time is still of the essence.
>
> Particularly when the scientists involved have to use slide-rules rather than computers.
>
> When you have to make do with second best electronic monitoring devices.
>
> And when you have to resort to economies like using garage petrol pumps for measuring your rocket fuel.
>
> For the truth of the matter is that where America uses dollars, France uses francs, and Japan uses yen, Britain falls back on good old ingenuity.
>
> But it is fast becoming apparent that Space projects can't live on ingenuity alone.

Both articles are, in different ways, making the same point. There is no such thing as a cut price space programme.

So much for the public perception of the British space effort. What was the Government's attitude? They were, after all, the customer.

With the advent of the Wilson Government in 1964, the Department of Economic Affairs (DEA) was set up as a counterbalance to the Treasury. One of its remits early in 1965 was to consider the UK space programme. To say that it was opposed to it in almost any form is no exaggeration. Thus one paper, when discussing the Small Satellite Launcher (Black Arrow) states: 'There may possibly be a long term interest in TV transmission by satellite, but this is never likely to be economic.' The first direct broadcast satellite was a Canadian satellite in 1972; nowadays, of course, Sky Television is ubiquitous. One of the problems with the Civil Service of the time, excellent though they may have been in many ways, is that they were not technically educated, nor had they any feeling for entrepreneurship. Even economists are seldom likely to spot the next future technology. This is re-inforced by a later paragraph in the paper:

> ... the fact remains that none of the applications of satellites at present even remotely in sight is likely to bring any economic return, either in terms of commercial profits to manufacturers, exploitation by HMG [Her Majesty's Government] as operator, or, through international contracts, across the exchanges.
> [This passage in the brief has been underlined and noted in the margin.]

One wonders how much consultation there had been with manufacturers, particularly those in the US. By 1964, two TELSTARs, two RELAYs, (medium orbit satellites) and two SYNCOMs (in geostationary orbit) had operated successfully in space. By the end of 1965, EARLY BIRD had provided 150 telephone 'half-circuits' and 80 hours of television service.

The paper then concluded: 'This proposal [Black Arrow] should be resisted as strongly as possible. Either it should be killed right away or remitted back to lower-level ...'

Another official in the same department as part of the same debate commented that competing with the US and USSR in space was 'a wanton waste of resources'. With regard to ELDO,

> ... unless Europe is to go on indefinitely squandering more and more resources in a field without significant economic return, some country sometime has got to take the lead in calling a halt, even at the cost of seeming opposed to European co-operation.

Implicit in this statement was the notion that the UK should be that country.

Then in 1965 came the first of the many disagreements in ELDO, followed later by the British reluctance to be further involved in the programme. It must have seemed odd to the remaining five members of the organisation to see the founder members, those who had pushed so hard for the organisation, to fall out in this fashion, and to lose enthusiasm for their own project.

A brief prepared for the Prime Minister, Harold Wilson, by the DEA on ELDO noted: '... the ELDO programme in general and our own proposed satellite launcher and satellite development programme are of low economic priority and cannot be justified on economic grounds.' And on the ELDO B launcher proposal: '... For much less money we could do more work than we do now in a field which is of direct concern to us and where we can make new technological contributions'. But these fields never seem to be specified in any of the documents.

So what proportion of the space budget was taken up by ELDO? A policy paper written in 1966 estimated that Britain would spend £20.65 million on space that year, of which ELDO's share would be £13.5 million, or more than 65%. Black Arrow, on the other hand, would come to around £1.7 million – or 8%!

Was the space budget exorbitant? Another paper of the period gives these figures for the estimated expenditure on civil Research and Development for 1966–1967[2]:

	£ millions	%
Space programme	18.8	6.4
Atomic Energy Authority	51.1	17.5
Research councils*	56.7	19.5
Universities	48.0	16.5
Ministry of Technology**	46.8	16.1
Other Government departments	23.1	7.9
Post Office*	5.9	2.0
*excluding space	** excluding atomic energy	

Take ELDO out of the space programme, and its fraction of research and development expenditure shrinks to around 2%!

All the documents disparage ELDO A (Europa) on the grounds of 'obsolescence'. This is a half-truth. What, in this context, does obsolete mean? The purpose of a satellite launcher is to launch satellites, and ELDO A would be a very good medium-sized launcher in the 1960s. Technically, the US was moving forward into solid fuel boosters and liquid hydrogen, but essentially, the technology has changed hardly at all in the half century from the advent of Atlas and Blue Streak. Launchers have grown bigger, but the latest designs would be quite comprehensible to the pioneers of the 1930s. The problem was rather different: there was little European demand for a medium-sized satellite launcher. Given in addition its increased cost compared with US launchers, then demand would indeed appear to be minimal (although it would have been as effective or more effective than the Delta rocket used to launch the UK Skynet satellites – had it been available on time).

ELDO B was written off by the same officials as still being smaller than some US launchers (Titan III). They commented: 'Even the advanced ELDO B launcher cannot be expected to be technically or financially competitive with American launchers'. This again misses the point, which is whether it would have been capable of launching the geosynchronous satellites for which there would be a market, and a market that exists today and is growing ever greater.

It is interesting to read Tony Benn's (Minister of Technology, 1966–1970) summation of ELDO, published in the New Scientist magazine in February 1971:

> ... The foundation of ELDO was, in fact, the first offloading by Britain of a high technology budget that its own industrial weakness no longer permitted it to carry. France identified it as a chance to build an alternative space programme to the American one, and de Gaulle dreamed of carrying the French language and culture to French Africa and possibly even to Quebec. Germany saw it as a first foot back into rocketry which helped to compensate for the loss of Werner von Braun, and Italy as a place for her in the big league plus contracts for Fiat. For European ministers of science it offered new ways to win a national reputation that would be popular with a public for whom the technological unity of Europe was slowly beginning to be real, even if only through the televising of the European cup and the Eurovision Song Contest.

And further: 'ELDO ... suffered from the fatal defect of being a hardware system in search of an application which was in any way economic.'

The counter argument was put forward by the Ministry of Aviation that unless the UK or Europe had its own capability, the US would have had a monopoly. Without Ariane, and without the availability of Russian launchers from the early 1990s, that would have been true. It is also probable that despite the wishful thinking of civil servants, the US would have charged a very great deal for launching satellites that would have competed with its own in the lucrative communications market. Indeed, it could have charged almost what it wanted to, or alternatively have retained its monopoly in the communications satellite area.

Lest it should seem that all the preceding quotes have been taken out of context, or that the quotes have been carefully selected to provide a one-sided argument, it is well nigh impossible to find any quotes in favour of pushing forward any space programme outside the Ministry of Aviation and the Foreign Office – and indeed after around 1966 even the Ministry of Aviation, or Technology as it had become, accepted the demise of the British effort. The Foreign Office was not concerned with the technology at all – merely the political implications of withdrawal from present programmes with regard to Europe and Australia.

Be all that as it may, British withdrawal, painful though it was, took place. The British experience with ELDO bit deep: so much so that the UK has never again become involved with any launcher programme. Even the European Space Agency, ESA, was described by one minister in the 1980s, Kenneth Clark, as 'an exclusive club designed principally to put a Frenchman into space'. Such wilful disparaging of one of the most commercially and scientifically successful space agencies is astonishing. The same minister also refused to fund the innovative British Hotol design, whilst at the same time declaring the engine design classified, and thus preventing development elsewhere.

Could commercial firms have carried on some of these projects? The companies involved in aerospace in 1971 were Hawker Siddeley Dynamics, successor to de Havilland, which itself had been swallowed up in British Aerospace, Rolls Royce and Saunders Roe, who by then were part of Westland (at this period they were working under the name of the British Hovercraft Corporation. Like too many of the Saunders Roe programmes, hovercraft seemed initially to have had a bright future but have turned out to be a dead end).

Saunders Roe, even as part of a larger organisation, were too small to be able to afford the capital investment needed for developing satellite launchers. Rolls Royce and British Aerospace could have worked together on further developments of Blue Streak, but this was only part of the problem.

As well as the rocket, launch sites are needed. Woomera was not suited to satellite launching, and, by this time, was near closure. Kourou, in French Guiana, did, however, have a Blue Streak launch pad. Ariane 1 was not launched until 1979; Ariane 4, which has been the mainstay of ArianeSpace, not until 1988. How prepared the French would have been to make the launch site available is, however, another question. In addition, NASA in its post-Apollo phase, was promoting the Shuttle very forcefully as the answer to satellite launching, with the re-usable nature of the craft. Indeed, it is arguable that part of the success of Ariane was the Challenger disaster of 1987, since by then NASA had almost halted its programme of satellites on conventional launchers.

However, Ariane was a more powerful launcher than all but the most sophisticated Blue Streak derivatives. Ariane 1 was optimised to put 1,750 kg into a Geo Transfer Orbit (GTO). A Blue Streak/Black Arrow combination, even supplemented with strap-on boosters or liquid hydrogen stages, would not have matched that performance. Ariane 4 and Ariane 5 are even more powerful.

But even if the three firms had joined forces to produce a launcher, who would have been their customers? The UK Government has launched some military communications satellites under the Skynet programme, but not on a scale large enough to justify such investment. The European Space Research Organisation (ESRO), ELDO's sister organisation, and other European countries might have been customers, but the market in 1971 was still very thin.

There is also a further political dimension: the UK aircraft industry was not in good shape in the 1970s; indeed, it was nationalised by the Wilson Government during that period. It was certainly in no position to undertake large speculative projects of this sort.

Could the UK Once Again Become Involved in a Launcher Programme?

In a word, no. There are many reasons for this.

Firstly, there is no infrastructure left. After the Black Arrow cancellation, the facilities at Cowes, at High Down and at Ansty were closed down. With the demise of ELDO, the Rolls Royce facilities were closed, and Spadeadam handed over to the RAF. There was still some work going on with the Falstaff programme designed to assist the Chevaline upgrade to the Polaris system (Chevaline itself also required some rocketry development). Some facilities were kept available until the mid-1990s as a 'strategic asset', but even those have gone. A little development work continues at Westcott related to satellite work.

Building new infrastructure would be very difficult. Gone are the days when redundant War Office sites or disused airfield could be pressed into service. The idea of building a rocket test facility at somewhere like High Down is laughable in today's Britain.

Secondly, the skills have gone. Whilst there may well be plenty of competent engineers available, none will have worked on rocketry systems. Such systems have their own peculiarities. If you have worked on them in the past, you are aware of the pitfalls to be avoided. This is sometimes described as 'tacit knowledge' – knowledge you have gained by experience, but which is very difficult to describe. But all those who worked on Blue Streak or Black Knight or Black Arrow have long since retired, and newcomers would have to learn many lessons which once were well known, but that knowledge has gone with the engineers of the past.

Thirdly, there is money. Rocketry demands a lot of money. The folk memory of the Treasury is long, and the experience of ELDO is burned into the collective subconscious of Whitehall. Never mind that it was the Government's fault in the first place – the money wasted serves as a stark reminder to anyone trying to resurrect the programmes.

On the other hand, space is not all about rockets. Rockets are but a means to an end, and that end is to launch satellites. Part of what is left of de Havilland's site at Stevenage has, by a long and tortuous path, ended as part of EADS Astrium, and still manufactures satellites, as does another site at Portsmouth, which in 2011 employed 1,400 people. Similarly, Surrey Satellite Technology (SSTL) is an example of what the Treasury was talking about when it insisted that if space was a profitable business, then private business should get on and do it.

What was left of the Ministry of Aviation became subsumed into the Ministry of Technology, then the Department of Trade and Industry. Now there is a new UK Space Agency, created in April 2010, replacing the British National Space Centre (BNSC) which was an umbrella organisation of ten Government departments, research councils and non-departmental public bodies. The UK civil space programme budget was at that time in the order of £270 million per year – about 76% of which is the UK's contribution to ESA projects.

There may have been relief in the Treasury and in the Government when the programmes were finally cancelled, but there was a great deal of bitterness among those who had worked on the projects. Let them have the last word: they built rockets with a success rate almost second to none on shoe string resources, and then retired into obscurity.

---

[1] TNA: PRO AVIA 66/7. Blue Streak satellite launcher project: Pt B.
[2] TNA: PRO CAB 129/127. Space Policy. Report by the Official Committee on Science and Technology. November 1966.

Appendix A

Original Documents

History of the Saunders Roe SR53 and SR177

The SR177 is being built to OR337, issued by the Air Staff on 2nd December, 1955. This is a development of an earlier requirement, O.R.301, first issued in 1951. The aircraft being built to this latter requirement is the S.R.53.

SR53

May 1951. Particulars of a proposed requirement for a rocket propelled fighter were circulated to the Air Staff... Because of the limitations of the early warning system and the likely scale of enemy attack, it was thought that a large force of high performance day fighters would be required. The ability of the fighters then being developed to deal with the very high altitude raider was doubted. The aircraft proposed was intended to fill the gap until effective Guided Weapons became available and to provide a strong backing for the day fighter force against mass daylight raids of B.29 type bombers. The operational role of the aircraft was to be based on an exceptional rate of climb, probably obtainable only by rocket propulsion. Target date for the first production aircraft was Spring, 1954. The aim was to combine simplicity and ease of manufacture with operational efficiency. Certain operational refinements were therefore to be sacrificed.

August. O.R.301 was issued for a rocket fighter with the following main features:
(a) Climb 60,000 ft. in 2 ½ mins.
(b) Speed. Aircraft of this type were required ultimately to be supersonic above 30,000ft. In the first instance, a maximum speed of $M = 0.95$ would be acceptable if this would shorten development time substantially.
(c) Landing speed. A low landing speed–this was more important than supersonic speed since landings would have to be made from the glide.
(d) Armament: Battery of 2" air-to-air rockets, with provision for fitting direct hitting air-to-air guided Weapon as an alternative.
November. Ministry of Supply accepted O.R.301.

1952

January. Ministry of Supply issued Specification (F.124T). This enlarged on O.R.301 by specifying that provision should be made for carrying Blue Jay [an air to air infra-red homing guided missile].

February. Ministry of Supply circulated the specification widely to aircraft firms … Tenders were submitted by Bristol, Fairey, Blackburn, A.V. Roe, and by Westland and Saunders Roe.

While firms were preparing designs, the Air Staff decided to ask for an ancillary jet engine to assist the return to base phase.

July. The Tender Design. Conference decided to recommend to C.A. that three prototypes each of the Avro and Saunders Roe aircraft should be ordered.

October. Ministry of Supply raised a Technical Requisition to initiate contract action.

1953

May. Ministry of Supply awarded a contract for three aircraft to Saunders Roe. The history of the Avro design is not followed in detail hereafter.

June. Ministry of Supply issued Specification (F138D) calling for Spectre, (rocket) and Viper (jet) engines, supersonic performance above 40,000 ft. and a subsonic cruising ceiling of not less than 70,000 ft …

August. … The target date for the aircraft to be in service was 1957.

1954

January. For reasons of economy, the Ministry of Supply order was reduced from three prototypes each from Saunders Roe and Avro to two prototypes each.

June. The Ministry of Supply forecast the first flight of the first Saunders Roe prototype for July 1955.

1955

January. The D.R.P.C. decided that for reasons of economy, either the Avro or the Saunders Roe development should be stopped. The Ministry of Supply made a study of the relative merits of each aircraft and its development potential.

March. D.M.A.R.D.(RAF) concluded that the Saunders Roe aircraft was likely to be more successful and would have an attractive performance in its developed form.

July. A.C.A.S.(O.R.) recommended to D.C.A.S. that the Air Staff should support the Ministry of Supply's proposal to abandon the Avro aircraft.

1956

The first prototype SR53 is expected to fly in July, 1956.

March. Delays have been due to two main reasons, each of which would have held up the first flight date.

(a) The fuel and designing a HTP system were more difficult than was first realised and required a large amount of testing.

(b) Development of the Spectre rocket has slipped and the engine has not yet been airtested. Tests with a Canberra are expected to begin in March, 1956.

S.R.177

1954

January. The Air Staff considered the further development of the aircraft to O.R.301. A.C.A.S.(O.R.) suggested that the O.R.301 prototypes might be used to provide early technical information for building a more advanced aircraft on similar principles.

February. Saunders Roe submitted a brochure to the Ministry of Supply proposing that a jet engine of similar thrust to that of the rocket be fitted to the aircraft being built to O.R.301.

June. Ministry of Supply asked R.A.E to assess the performance of the aircraft proposed by Saunders Roe when fitted with a Gyron Junior engine.

1955.

February. Ministry of Supply raised a Technical Requisition for design studies of the possibility of using an engine of 7,000 to 8,000 lb. thrust in the P.138D.

August. Air Staff circulated Draft O.R.

September. Ministry of Supply issued a further contract instructing the company to proceed with fullscale design, pending a main contract, on the basis of the Draft O.R.

December. The Air Staff issued O.R.337. The preamble stated that the main threat to the country was still subsonic, but attacks by aircraft capable of speeds up to M = 1.3 at heights up to 55,000 ft. might be expected in 1960/62 …

The flexibility given by A.I. [Airborne Interception], navigation aids and auto-pilot facilities was essential.

The aircraft was required in service as soon as possible and not later than July, 1959.

1956.

January. The Ministry of Supply accepted the O.R. …

February. D.R.P.C. accepted the S.R.177 as a development project for RAF and Navy. Ministry of Supply sought Treasury approval to place an order for a development batch of 27 aircraft. As this was not readily forthcoming, in April the Firm was authorised the expenditure of a further £100,000 to maintain continuity.

February. The two S.R.53 prototypes are now regarded primarily as a lead in to the F.177, rather than as a research project.

July. Specification [handwritten: F177 to meet OR337] issued by Ministry of Supply.

Treasury agreed to a development batch of 27 aircraft, but authorised the build of only 9 aircraft with long dated materials being allocated to support the remaining 18 aircraft. The delay in Treasury approval being granted was due to reviews of patterns of fighter defences of the future, and the atmosphere of financial stringency and economy generally.

The S.R.53 has not yet made its first flight. The first F177 (SR177) is scheduled to make its first flight in April 1958, but this is likely to slip by 6 months.

September. Ministerial approval having been granted, O.R.337 is formally accepted for action by the Ministry of Supply. Design work has however been proceeding since September 1955. The main adverse effect of the delay in placing the final contract has been that it has prevented Saunders Roe placing sub-contract orders.

1957

March. The first flight of the S.R.53 remained "imminent" until the end of 1956, but it has not yet flown and is scheduled for mid-April 1957. There have been troubles with the Spectre engine, but the airframe also is not fully ready.

[handwritten] 29th March. Air Staff cancellation of OR337 was formally sent to the M of S [Ministry of Supply] on the 29th March.

K11 Underground Launcher

The paper that follows is the Air Staff description of the prototype Blue Streak underground launcher. The prototype was known as K11.
   A drawing showing a full reconstruction of the launcher can be found in Chapter 6 (Figure 50).

K.11 prototype underground emplacement

(1) The potential attacker is believed to have the capability to produce an explosion of 1 megaton yield on the ground or in the air with an accuracy of ½ nautical mile from his target. The launcher must be able to withstand such an explosion and successfully fire its own missile without outside assistance within 24 hours.

(2) The emplacement must be able to fire the missile in all weathers.

(3) The emplacement must contain the missile and the necessary facilities for operating and servicing it and for messing and accommodating the concerned personnel. Since an alert may be sounded when the outgoing shift is handing over to the incoming shift, messing facilities must be adequate for two shifts.

(4) Storage space for the missile propellant fuels, food and other stores and equipment must be provided.

(5) Adequate ventilation including the efficient and speedy expulsion of missile exhaust after firing, must be provided together with facilities for conditioning, purifying and circulating air.

(6) Insulation against the electro-magnetic effects associated with a nuclear explosion.

(7) The emplacement must be self-contained for an emergency period of four days (covering three days before an attack is expected and one day afterwards).

II. SITE CRITERIA

1. Rock mass (hard chalk, limestone or better) not less than 300 ft thick and preferably with no overburden. But if overburden is present, it must be soft and not more than 25 ft thick.

2. Easy and firm access from main road to emplacement for transport of missile, equipment and stores.

3. Ease of guarding.

4. Neighbouring inhabited property must be more than 3,000 feet from the emplacement (this may be reduced as experience is gained in K.11).

III. DESIGN OF EMPLACEMENT

1. Basically, the emplacement consists of a hollow re-inforced concrete cylinder, 66 feet internal diameter, extending downwards from ground level to a depth of 134 feet and divided internally into two main sections by a vertical concrete wall. One section houses a U-shaped tube, the arms of which are separated by a concrete wall and are, respectively, the missile shaft and its efflux duct. The surface apertures of this U-tube are covered by a lid which can move horizontally on guide tracks. The other main section within the cylinder is divided into seven compartments, each with concrete floor and ceiling, for the various storage, operating, technical and domestic functions.

2. The internal diameter (66 feet) of the concrete cylinder is determined solely by what is to be accommodated. Protection against an explosion as … above is given by the lid and by the re-inforced concrete roof walls and foundations. The wall thickness will depend on the geological characteristics of the surrounding rock and may well be of the order of 6 feet. The depth of 134 feet is arrived at primarily to give sufficient clearance below the missile (itself 79 feet long) to allow for de-fuelling and re-fuelling the missile into and from the liquid oxygen and kerosene storage tanks located on the 7th floor.

3. A nuclear explosion produces certain electro-magnetic effects which could gravely injure the electronic systems built into the missile and on which its efficient functioning depends. To screen the emplacement from these effects the concrete cylinder will be wholly encased in ½" thick mild steel plate.

Missile shaft

4. The shaft is octagonal in section, 25 feet across and has an acoustic lining. The octagonal shape, which has been proved by tests, will facilitate the mounting of

the acoustic lining and of the four hinged platforms which are spaced at intervals down the shaft.

5. The purpose of the acoustic lining is to prevent damage to the missile from the extremely high noise level produced by the main thrust chambers in the confines of the missile shaft.

6. The missile rests vertically in the shaft on a launcher supported by four suspension limbs attached to the wall of the shaft.

7. Access to the shaft for servicing purposes is through blast-proof doors opening on to the second and sixth floor.

Efflux duct

8. This has an area approximately 60% of that of the missile shaft and in section is half-octagonal in shape. This gives symmetry in the structure and at the surface aperture. A series of deflector plates at the exit will take the exhaust gases away from the missile as it leaves its own shaft.

Storage, Operating, Technical and Domestic Section.

9. This is divided into seven floors, as below, connected by a lift and staircase running from the first floor (at the top) down to the sixth floor:

First Floor

This floor contains:

(a) lid operating mechanism

(b) generating equipment

(c) air conditioning equipment

(d) blast valves for all intakes and exhaust ducts.

All this equipment has been centred as far as possible on this floor to avoid large air trunking systems being provided throughout the site. In the event of contaminated air being taken in, arrangements will be made to close off this floor (other than the general access facilities) thus allowing the generating and air conditioning plant to continue to operate without risk of contamination of the rest of the site.

Second Floor

This floor contains:

(a) Upper storage and maintenance area for the missile, together with two magazine type stores for the payload and the pyrotechnic equipment of the missile, i.e. retro rockets, head propulsion rockets, etc.

(b) Certain items of heating and ventilating equipment for which space is not available on the first floor.

(c) The refrigeration supply for the missile guidance equipment

(d) Blast proof access doors to the upper portion of the missile shaft.

Third Floor

This floor contains:

(a) Auto-collimator equipment

(b) Radio and communications equipment

(c) Site and missile control and checkout equipment

(d) Azimuth bearing and general purpose telescopes.

This floor level is controlled by the relationship required between the auto-collimator and the inertial guidance unit in the missile.

Fourth Floor

This floor contains all the general domestic accommodation including kitchen, recreation and sleeping facilities, etc., together with a small battery room and a switch room.

Fifth Floor

This is intended as the main storage area for the site generally. It also contains one or two tanks which it is not practical to put in the tank room on the seventh floor.

Sixth Floor

This floor is the lower maintenance area and contains the blast proof access door to the lower portion of the weapon shaft. Small hydraulic units are installed on this floor to supply the auto-pilot and launcher services. A small mono rail is provided that can be extended into the missile shaft for maintenance purposes. Access is also provided into the lox and kerosene ['and water systems' crossed out in original and 'rooms on the seventh floor' handwritten in]

Seventh Floor

This floor is divided in two by a structural wall to separate the liquid oxygen and nitrogen systems from the kerosene and water systems.

The Lox room contains:

(a) Main Lox storage tank
(b) Main liquid nitrogen tank
(c) High pressure gaseous nitrogen storage bottles
(d) The Lox start tank
(e) Liquid oxygen topping up pump

Subsidiary rooms contain:

(a) Liquid oxygen recondensing units
(b) Liquid nitrogen recondensing unit
(c) Liquid nitrogen topping up pump
(d) Liquid nitrogen evaporating plant

The kerosene room contains:

(a) The main kerosene storage tank
(b) The main water storage tank
(c) The kerosene recirculating pump
(d) The kerosene start tank

The access doors from the sixth floor will normally be kept closed and ventilation shafts are provided from these two rooms through the main structure to the surface pipe systems are also provided in these vent shafts for filling these systems from the surface,

IV. DESIGN OF LID

1. The detailed design of the lid is about to form the subject of a special design study by selected firms.

2. The purpose of the lid is to protect the missile from the effects of attack and to remain fully serviceable itself after such attack. Since the missile is completely unprotected when the lid is open, the time allowed immediately prior to firing the missile for opening the lid must be kept to a minimum and has been put at 17 seconds.

V. DESIGN OF SITE

1. The main requirement is to achieve maximum security and this calls for both the site and its immediate surrounds to be enclosed by a security fence and to be clear of obstructions to visions. The cleared area extends also beyond the site perimeter. The need to camouflage the site is at present being considered. A simple road system with associated hardstandings must be provided within the site.

2. The site will be about 3 acres in extent.

3. The site includes the main entrance to the emplacement, consisting of three flights of steps, protected only against weather and leading down to a cylindrical air lock giving access to the first floor of the emplacement.

Use of Spadeadam for Space Firings

The document below is the full text of the note written by Sir Steuart Mitchell, CGWL, to the Minister of Aviation, Peter Thorneycroft, concerning the possibility of using Spadeadam to launch Blue Streak.

The following are hastily prepared views on the above.

Technically

Spadeadam certainly would be feasible, and in nearly every way technically would be better than anywhere else.

2. <u>Costwise</u>. (Capitol plus operating) Spadeadam would be as cheap as Woomera, and cheaper than anywhere else.

3. Method

Trajectory about north 35 east.

Overflies Kelso. Crosses coast ever Eyemouth.

Passes 25 miles east of Aberdeen, and 100 miles west of the Norwegian coast.

Down range station, very well placed, would be in Spitsbergen (Norway, open all year round). Alternative would be at Tromso, which is possible but not so well placed.

First stage impact 200 miles off Norwegian coast abreast Namsos. (This is west of the Narvik – North Sea ore traffic lane).

Second stage impact on the polar ice cap.

4. Risks

Quick and Rough estimates are:-

1. The chance of having to cut down the missile on to UK territory beyond the Spadeadam Range Area is approximately:-

½% to 2% during the development period.

2. The chance of a missile having to be cut down and then landing in a "populated area" is approximately:-

1/50% to 1/5% during the development period

3. The chance of killing a person is approximately 1 in 10,000 per round fired in the development phase.

4. The risk to Norwegian territory is negligible.

5. The risk to shipping, is negligible.

5. Nature of the Risk

The cut-down risk is numerically greatest within the first mile. The Spadeadam Range area extends to just over a mile from the launch.

A missile cut down within five miles would have a considerable fire risk from fuel and oxidant. Outside five miles a cut-down missile is primarily a fragment risk not a fuel oxidant risk.

6. Comparison with Aircraft Risks

Aircraft crashes over the last 5 years in the UK average about 90 per annum. The probable total damage to lives and property of persons on the ground per annum from firings from Spadeadam is estimated to be about 1/10 of that from aircraft crashes in the UK per annum.

One Boeing 707 crashing near take off from London airport with full tanks and 128 passengers, or two Boeings colliding over London, would be far more serious than any conceivable accident with a space launcher.

7. Black Knight experience

None of the total of eleven firings so far done with Black Knight would have landed in UK territory if they had been fired from Spadeadam.

8. Conclusion

Spadeadam is technically both feasible and attractive. From the cost point of view, it is approximately the same as Woomera, and is much cheaper than any alternative.

It must be accepted, however, that some cut-downs on to UK territory would inevitably occur if we fire from Spadeadam. The chance of serious damage to life and property from such cut-downs are numerically small.

The risk of damage to foreign countries, or to shipping, is negligible.

The crucial point is the political acceptability of the risk in the UK Hitherto this has been regarded as unacceptable, and it would be no less now than when previously considered. My advice is that the risk is appreciable and should not be accepted.

S.S.C.M. C.G.W.L.   27th October 1961
[Taken from TNA: PRO AVIA 66/7]

Space and the Future Of The European Launcher Development Organisation (ELDO)

[Handwritten on the top of the document: 'This is one of the background of briefs prepared for new ministers in October 1964.']

Recommendation.

Some big decisions will be needed very soon about UK space policy. There is to be a meeting of the ELDO Council around the end of the year (it was fixed for mid-December but is now likely to be deferred), at which the UK will be expected to show its hand. All this will be submitted to the Cabinet in due course. In the meantime there is an urgent need to put into cold storage a project authorised at the beginning of September by the last government–namely the development of a small satellite launcher based on Black Knight.

The four aspects of space.

2. The problems of UK space policy fall under four heads:
    (i) The scientific space research programme.

(ii) UK participation in European efforts on launchers and communications satellites.

(iii) Military space.

(iv) The so-called "national programme"

To take each of these in turn:

The scientific research programme.

3. This is the programme of scientific investigation into what happens in space: expenditure on it is financed from the so-called "space research budget", the level of which is now £3¼ m a year and is expected to rise to £6m towards the end of the decade. The biggest single item in this is the UK contribution to the European Space Research Organisation (ESRO) and the development of the UK 3 space research satellite (to be put up by an American launcher). The Treasury is not at this time pressing for the termination of this programme, taken as a whole, or our withdrawal from ESRO.

The European effort on launchers and communications satellites.

4. The UK is a member of ELDO (along with France, Germany, Italy, the Netherlands, Belgium and Austria). The ELDO launcher to be developed in accordance with the Lancaster House agreement of 1961 will consist of a UK first stage (the old Blue Streak), a French second stage and a German third stage. The whole launcher was originally estimated to cost £70m. It is now estimated at least £90 m. (of which the UK pay about 40 per cent) and will almost certainly rise further. For all this money (now around £6m a year on the Ministry of Aviation vote) we will get a launcher which is likely to have little, if any, practical use. What has gone wrong, we understand, is that the engineering of the launcher has not been up to the standard of the original design (which was largely cribbed from the US anyway). This means that the ELDO launcher when it is ready to be fired (probably in 1967/68) will not be powerful enough to put a satellite of any size in to a high orbit. Since communication satellites tend to be big and to operate most satisfactorily in high orbits, the launcher as developed under the initial ELDO programme ELDO-A) will not be able to be used to put communications satellites into orbit. (Whether ELDO-A could have any other, not communications, uses is quite uncertain.)

5. A proposal has therefore now been canvassed (it will be put before the meeting of the ELDO council around the end of the year) to develop an apogee motor (a device to give a satellite an extra "kick" into orbit) so as to enable the launcher to put communications satellites into high orbit. The cost of this is estimated at up to £20m (our share, say, £7m). Beyond this the clouds of very much larger

projects are already looming. The first such project would be the development of the so-called ELDO-B launcher; this would involve the introduction of liquid hydrogen propellant for the upper stages of the launcher, and development might start around the end of 1965. The cost of this might be of the order of £50-100m. (UK share say around £20–40m.). This could be followed – towards the end of the decade – by ELDO-C, which would be a completely new launcher using liquid hydrogen as a propellant and the cost of which can only be guessed at now (£100–200m.).

6. There is also the prospect of a collaborative European effort on developing and constructing communications satellites to be sold to the world communications organisation – if that organisation will buy them – and the question of whether the UK should participate in this programme will arise.

7. By the beginning of the 1970s, we could be spending up to £30–40m. a year on all these "European" activities, and on the associated "national programme" referred to below. We should have established a big new growing point, both the public expenditure and for the use of fallible human resources (scientists and technologists).

8. The Treasury view is: -
   (i) This is not expenditure which we are likely to be able to afford or which has any prospect of bringing us a commensurate economic return.
   (ii) We should therefore contract out of all these activities; if Europeans wished to go ahead wasting money on space, that is their affair.
   (iii) The crucial decision in our view is on the apogee motor; if we go on with this – the first addition to the initial ELDO programme – we shall find that our room for manoeuvre at later stages will be drastically reduced. At every stage it will be argued–as it was argued when was first set up – that, although large sums and money had been admittedly spent to no purpose in the past, all will be well if we will only spend some more. What is needed in the Treasury view is the decision to go no further in the construction of launchers and communications satellites. But there are political, and perhaps also defence, implications which Ministers have to consider in due course.

Military space.
9. There are some big unresolved questions here. The military are hankering after a defence satellite communications system. The Treasury believe that this is likely to be a great deal more expensive than can be afforded out of a tolerable defence budget, and that, if we were in fact to maintain our military posture east of Suez at all, the military will have to do make do with some relatively

inexpensive improvements to the existing means of communication with Singapore etc. But the issues here have not yet been fully explored. At the moment the military are contemplating two possible alternative lines of approach – one to go in with the US on the joint defence satellite communications system which the Americans are likely to set up; and the other to construct some limited defence satellite communication capability of our own. The former course would not be cheap, but the latter would almost certainly be fearfully expensive.

The so-called "national space programme".

10. Apart from some effort within Ministry of Aviation establishments, and a limited amount of research by industry (at government expense) into the basic technology of building satellites, the principal commitment of the UK "national civil space programme" is the development [handwritten: 'at an estimated cost of £10m.'] of the small satellite launcher based on the military research rocket Black Knight. The decision to go ahead with this development was taken at the beginning of September by the late government. It was justified [handwritten: 'by the protagonists'] on the grounds that, if Britain is to go in with Europe in collaborative launcher and satellite development programmes, it was necessary for Britain to have her own "research tool" which would provide her with the necessary technological know-how and an opportunity to test in flight bits of the communications satellites which she will be contributing to the European programme. But no decision has yet been taken on whether the UK should in fact participate in any further European launcher or communications satellite programmes – as stated above, the Treasury would be opposed since to such participation. Until the decisions on this have been taken, it is in the Treasury's view wholly premature to embark on the development of the United Kingdom small satellite launcher. We recommend therefore that until Ministers have had the opportunity to consider collectively the future course of UK space policy, no work should proceed on a small satellite launcher, and that the project should be regarded as in abeyance. If the Chancellor agrees, we will submit a draft minute for him to send to those of his colleagues who are concerned with these matters.

(F.A. Barrett)   19th October 1964

[From TNA: PRO T 225/2765. Ministry of Aviation space programme: future policy.]

Appendix B

A Black Knight Launcher?

There is a reference made in a file in the Public Record Office[1] to a telephone conversation in which the RAE was asked by an official at the Ministry of Aviation in October 1961:

> What has the RAE done on solid propulsion and what has been done on smaller rockets? (Opinion and any possible objections).

The context is not entirely clear, but Dr Schirrmacher of RAE sent a fairly lengthy reply, and part of this reads:

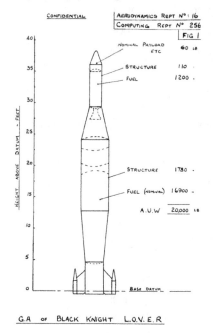

G.A of BLACK KNIGHT L.O.V.E.R

Figure 120. LOVER

(b) For smaller satellites two solutions are worth mentioning:

Black Knight boosted by two Ravens and carrying a Rook as 2nd stage with a Cuckoo as 3rd stage. One will get 100lb into a 200mile orbit. Present development at Westcott for an improvement of the mass fracture [*sic* – probably 'fraction'] indicates that the payload could probably be increased by a factor 2 [*sic*].'

Dr Schirrmacher was referring to a report which was written at RAE in January 1961 as part of a more general paper on the possibilities of using Black Knight as a satellite launcher, and the study was based around what might be called the 'Dazzle' Black Knight – i.e., the 36-inch vehicle with a lift-off thrust of 21,600lb.[2]

It is also interesting to note, in the Black Knight context, that Saunders Roe's Aerodynamics Department and Computing Department had produced a joint report of a vehicle which they christened LOVER, standing for Low Orbit Vehicle for Experimental Research. This was nothing more than the 54-inch Black Knight together with the Kestrel motor (but pointing upwards). Somewhat optimistically, it concludes that

> The proposed satellite launcher using a 4' 6" diameter Black Knight as a first stage and a "Kestrel" powered second stage can be used to inject a small satellite into orbit at an altitude of between 100 and 200 n.miles…[3]

The 54-inch Black Knight is a rather more plausible contender for a satellite launcher than original 36-inch version, and in the context of what might have been possible in 1963, some performance figures would be helpful:

	Thrust (lb)	Burn time (seconds)	Total impulse (lb.seconds)	Weight (lb)	Diameter (inches)
36-inch Black Knight	21,600	120	2,600,000	13,000	36
54-inch Black Knight	25,000	140	3,500,000	17,000	54
Raven VI	15,000	30	450,000	2,540	17
Rook IV	66,700	6.65	480,000	2,370	17
Cuckoo II	8,200	10	82,000	500	17
Kestrel	27,000	10	276,000	1,600	24
Waxwing	3,500*	55	195,000	760	30

* in vacuum

Why 1963? This was the time when the design which would become Black Arrow was evolving. Instead of Black Arrow, this section will consider the possibilities of a launcher using Black Knight as a central core and adding assorted solid motors, both as strap-on boosters and as upper stages.

Adding solid fuel boosters to Black Knight:

	Initial Thrust (lb)	Total impulse (lb.seconds)
36-inch Black Knight + 2 Ravens	51,600	3,500,000
36-inch Black Knight + 4 Ravens	81,600	4,400,000
54-inch Black Knight + 2 Ravens	55,000	4,400,000
54-inch Black Knight + 4 Ravens	85,000	5,300,000
Black Arrow	50,000	6,350,000

The total impulse could be increased even further by 'stretching' the Black Knight stage – that is, increasing the size of the fuel tanks. The Raven is the most suitable, since it has a relatively long burn time, so that the central core has had time to burn off some of its fuel. Total impulse is, of course, not the only criterion by which a rocket motor should be judged, but it is useful in this context as a rough and ready guide.

There would be very little extra development work needed to be done, since by 1961, the Raven, Rook and Cuckoo were fully developed – the Raven and Cuckoo were in use on the Skylark rocket, and the Cuckoo II was also the second stage for the 36-inch Black Knight. The Kestrel was to have been the second stage for the 54-inch Black Knight, and so if we have the 54-inch Black Knight, we also have the Kestrel. Waxwing is the only rocket not available in 1963, since it was developed specifically as the apogee stage for Black Arrow.

In a launcher which has several stages, it helps if they are well matched in weight. For reasonable performance, very roughly the first stage might be around 60% of the all up weight, the second stage perhaps 30%, and the third stage 10% or so. They also need to be matched in thrust: if the first stage only has a limited thrust, it cannot lift the upper stages. One of the points of the strap-on boosters is to give that extra thrust, so that when they have burnt out, the main stage has used up enough fuel to carry on by itself.

Another consideration is geometry, and this is where the idea of using the Rook and Cuckoo as second and third stages falls down. The Rook had a diameter of only 17 inches (at this time, 17 inches was the largest solid motor made in the UK). The Kestrel would have had a 24-inch diameter. The largest motor made in the UK was the Stonechat at 36-inch, but this would have been too heavy to use as a second stage. Larger diameter solid fuel motors were produced on an experimental basis at Westcott in the 1960s. Bristol Aerojet Ltd (BAJ) produced high strength steel tubes using the helical welding process, and Westcott developed the propellant charge design. Tubes up to 54 inches in diameter were produced, and one such tube was filled and fired at Westcott.[4]

Fitting a 17-inch motor onto a 54-inch stage would have been awkward, but the bigger limitation would have been in the size of satellite carried. Even the smallest satellite would have been more than 17 inches across, and so an extremely bulbous fairing would have to be fitted on top. Even with a Kestrel, the geometry looks awkward, as the diagrams in Figure 121 demonstrate.

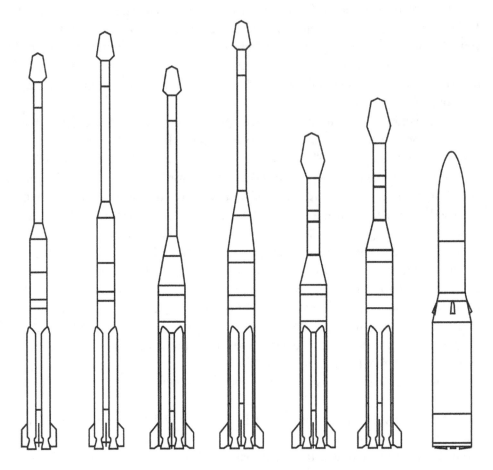

Figure 121. Various possibilities for satellite launchers based on Black Knight.

From left to right:
1: 36-inch Black Knight + Rook + Cuckoo;
2: Stretched 36-inch Black Knight + Rook + Cuckoo;
3: 54-inch Black Knight + Rook + Cuckoo;
4: Stretched 54-inch Black Knight + Rook + Raven;
5: 54-inch Black Knight + Kestrel + Kestrel + Waxwing (under shroud);
6: Stretched 54-inch + Kestrel + Kestrel + Waxwing (under shroud)
7: Black Arrow.

Each of these should have been capable of putting some payload into orbit; option 6 might have been a good deal more effective than Black Arrow!

The Rook would seem to be better than the Kestrel on grounds of total impulse, but was less structurally efficient. On the other hand, there is no reason why two or even three Kestrels could be used – two Kestrels give nearly the same impulse as the Rook, they are structurally more efficient, and staging improves that efficiency.

In either case, the payload would be in the region of 200 lb or so – possibly more. In fact, the payload could be increased further: the total mass of the launcher would have been in the region of 29,000 lb or so, whereas the lift-off thrust was 75,000 lb. As mentioned, this gives the possibility of 'stretching' the Black Knight stage by lengthening the tanks to carry more fuel. By the time the boosters have burned out, the extra fuel will also have been burned off. This is quite efficient – the extra structural weight added is merely that of the thin walled tank.

Why is this a good deal cheaper than developing Black Arrow? There is very little in the way of development costs. By the time of the cancellation of the re-entry experiments, BK26, the first 54-inch Black Knight, was quite well advanced in construction. The development costs of the Black Knight and the Kestrel would already have been absorbed by the re-entry programme. One of the major development costs for Black Arrow was the design and testing of the Gamma 8 and Gamma 2 motors. Developing more capable solid fuel motors is a good deal cheaper.

So, it might well have been possible to develop a Black Knight launcher costing not much than around £2 million which might have been able to put 200–300 lb into orbit, compared with Black Arrow costing £10 million ... and which could put around 150 lb into orbit.

---

[1] TNA: PRO AVIA 65/1567. ELDO satellite launcher system.
[2] Black Knight as a Satellite Launcher. AP Waterfall, Guided Weapons Department, RAE. 18 January 1961 (Science Museum Library and Archive).
[3] Black Knight LOVER, Saunders Roe, May 1963.
[4] Moore PO, 1992. *Stonechat*, JBIS, Vol 45, pp.145–148.

Appendix C - Timeline

1957
13 February		First launch of Skylark
16 May		First flight of Saunders Roe 53 (XD 145)
4 October		Sputnik 1 launched into orbit
18 December		First flight of XD 151

1958
5 June		Crash of Saunders Roe 53 (XD 151)
7 September	BK01	First Black Knight launch, explodes near end of flight

1959
12 March	BK03	Repeat of BK01
11 June	BK04	First successful re-entry experiment
29 June	BK05	
30 October	BK06	

1960
13 April		Cancellation of Blue Streak as military weapon
24 May	BK08	First two stage vehicle
21 June	BK09	
25 July	BK07	

1961
7 February	BK13	
12 April		Yuri Gagarin is first man in space
9 May	BK14	
7 June	BK17	

1962

20 February		John Glenn is first American to orbit the Earth
1 May	BK15	
24 August	BK16	
30 November	BK18	

1963

17 October	BK11	Testing range safety equipment for ELDO

1964

5 June	F1	First Blue Streak launch. Cut out prematurely
6 August	BK19	
20 October	F2	Completely successful
6 November	BK20	

1965

22 March	F3	Completely successful
24 April	BK21	
27 July	BK23	
29 September	BK24	
25 November	BK25	Last Black Knight firing

1966

24 May	F4	Dummy upper stages. Destroyed by Australian Range Safety Officer
15 November	F5	Dummy upper stages

1967

4 August	F6	Live second stage
5 December	F7/1	Live second stage

1968

30 November	F7/2	All stages live, third stage exploded

1969

28 June	R0	First Black Arrow launch. Control failure, vehicle destroyed
20 July		Apollo 11 lands on Moon
31 July	F8	All stages live, third stage exploded

1970

4 March	R1	Two live stages, success
11 June	F9	All stages live, fairings did not separate
2 September	R2	Orbital attempt, second stage shut down prematurely

1971

| 28 October | R3 | Prospero launched into orbit |
| 5 November | F11 | Kourou, satellite attempt. Guidance system damaged; vehicle goes out of control and breaks up. |

Further Reading

Ashford, D.
Spaceflight Revolution (London: Imperial College Press, 2003) ISBN 186094325X.

Budd, R and Gummett, P (*et al.*).
Cold War, Hot Science: Applied Research in Britain's Defence Laboratories 1945–1990 (London: Science Museum, 2002) ISBN 1900747472.
Also useful for a more general perspective of British technological projects of this period.

Cocroft, WD and Thomas, RJC.
Cold War: Building for Nuclear Confrontation 1946–89 (Swindon: English Heritage Publications, 2005) ISBN 1873592817.
This covers the infrastructure behind both Blue Streak and Black Knight, as well as the V bombers and Blue Steel. There is an excellent description of the proposed Blue Streak underground launcher.

Gibson, C and Buttler T.
British Secret Projects: Hypersonics, Ramjets and Missiles (Surrey: Midland Publishing, 2007) ISBN 1857802586.
Covers Blue Steel, Black Knight and Blue Streak among many other projects.

Madders, K.
A New Force at a New Frontier (New York: Cambridge University Press, 1997) ISBN 0 521 57096 4.
Covers space science and policy in Europe generally, and ELDO specifically. Some rather unfortunate factual errors.

Martin, CH.
De Havilland Blue Streak (London: British Interplanetary Society, 2004) ISBN 0950659770.
Written very much from an engineer's point of view – Charles Martin was a member of the HSD team working on the civilian applications of Blue Streak.

Millard, D.
Black Arrow Rocket: A History of a Satellite Launch Vehicle and Its Engines. (London: Science Museum, 2001) ISBN 1900747413.
A rather brief but useful account of the use of HTP and its application to Black Arrow.

Morton, P.
Fire Across the Desert: Woomera and the Anglo-Australian Joint Project 1946–1980. (Canberra: Australian Government Publishing Service, 1989) ISBN 0644060689
An account of the work carried out at Woomera, South Australia, as part of the Anglo-Australian Joint Project. Concentrates on people rather than hardware.

Southall, I.
Woomera. (Sydney: Angus and Robertson, 1962) ISBN B0000CLLIO
A slightly fanciful and now rather dated description of the work being done at Woomera, but with some interesting insights.

Spufford, F.
The Backroom Boys: The Secret Return of the British Boffin (London: Faber and Faber, 2004) ISBN 0571214975
The first chapter is an excellent account of Black Arrow described from the point of view of those who worked on the project.

Tagg, AE and Wheeler, RL.
From Sea to Air (Newport: Crossprint, 1989) ISBN 0 9509739 3 9
A useful and well illustrated account of the work of Saunders Roe.

Wynn, H.
RAF Nuclear Deterrent Forces (London: The Stationery Office, 1994) ISBN 0 11 772833 0
Blue Steel and Blue Streak are covered comprehensively, with attention given to development, deployment, and operations.

Archives

The National Archives at Kew hold the government policy papers of the Cabinet Office and the Treasury. There are more specific policy papers in the files of the Air Ministry, Ministry of Supply/Aviation, and Ministry of Defence. Technical papers relating to many of the projects can also be found here.

Many of the brochures and photographs from the Bristol Siddeley establishment at Anstey are now in an archive in the Herbert Art Gallery and Museum, Coventry.

When Space Department at RAE was closed down, many of its papers were rescued and are now housed the Science Museum Library and Archives at Wroughton, Wiltshire. These relate in the main to Black Knight, Black Arrow and the re-entry experiments at Woomera.

The ESRO, ELDO and ESA archives are housed at the Historical Archives of the European Community at the European University Institute in Florence, Italy.

Index